I0036416

Faneuil D. Weisse

Practical Human Anatomy

working-guide for students of medicine and a ready-reference for surgeons and physicians - Vol. 2

Faneuil D. Weisse

Practical Human Anatomy
working-guide for students of medicine and a ready-reference for surgeons and physicians - Vol. 2

ISBN/EAN: 9783337369798

Printed in Europe, USA, Canada, Australia, Japan

Cover: Foto ©berggeist007 / pixelio.de

More available books at **www.hansebooks.com**

PRACTICAL

HUMAN ANATOMY

WORKING-GUIDE FOR STUDENTS OF MEDICINE

AND A

READY-REFERENCE FOR SURGEONS AND PHYSICIANS

FOURTH EDITION, "ATLAS" FORM

BY

FANEUIL D. WEISSE, M.D.

PROSECTOR (1863 TO 1865) TO THE LATE VALENTINE MOTT, M.D., LL.D., EMERITUS PROFESSOR OF SURGERY AND SURGICAL ANATOMY, PROFESSOR OF SURGICAL PATHOLOGY (1874-1875), PROFESSOR OF SURGICAL ANATOMY (1875-1876) PROFESSOR OF PRACTICAL AND SURGICAL ANATOMY (1876 TO 1889), MEDICAL DEPARTMENT OF THE UNIVERSITY OF THE CITY OF NEW-YORK

PROFESSOR OF ANATOMY, NEW YORK COLLEGE OF DENTISTRY, SINCE 1865

ILLUSTRATED BY 222 LETTERED PLATES CONTAINING 321 FIGURES

NEW YORK

JAMES T. DOUGHERTY

409 & 411 WEST 59TH STREET

1899

Copyright, 1896, by
WILLIAM WOOD & COMPANY

Copyright Assigned, 1898, by
William Wood & Co. to
Dr. FANEUIL D. WEISSE

CHAUNCEY HOLT, PRINTER,
27 ROSE STREET, NEW YORK.

To

JOHN A. WEISSE, M.D.

MY FATHER,

THIS WORK IS DEDICATED

WITH THE WARMEST FILIAL AFFECTION, AND A DEEP SENSE OF GRATITUDE

FOR HIS PERSONAL GUIDANCE OF MY EDUCATION.

PREFACE TO FIRST EDITION.

This work was commenced with a desire—after an experience of nearly twenty years in study, actual dissections, and the teaching of anatomy—to produce a practical working-guide for the student at the cadaver, and a ready-reference book, which would take the place of the cadaver, for practitioners of surgery and medicine.

The plan of the work embraces the following points:

1st, the division of the body into *practical dissections;*

2d, the giving, in *dissection-paragraphs*, the progressive steps by which the several parts, involved in a dissection, are to be systematically displayed;

3d, the guidance, by lines across the parts in the plates—called *section-lines*—to the points where they are to be cut, for their reflection, in order to advance to a succeeding stage of the dissection;

4th, the indication, by *numbering the parts of the dissections*, of the order in which they are exposed;

5th, the description of the parts, in *descriptive-anatomy paragraphs*, as they are brought into view;

6th, the adherence, in expressing the relations of parts, to a well-defined nomenclature of *general* and *special anatomical terms;*

7th, the illustration of the anatomy of the regions and viscera of the body by plates, *with the names of the parts printed upon them or at the sides of the figures with indicating lines to them*—the *dead-anatomy* is thus presented to the student, and the *living-anatomy* to the surgeon and physician.

The dissections have been planned, as far as practicable, to give all the attachments of included muscles.

Each dissection has been completely described and illustrated, irrespective of the contiguous dissections upon which it may encroach.

The bones of the body have not been described in detail; their practical relations to the anatomical elements of the several dissections are shown in the plate illustrations of their surfaces with the areas they afford for muscle attachments; together with their presentation in the progressive steps of the dissections.

The attachments of muscles are given without any distinction as to their being the *origin* and *insertion* of the same; it is more natural and clear to enumerate the attachments only.

The illustrations include the following:

153 full-page, original plates;
51 plates of 132 original figures;
6 full-page plates selected from standard authors;
9 plates of 25 figures, selected from standard authors;
Plate 56, of one original and two selected figures;
Plate 66, of one original and one selected figure;
15 original text figures.

The original plates and figures were sketched and drawn by Mr. Maximilian Cohn from dissections by the author.

The illustration and description of the complete distribution of the sympathetic nervous system and of the lymphatic system have not been given, as they do not come within the scope of the work. The dissection of the globe of the eye, and of the auditory apparatus have been omitted.

No reference has been made to surgical and medical or applied anatomy, as it is deemed best to do *one thing at a time*. Moreover, the systematic mastery of the anatomy of the body places one in a position to make practical applications of the same, to the elucidation of surgical, medical, and obstetrical problems; and to read, appreciatively, the special works on surgical and medical anatomy.

That which can be demonstrated to the unaided eye is the province of anatomy; therefore, the structure of parts, requiring the aid of the microscope, has not been alluded to, as it belongs to Histology.

The practitioner of surgery and medicine, in the absence of a cadaver to refer to, has been much in need of illustrations of anatomy, that would present him a progressive series of dissections of a region or organ about which he desired information. To meet this want has been one of the aims in the preparation of this work, special attention having been given to those regions and organs which claim frequent surgical and medical care. To this end also the contents and index have been so elaborated as to facilitate the finding of the plate illustrations and text descriptions of any part.

Too much cannot be said in praise of the comparatively wonderful artistic skill displayed by Mr. Maximilian Cohn, in his faithful reproductions of nature, given us in the plates and plate figures; and the clearness of his lettering of the same. The illustrations are photographic in their representations of nature and are works of art in themselves.

Thanks are due to the Moss Engraving Co. for the care which has been bestowed upon the photo-engraving and the printing of the plates.

Nature has been the text-book to which reference has always been made; but, a due respect for the labors of our fathers and of our contemporaries has been kept in view.

The original intention was to illustrate this work by selections from the illustrations of standard authors, simply adapting and lettering the same to fulfil our purpose. After having carried out this scheme to the extent of some 300 photo-plates and 150 electrotypes, the plan was abandoned because it was found impracticable to accomplish the object of the work therewith, and *the cadaver was resorted to.*

Errors and omissions, no doubt, may be found in the text and plates, but should any, *of whatever nature*, appear to a reader, his personal communication of the same to the author will be regarded as a kindness.

The manifold difficulties and disappointments that have been met with, in the evolution of this volume, are only for a private ear. The desired end will have been attained, if the fruits borne of years of time, thought, and labor are: that it enables the student of medicine to acquire, more readily and thoroughly, a knowledge of human anatomy; and proves a companion to the surgeon and the physician, to keep them in mind of the practical facts of anatomy.

51 West 22d Street,
New York City, December, 1885.

PREFACE TO FOURTH EDITION, "ATLAS" FORM.

The solicitation of many who have used "Practical Human Anatomy," now out of print for several years; the great demand for second hand copies—which are found almost impossible to obtain; the constantly recurring statements of *dissectors*, that the plates are, in themselves, all they need to guide them in their work; and of *surgeons*, that they had learned, by experience, to rely upon a review of the plates of a region to guide them in a given operation; the favorable reception of the three first editions, as evidenced by letters and personal opinions from accredited authorities on anatomy, from surgeons and from students, have decided the author to present a Fourth Edition of his work, in "Atlas" form, to medical students, surgeons and physicians.

FANEUIL D. WEISSE, M.D.

46 West 20th Street, New York, June, 1899.

CONTENTS.

INTRODUCTION.

FIRST DISSECTION.

SECOND DISSECTION.

THIRD DISSECTION.

FOURTH DISSECTION.

FIFTH DISSECTION.

SIXTH DISSECTION.

SEVENTH DISSECTION.

EIGHTH DISSECTION.

NINTH DISSECTION.

TENTH DISSECTION.

ELEVENTH DISSECTION.

TWELFTH DISSECTION.

THIRTEENTH DISSECTION.

FOURTEENTH DISSECTION.

FIFTEENTH DISSECTION.

SIXTEENTH DISSECTION.

SEVENTEENTH DISSECTION.

EIGHTEENTH DISSECTION.

NINETEENTH DISSECTION.

TWENTIETH DISSECTION.

TWENTY-FIRST DISSECTION.

TWENTY-SECOND DISSECTION.

TWENTY-THIRD DISSECTION

TWENTY-FOURTH DISSECTION.

TWENTY-FIFTH DISSECTION.

TWENTY-SIXTH DISSECTION.

TWENTY-SEVENTH DISSECTION.

SOURCE OF SELECTED ILLUSTRATIONS.

PRACTICAL HUMAN ANATOMY.

INTRODUCTION.

It is, for the following reasons, of the greatest importance that the dissector read this introduction, carefully, before commencing his dissections; first, that he may be prepared to use this work to the greatest possible advantage; second, that he may have a general appreciation of the anatomical elements that he is to meet with; third, that he may obtain a general idea of how to dissect the several parts which enter into the make-up of the body; fourth, that he may know the relative importance of the parts exposed in a given dissection, and, thereby be guided, as to which to preserve and which to cut away, as he progresses.

WORKING PLAN.

Dissectors Use of the Work.—The plates illustrating a given dissection are to be carefully reviewed, before commencing the dissection, as follows:

The bone areas involved are to be noted, then the consecutive order in which the parts come into view in the plates presenting the several anatomical elements. By carrying forward the dissection with strict adherence to the sequence presented by the plates, when the bone surfaces are reached, the dissector will have obtained a complete practical knowledge of the anatomy of the region or part.

In pursuing the dissection, by the plates, the dissector will notice *black lines crossing parts in the plates* (muscles, vessels, etc.). They indicate the points *where* and the stage of the dissection *when* the parts lined are to be sectioned, that the same may be reflected in order to expose to advantage subjacent parts. In some of the plate series, after the section lines have presented on parts, those parts do not appear in subsequent plates of the series—the dissector need not cut away the parts sectioned, but simply reflect them off, end for end, for future replacement and removal, as he would turn the leaves of a book.

Surgeons Use of the Work.—With a given operation in view, the surgeon may, by a review of the plates of the region of the operation, refresh his mind on the relations of the anatomical elements involved.

Physicians Use of the Work.—With a case of disease of a given organ, a physician may, by a review of the plates displaying the organ and its relations, renew his knowledge of the part.

Contents and Index.—The Contents groups the plates of a given region or of a given organ or organs; while the Index—exhaustive as it is—affords a rapid guide to the finding of any given anatomical element, in all the plates in which it is displayed.

Of Right and Left Parts.—Of necessity the plates present in some instances a view of the right side, in others of the left side; the student or reviewer must therefore take his bearings accordingly.

Special Plates to be Noted by the Dissector.—In Plate 38, section lines, are indicated on abdominal aorta and inferior vena cava, which will enable the kidneys and supra-renal capsules with their vessels, included portions of abdominal aorta and inferior vena cava, and the ureters and spermatic or ovarian vessels to be dissected off and carried inferiorly into the pelvis, from which they can be removed from the body collectively, through the outlet of the pelvis, with the bladder, rectum and male or female internal genitalis, for subsequent special dissection.

In Plate 138, section lines indicate where parts at the base of the neck are to be cut so as to enable the viscera to be collectively taken out of the thorax, as follows: free the right and left lung from parietal pleuritic adhesions, cut the pericardium at the circumference of its diaphragmatic attachment, then grasp firmly the inferior ends of the sectioned parts at the base of the neck, and drag the organs forwards and downwards from the thoracic cavity, freeing them below by cutting the inferior vena cava as it comes through the diaphragm, and the phrenic nerves and vessels where they enter the superior surface of the diaphragm.

In Plate 197, section lines indicate where parts are to be cut at the base of the neck, previous to separation the œsophagus and pharynx with the collateral vessels and nerves, from the prevertebral muscles of the neck; by reflecting them in a direction forwards and upwards, *until* a saw can be passed between the pharynx and the prevertebral muscles, that the base of the cranium may be sawed upwards, thereby removing, in mass, the viscera of the neck, the face and anterior half of the cranium, from the prevertebral muscles of the neck and the posterior half of the cranium.

In Plates 153 and 154 the laminæ of the vertebræ and of the sacrum are to be sawn through on both sides of the median line and the included portions of the posterior wall of the spinal canal removed, in sections, to expose the spinal cord as displayed.

DISSECTING INSTRUMENTS.

Dissecting Instruments, Plate 1.—Experience has proved, that the following instruments, etc., will meet all the wants of the dissector, and in some respects, better than the contents of the conventional dissecting case; a pair of modified Coxeter forceps, two scalpels, a pair of sharp-pointed curved scissors, chain-hooks, thumb-pins, clasp-pins, two probes, and a scalpel strop. As contained in their case (1), they fulfil the ends of *compactness, usefulness* and *economy.*

Forceps.—The forceps (2) is the "Coxeter forceps," modified by having its head cut square across, so that, when the closed blades are held in the hand, it may be used to strip muscles apart, etc.

Scalpels.—Each scalpel has a thick handle; a portion of the blade forms a shank to the knife, which strengthens the instrument, facilitates its manipulation, and limits its cutting portion to its available edge. One of the scalpels (3) has a rounded end, and a very convex cutting edge, which continues to the end of the blade; this scalpel is adapted for coarse work, such as reflecting flaps of skin, etc., and clearing fasciæ and muscles. The pointed scalpel (4) has a very convex cutting edge, and is intended for delicate work on nerves and vessels. These two knives will fulfill all the requirements of dissection.

Curved Scissors.—The sharp-pointed curved scissors (5) presents nothing peculiar. As an instrument for the dissector, it should be much more used than it is; after a little practice, it can with great advantage, in many instances, be made to take the place of the scalpel. As a scissors, it answers all the needs of a straight pair.

Chain-Hooks.—The chain-hooks (6) have blunt points and the chain is very strong.

Thread.—The dissector should provide himself with *coarse linen thread*, and a needle for the same. Flaps may be advantageously reflected by threads tied into perforations at their borders. Loops of thread passed around vessels, nerves, etc.—with their ends tied—will be found useful in holding them off.

Probes.—These (7) are used to demonstrate ducts, vessels, sheaths of tendons, etc.

Scalpel Strop.—This (8) will be found very convenient; one's knife is continually dulled, while dissecting; but a few passes of the scalpel over the strop will sharpen it.

A saw, chisel, hammer and hook, costotome, intestine scissors, etc., required by the dissector, constitute a part of the furniture of a practical anatomy room.

GENERAL RULES FOR DISSECTION.

Division of a Cadaver.—A cadaver may be divided into *sections* to be worked by five, six, or eight dissectors. A body assigned to five, one takes the head and neck, two take the upper extremities and thorax, and two the lower extremities, the pelvis and the abdomen. A subject, apportioned to six, one works the head and neck, one the trunk (thorax, abdomen, and pelvis), two work the upper extremities, and two the lower extremities. A body, dissected by eight, there will be two to the head and neck, two to the trunk (thorax, abdomen and pelvis), two to the upper extremities and two to the lower extremities.

Object of Dissection.—*The object of dissection is to separate parts, not to cut them.* With the separation of parts is included the removal from their surfaces of fibrous tissue of investiture—as membranes, areolar tissue, fasciæ, inter-muscular septa and vessel-sheaths.

Rules for Dissection.—There are three rules to be followed to make a good dissector:

First.—*Know what you are to look for.* This knowledge is attained by having previously read a description of the parts to be found in a given region.

Second.—*Work slowly and thoroughly.* To fulfil this rule, do not allow yourself to work without system or method, but follow the progressive steps of a given dissection, as laid down in the book, which you have selected to guide you.

Third.—*Never let your knife cut when you do not know what it is about to divide.*

Review of a Dissection.—When finishing work replace parts, as nearly as possible in their normal relations, so that on resuming work they may be removed in their relative order; this affords repeated reviews of a dissection.

How to Keep a Dissection.—Re-apply skin flaps; *lay on the part the refuse tissue from the dissection;* cover with a *dampened* piece of muslin; and, outside of all, wrap a piece of dry muslin or oil-silk.

SPECIAL RULES OF DISSECTION.

Anatomical Elements.—By an anatomical element, borrowing the term from chemical nomenclature, is to be understood a structural part of the body, such as the skin, a muscle, etc. The anatomical elements entering into the construction of the regions of the body are: *epidermis, skin, subcutaneous tissue, superficial fascia, intermuscular septa, muscles, bursæ, synovial sheaths of tendons, deep fasciæ, arteries, veins, lymphatic vessels, lymphatic glands, nerves, viscera, ducts, mucous membranes, serous membranes, ligaments, fibro-cartilages, cartilages* and *bones.* All these elements are not present in every region of the body and some are found only in special regions.

Epidermis.—The epidermis or cuticle claims the respect of the dissector as a useful portion of the skin to him: where the epidermis is removed, the evaporation that takes place causes the skin and subjacent tissues to become hard, dry and matted together, so as to interfere materially with dissection.

Skin.—The skin or derma varies very much as to thickness, in the different regions. In making skin incisions, care should be taken, that the skin alone is cut through, as subcutaneous vessels, nerves and even the muscles may be divided, and mar, in consequence, subsequent dissection.

Dissection.—The position of the knife, in making an incision through the skin or any other membrane, should be vertical to the surface (Fig. 1, Plate 2); in this position, the knife should be steadied, by the little finger resting upon the surface, and driven by the index finger at its shank; it should be drawn, as thus held, from the initial point to the terminus of the incision, the point only of the knife cutting the tissue. In reflecting flaps of skin, or other membranes (fasciæ, etc.), they should be commenced (Fig. 2, Plate 2) by pinching up the initial flap with the forceps, and incising the subcutaneous tissue, so as to raise the skin alone; as soon as there is sufficient flap to enable it to be grasped, it should be held taut in the fingers (Fig. 3, Plate 2) at about two inches from its attached margin; the scalpel should be held lightly with its blade flat on the subcutaneous tissue, its cutting edge at a right angle to the skin; in this position the strokes should be made in long sweeps, never allowing the cutting edge to actually touch the skin. It is not a cutting that is effected, but a scratching with the edge of the knife, which parts the taut fibrous framework of the subcutaneous tissue, thus allowing the skin to be raised from the surface beneath. Never cut away a portion of reflected skin, as it is the best possible protective covering to dissected parts.

Subcutaneous Tissue.—The subcutaneous tissue is more or less loaded with fat and has embedded in it vessels and nerves. In some regions it can be split into two layers, which may be designated as the *superficial and the deep layer of the subcutaneous tissue.* This nomenclature avoids the confusion, which arises, if the subcutaneous tissue is called superficial fascia, and its layers superficial and deep fascia or superficial and deep layers of the superficial fascia.

Subcutaneous Veins.—The veins, found in the subcutaneous tissue, occupy a superficial plane, and are distinguished because of the dark color, imparted to them by blood clot within.

DISSECTION.—The subcutaneous veins should be raised free from the tissue in which they are embedded, so as to lie loosely thereon.

Subcutaneous Nerves.—The subcutaneous nerves lie in a plane beneath the veins, and, as a rule, contiguous to and parallel with them.

DISSECTION.—They may be found by scratching through the subcutaneous tissue, at a right angle to the course of the nerves (Fig. 4, Plate 2); thus the subcutaneous tissue is displaced and the resistant nerve-cord becomes apparent; once recognized, at a given point, the nerve may be raised with the forceps and stripped out from its bedding with the point of the scalpel (Fig. 5, Plate 2), or with scissors. Having recognized the subcutaneous veins and nerves, the subcutaneous tissue, as flap or otherwise, may be cut away from the area of the dissection region. The subcutaneous veins and nerves may be divided and reflected as may be directed.

Superficial Fascia, Fig. 1, Plate 3.—The superficial fascia of a region is a sheet of fibrous tissue, which covers the superficies of the muscles; it is continuous over the whole body, and here and there will be seen to form special thickenings, annular ligaments, etc., to subserve the office of bands of protection and inclusion, to bind down tendons or insure firm packing of subjacent parts.

DISSECTION.—Incisions of fascia should be made (the same as skin incisions) parallel with the fibres of the subjacent muscles. Flaps of fascia should be reflected, the same as the skin, the strokes of the blade of the scalpel should be parallel with the muscle fibres (Fig. 6, Plate 2). In reflecting the fascia from off a group of muscles, it will be noted that, at each intermuscular space, a continuity of the fascia with fibrous tissue in the intermuscular space exists; it is therefore necessary to cut through this fibrous tissue septum, in order to expose the adjacent muscle. In certain regions, the subjacent muscles are attached to the under surface of the fascia; at these areas, no attempts should be made to raise the fascia, but the same may be left upon the muscle, by cutting the fascia at the circumference of the attached portion. *Reflections of fascia should not be cut away.*

Intermuscular Septa, Figs. 2 and 3, Plate 3.—The intermuscular septa are fibrous tissue walls, recognized above, in continuity with the deep surface of the superficial fascia. They occupy the interspaces between muscles, forming compartments (Fig. 2) for their lodgement, and completely isolating each from the other. In Fig. 3, the intermuscular septa are shown in a transverse section of a limb.

Muscles.—The voluntary muscles invest the bony framework, being attached to the bones, at both ends, so as to produce movements at their articulations, or as in the case of the muscles of the face—attached to bone at one end and the skin at the other—to produce the facial expressions. Structurally, a muscle consists of a framework of fibrous tissue, which is continuous between its attachments; the tendon and the aponeurosis (flat tendon of a broad muscle) are respectively the fibrous framework of the muscle continued by itself; the fleshy portion of the muscle has, in addition, the muscle structure, lodged in the interstices of the fibrous framework. Every muscle is supplied with arteries, veins, lymphatics and nerves; the arteries and nerve or nerves are derived from contiguous trunks. The arterial and nerve supply should always be recognized and the same traced to where they enter the muscle; their entrance will usually be found at the protected surfaces of the muscle.

DISSECTION.—In cleaning a muscle, never grasp it with the forceps, but let the tissue, to be removed from the muscle, be held off; the scalpel should be guided parallel with the muscle fibres (Fig. 6, Plate 2); the handle of the scalpel and the head of the forceps are excellent instruments to strip the sides of a muscle free from contiguous parts; but, in so doing, care must be taken not to break off nerves and vessels at their points of penetration into the muscle. In unpacking or separating muscles, lying in different planes or in contiguity in the same plane, they cannot be regarded as cleaned, until all their surfaces are freed from fibrous tissue.

Bursæ, Fig. 1, Plate 4; Plate 60.—A bursa is a fibrous-tissue bag, containing fluid, which is lodged upon a bony prominence, upon which skin or muscle plays; its object is to obviate undue irritation of the skin or muscle by pressure. The subcutaneous bursæ over the patella (Plate 60), and the submuscular bursæ of the gluteus maximus and obturator internus (Fig. 1, Plate 4) are examples.

DISSECTION.—After recognition, a bursa may be cut away and its relations appreciated.

Sheaths of Tendons, Fig. 2, Plate 4.—Sheaths invest the long tendons of the limbs, for the isolation of the same, and the play of the tendons is facilitated by the sheaths being lined by serous membrane, which secretes synovia upon them to lubricate their surfaces. Examples of the synovial sheaths are well seen at the wrist and palm.

DISSECTION.—In the appreciation of the synovial sheaths of tendons, they should be opened at a given point and a probe inserted into them, along the tendon in both directions, to determine the extent of their investiture. After recognition, they should be stripped from the tendons with the curved scissors or scalpel.

Deep Fasciæ.—The deep fasciæ are specially thick septa or fasciæ, prolonged between the anatomical elements of a region, which form compartments for the grouping of muscles or the isolation of contiguous parts. They are also found lining the interior of cavities.

DISSECTION.—Their extent and points of attachment should be appreciated, after which, *if they interfere with subsequent dissection, they should be cut away.*

Arteries.—The arteries, when injected with substances such as plaster, wax, or rubber, are readily recognized, but when not so injected, they appear as flattened empty tubes. The smaller arteries are, as a rule, accompanied by two veins, called venæ comites, the larger ones by a single venous trunk.

DISSECTION.—The arterial trunks of a region should be first cleared of areolar tissue and sheath investitures; then their primary branches determined in the order of their size. All branches of distribution should be followed to the parts to which they distribute, and the anastomoses of branches should be recognized, where possible.

Veins.—The veins of the body are subcutaneous and comites. The subcutaneous veins are lodged, as their name indicates. The comites or deep veins accompany arteries; some of the deep veins are not comites. The venous channels (sinuses) within the cranium are not comites of arteries.

DISSECTION.—With a few exceptions, which will be noted, the subcutaneous veins do not require the dissector's attention; the venæ comites of the small arteries do not warrant preservation, as the recognition of the artery carries with it the appreciation of its companion veins; *the venæ comites of the small arteries may therefore be stripped away from them in the cleaning of the latter.* The large venous trunks run contiguous to the arteries, one to each, and their relations should be carefully noted; special directions will be given, when and how they are to be removed. *In dissecting arteries and veins the same rule holds as with muscles, viz.: that the vessels should not be raised by the forceps, but only the adventitious tissue and venæ comites (in the case of small arteries) are to be drawn away and cut from the surface of the vessel* (see the figure illustrating the cleaning of a muscle, Fig. 6, Plate 2), *the knife cutting parallel with the vessel, never across it.*

Nerves.—The nerve trunks of a region will be found to run parallel with its vessels. If not immediately contiguous they are not far removed. In appreciating a nerve it must be remembered that it is either motor, sensory or mixed (containing both motor and sensory fibres). If a nerve is motor, its distribution will be to muscle only; if sensory only, it ends at an organ of special sense or the skin; if a mixed nerve, it will be found to have a deep distribution of its motor filaments to the muscles and a superficial distribution of its sensory filaments to the skin. With the exception of a few nerves in the region of the head, which are either specially sensory or specially motor, the nerves of the body are mixed, having a deep muscle and a superficial skin distribution.

DISSECTION.—In exposing a nerve it should be first isolated for a short portion of its course, as with a subcutaneous nerve (see Fig. 4, Plate 2), then raise the nerve taut with the forceps (see Fig. 5, Plate 2), and with the sharp-pointed scalpel, held with its edge from you and cutting parallel with the nerve, strip out the nerve from the fibrous tissue, in which it is imbedded; as soon as enough of the nerve has been freed, it should be held in the fingers of the left hand and its stripping out continued. Observe the giving off of branches, those to the muscles, those communicating with contiguous nerves, and those to the skin, as the case may be. All the branches of a nerve should be traced to the parts, to which they distribute; the most delicate nerves, when once stripped out, will bear a great deal of handling, and the completeness of a dissection will more than reward the extra pains, taken to preserve them.

The dissector, in his general work, will be able to find, with but few exceptions, all the nerves shown in the plates. In a few of the plates, not original (Plate 33), the nerves were dissected out after the regional sections had been preserved, for a long time, in dilute nitric acid.

Lymphatic Glands.—Lymphatic glands are to be found in given situations in the body; when large they are evidence of a pathological condition, as in the perfect organism they may escape detection.

DISSECTION.—In the course of dissection it is sufficient to recognize the lymphatic glands as they may present, but it is needless to preserve them.

Lymphatic Vessels.—Lymphatic vessels are recognized only in special regions, viz.: the vicinity of the receptaculum chyli and along the left lymphatic or thoracic duct. They are, as a rule, so small elsewhere as to escape detection. Pictures illustrating them are drawn from specimens, where the lymphatic vessels have been injected by special methods necessary therefor.

Viscera.—The viscera of the body are certain organs, contained within the trunk and head; the direction for their dissection will be given in their proper places.

DISSECTION.—The anatomy of a viscus is complete in itself, therefore it may be removed from the body and kept in some preservative fluid—a solution of arsenite of soda or of chloride of zinc—for special dissection. Before removal of an organ from the body, its relations, the source of its arterial and nerve supply, the destination of its vein or veins, and its duct (if it has any) should be recognized, as far as practicable.

Ducts.—Ducts are the efferent canals from the secreting viscera, for the conveyance of the products of secretion · they are found only in the trunk and head regions of the body.

DISSECTION.—The relation of a duct, in situ, is to be first appreciated; then, its point of emergence from the viscus and its destination.

Mucous Membrane.—Mucous membrane lines all canals of the body with a surface outlet; at these points the membrane will be found continuous with the skin.

Serous Membrane.—A serous membrane invests all opposed surfaces, which move upon each other, the exception being the cartilage surfaces of joints; it covers all movable organs, and lines the interior of the cavities which contain them; it lines the ligaments of the movable joints, and forms the sheaths of tendons. It secretes a lubricating fluid to facilitate friction and prevent irritating effects therefrom.

DISSECTION.—Mucous and serous membranes should be dissected from subjacent tissue; a submucous and a subserous plane of areolar tissue may sometimes be demonstrated.

Bones.—The bones, which form the framework of a given dissection, should be considered, with reference: *first*, to the relations of their surfaces to the anatomical elements of the region; *second*, to the areas of attachment of muscles. The regional dissections have been planned so as to include all the attachments of a given muscle, thereby facilitating the appreciation of its function.

Dissection.—After a dissection has been completed, the muscles should be cut from the bones, one by one, noting the areas of their attachments. In so doing, not only the situation of a given muscle attachment is to be seen, but also its relations to contiguous areas of other muscle attachments.

Joints.—The joints of the body are the points where bone surfaces are in contiguity; they are complex in their construction, including: *bones*, *ligaments*, *cartilage*, *synovial* (serous) *membrane*, and *fibro-cartilage*. Joints are movable, partially movable, and immovable: a *movable joint* (elbow-joint) is constructed of bones covered with articular cartilage, and joined by ligaments lined by synovial membrane; if the joint is exposed to concussions (knee-joint), protection is afforded, from bone injury by the presence of movable plates of fibro-cartilage (interarticular), between the cartilage-covered articular surfaces of the bones. In other joints (hip-joint) the articular cavities are deepened by the rimming of their borders with immovable fibro-cartilage (circumferential). In the articulation of bones, which form partially movable joints (the vertebræ), cartilage and synovial membrane are wanting, and plates of immovable, fibro-cartilage (interosseous) are fixed between the articular surfaces (intervertebral discs); these latter points of bone articulation allow a slight motion of their surfaces upon each other. In the *immovable joints* (cranial and upper-jaw regions of the head) the articular surfaces of the bones are held in apposition: for the cranium, by the continuity of the exterior periosteum and the interior dura mater; for the upper-jaw region, by the continuity of the periosteum; ultimately, in both regions, the articulations between the bones are obliterated by the development of osseous tissue.

Dissection.—In the dissection of joints it is absolutely necessary that they should be moist and pliable; to keep them so, they should be kept covered by refuse tissue, with wrappings of wet muslin, and outside dry muslin or oil-silk. If they have dried from neglect, they should be soaked in water till the ligaments are again pliable. The cleaning of the surface of ligaments may be effected with curved scissors and subsequent scraping with a scalpel.

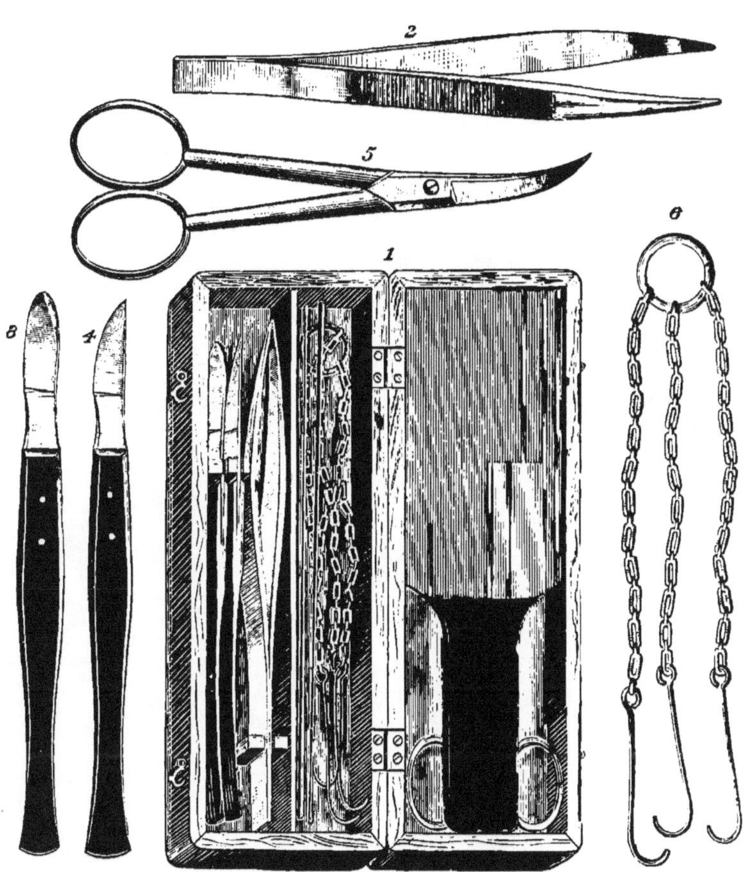
2
5
1
6
3
4

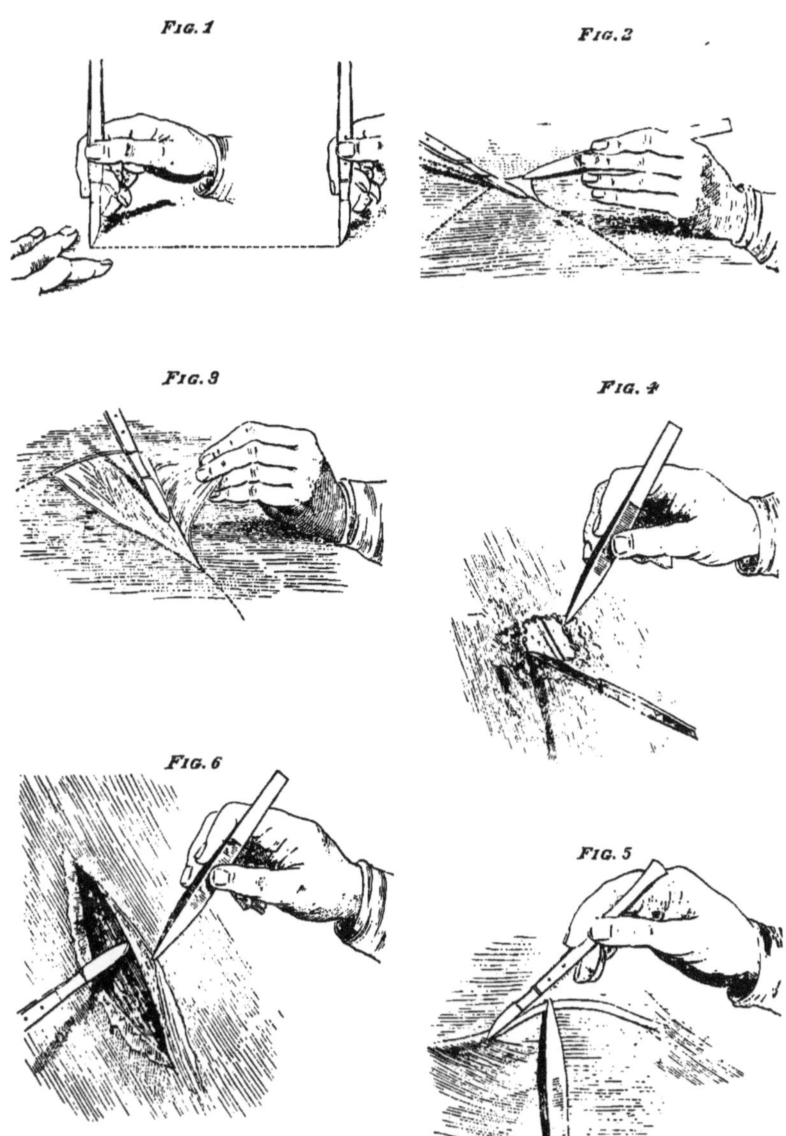
Fig. 1
Fig. 2
Fig. 3
Fig. 4
Fig. 6
Fig. 5
M. Cohn, ad naturam del.

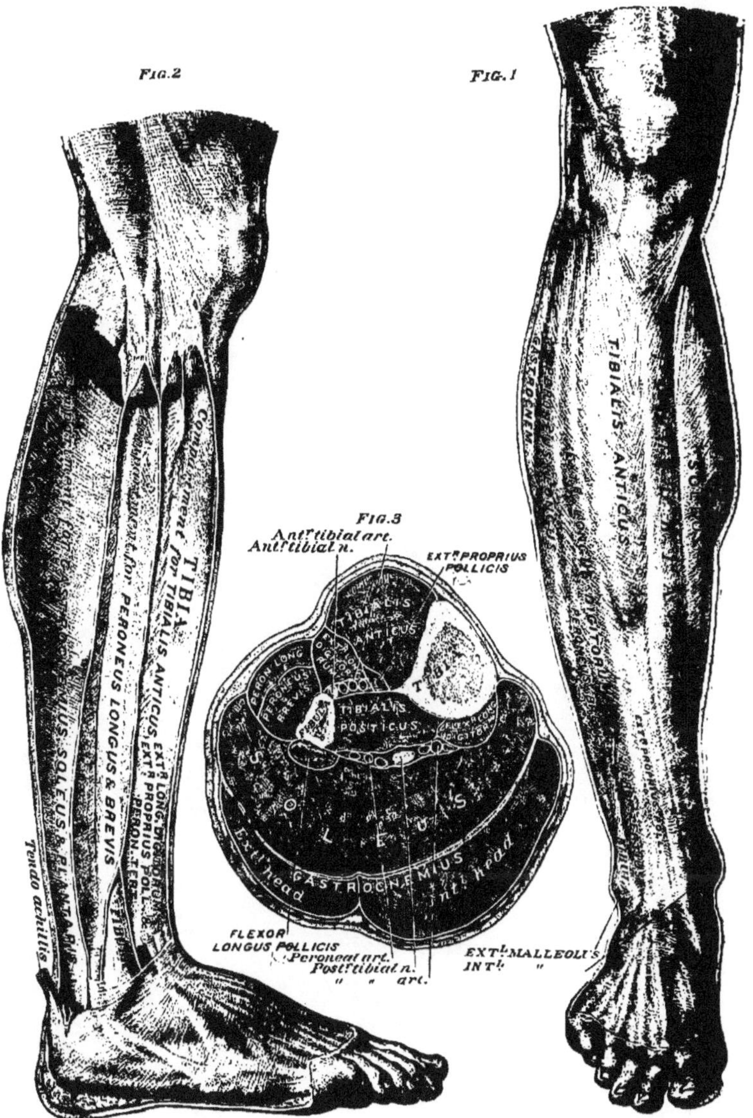
FIG. 2
FIG. 1
FIG. 3
Antʳ tibial art.
Antʳ tibial n.
EXTʳ PROPRIUS POLLICIS
TIBIALIS ANTICUS
PERON. LONG
PERONEUS BREVIS
TIBIALIS POSTICUS
SOLEUS
GASTROCNEMIUS
Extˡ head
intˡ head
FLEXOR LONGUS POLLICIS
Peroneal art.
Postʳ tibial n.
" " art.
EXTˡ MALLEOLUS
INTˡ "
TIBIA
Complement for TIBIALIS ANTICUS
Complement for PERONEUS LONGUS & BREVIS
EXTʳ LONG. DIGITORUM
EXTʳ PROPRIUS POLL.
SOLEUS & PLANTARIS
Tendo achillis
TIBIALIS ANTICUS
GASTROCNEMIUS
SOLEUS

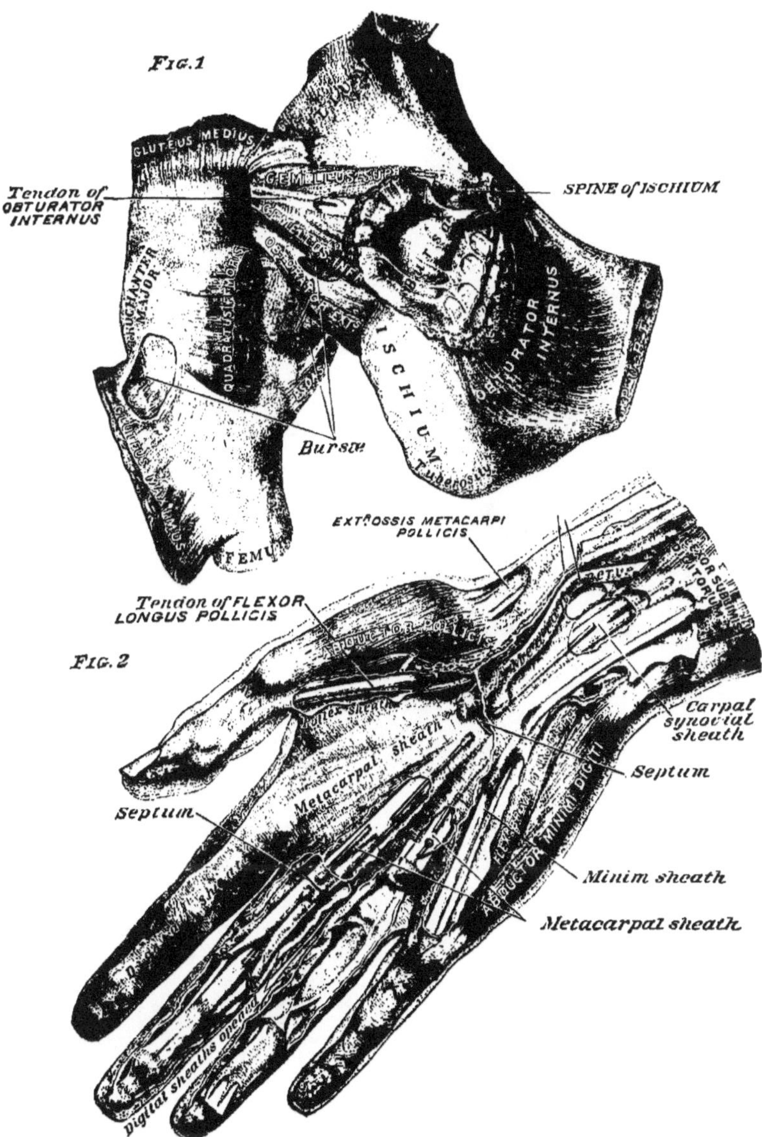
FIG. 1
GLUTEUS MEDIUS
GEMELLUS SUP.
Tendon of OBTURATOR INTERNUS
SPINE of ISCHIUM
TROCHANTER MAJOR
QUADRATUS FEMORIS
OBTURATOR INTERNUS
ISCHIUM
Tuberosity
Bursæ
FEMUR
EXT. OSSIS METACARPI POLLICIS
Tendon of FLEXOR LONGUS POLLICIS
FIG. 2
Carpal synovial sheath
Septum
Metacarpal sheath
Septum
ABDUCTOR MINIMI DIGITI
Minim sheath
Metacarpal sheath
Digital sheaths opened

PLATE 5

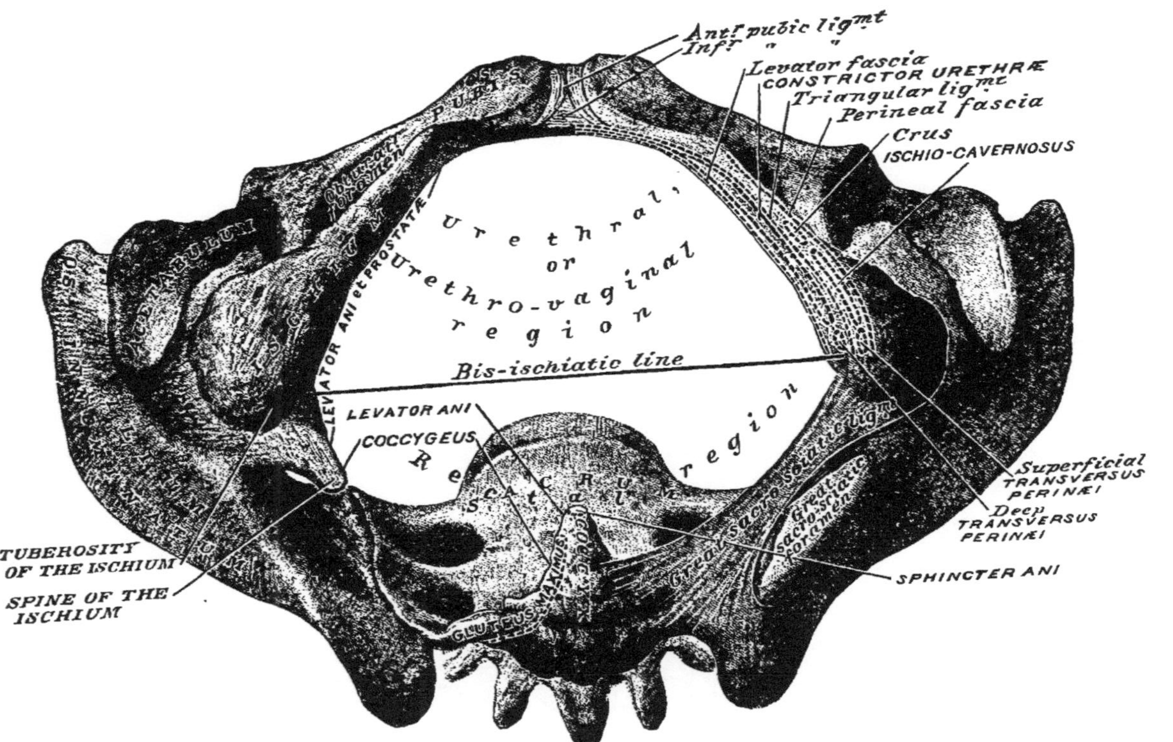

Antr pubic ligmt
Infr " "
Levator fascia
CONSTRICTOR URETHRÆ
Triangular ligmt
Perineal fascia
Crus
ISCHIO-CAVERNOSUS
Urethral, or Urethro-vaginal region
Bis-ischiatic line
Rectal region
LEVATOR ANI et PROSTATÆ
LEVATOR ANI
COCCYGEUS
ACETABULUM
Obturator foramen
SACRUM
COCCYX
GLUTEUS MAXIMUS
Great sacro-sciatic ligt
Great sacro-sciatic foramen
Superficial TRANSVERSUS PERINÆI
Deep TRANSVERSUS PERINÆI
SPHINCTER ANI
TUBEROSITY OF THE ISCHIUM
SPINE OF THE ISCHIUM

M. Cohn, ad naturam del.

PLATE 6

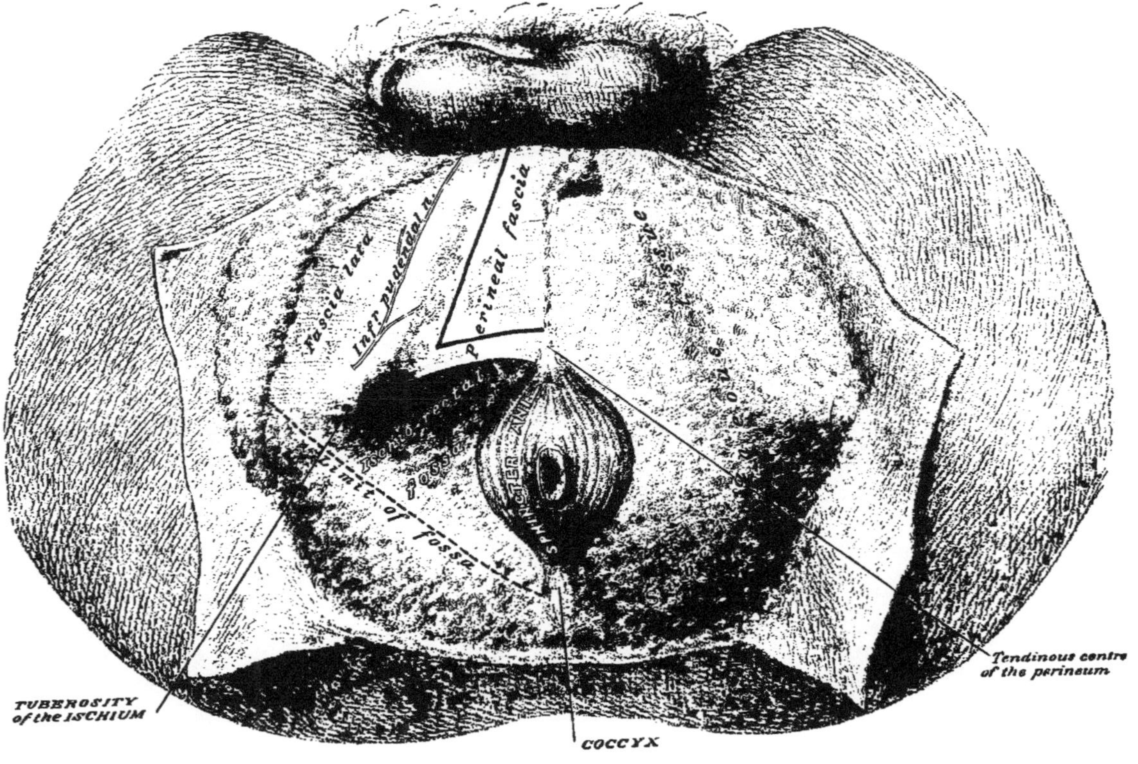

Fascia lata
Infr pudendal n.
Perineal fascia
Ischio-rectal fossa
Limit of fossa
SPHINCTER ANI
TUBEROSITY of the ISCHIUM
COCCYX
Tendinous centre of the perineum

PLATE 7

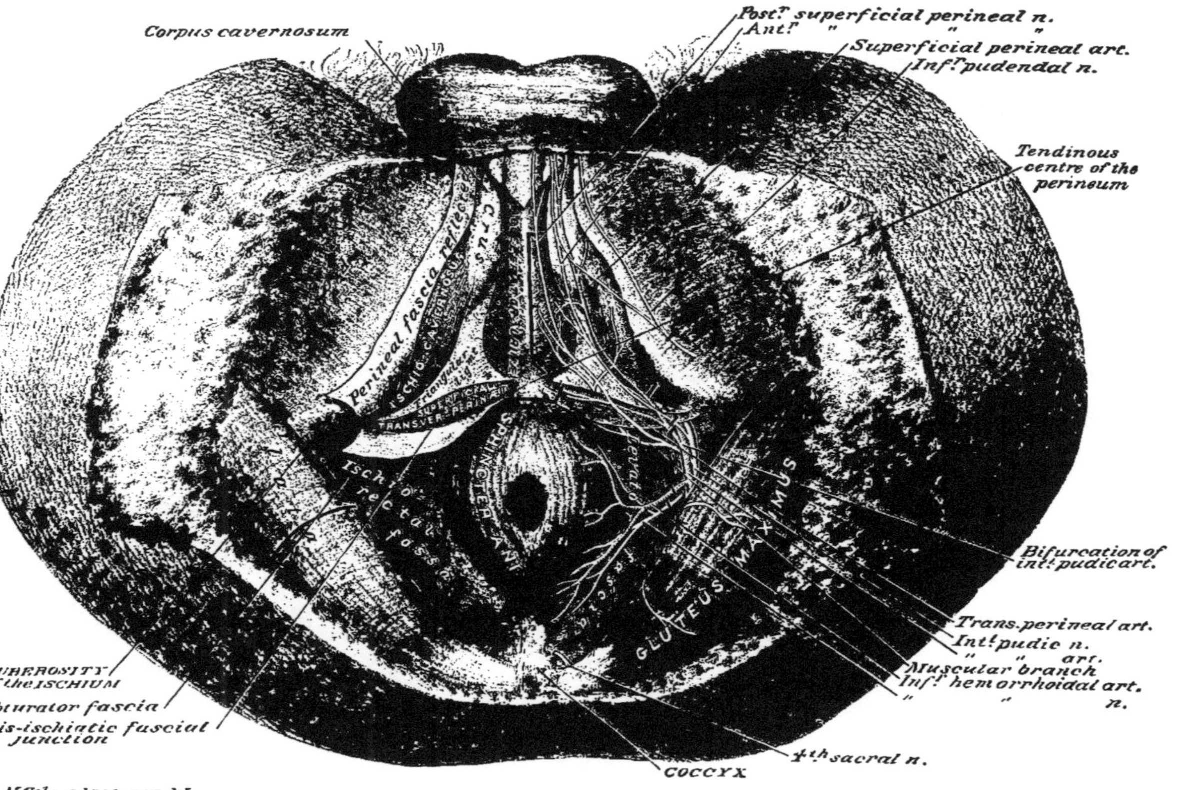
Corpus cavernosum
Post.r superficial perineal n.
Ant.r " " "
Superficial perineal art.
Inf.r pudendal n.
Tendinous centre of the perineum
Crus
Perineal fascia reflected
ISCHIO-CAVERNOSUS
Triangular lig.t
SUPERFICIAL TRANSVERSE PERINEI
SPHINCTER ANI
Levator
Ischio rectal fossa
fascia
lata
GLUTEUS MAXIMUS
Bifurcation of int.l pudic art.
Trans. perineal art.
Int.l pudic n.
" " art.
Muscular branch
Inf.r hemorrhoidal art.
" " n.
4th sacral n.
COCCYX
TUBEROSITY of the ISCHIUM
Obturator fascia
Bis-ischiatic fascial junction
M. Cohn, ad naturam del.

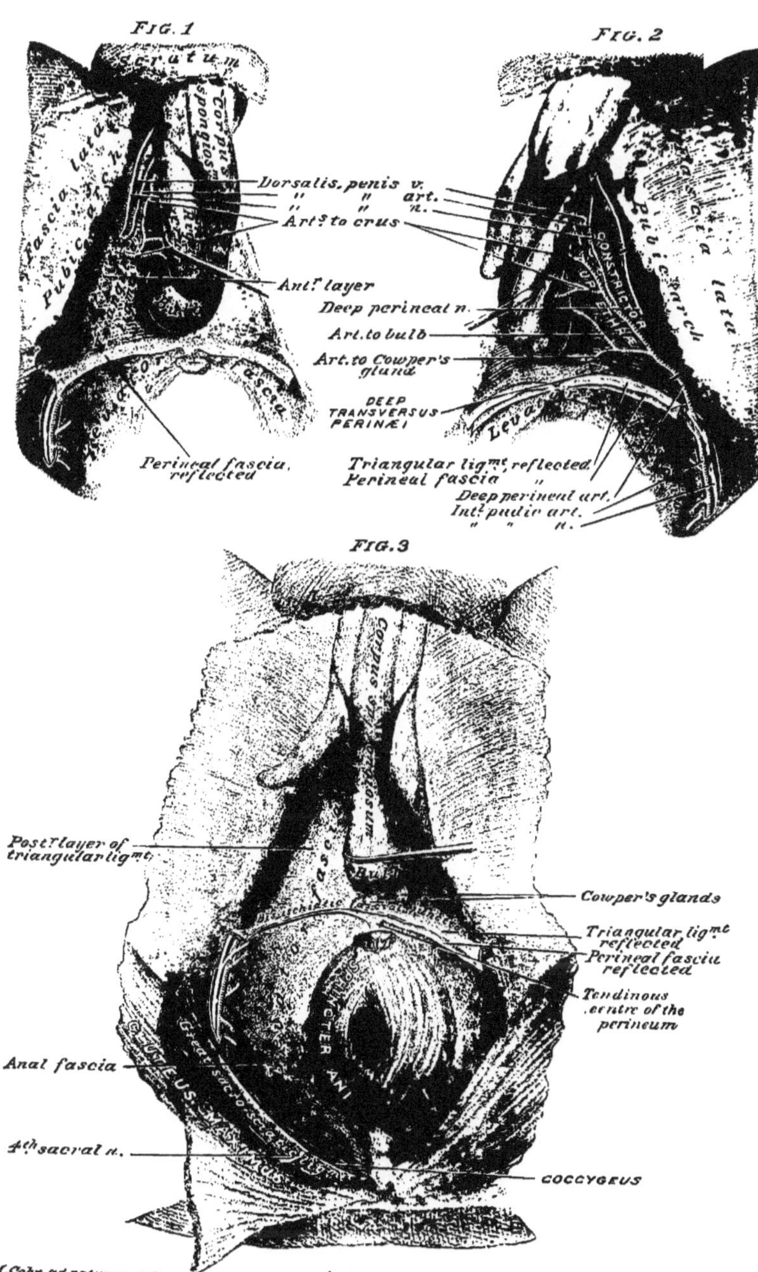
FIG. 1
Scrotum
Corpus spongiosm
Fascia lata
Pubic arch
Dorsalis, penis v.
" " art.
" " n.
Artˢ. to crus
Antʳ. layer
Perineal fascia, reflected
FIG. 2
Fascia lata
Pubic arch
CONSTRICTOR URETHRÆ
Deep perineal n.
Art. to bulb
Art. to Cowper's gland
DEEP TRANSVERSUS PERINÆI
Levator
Triangular ligᵐᵗ. reflected
Perineal fascia "
Deep perineal art.
Intˡ. pudic art.
" " n.
FIG. 3
Corpus spongiosum
Postʳ. layer of triangular ligᵐᵗ.
Cowper's glands
Triangular ligᵐᵗ. reflected
Perineal fascia reflected
Tendinous centre of the perineum
SPHINCTER ANI
Anal fascia
GLUTEUS MAXIMUS
Great sacro-sciatic lig.
4ᵗʰ sacral n.
COCCYGEUS
M. Cohn, ad naturam del.

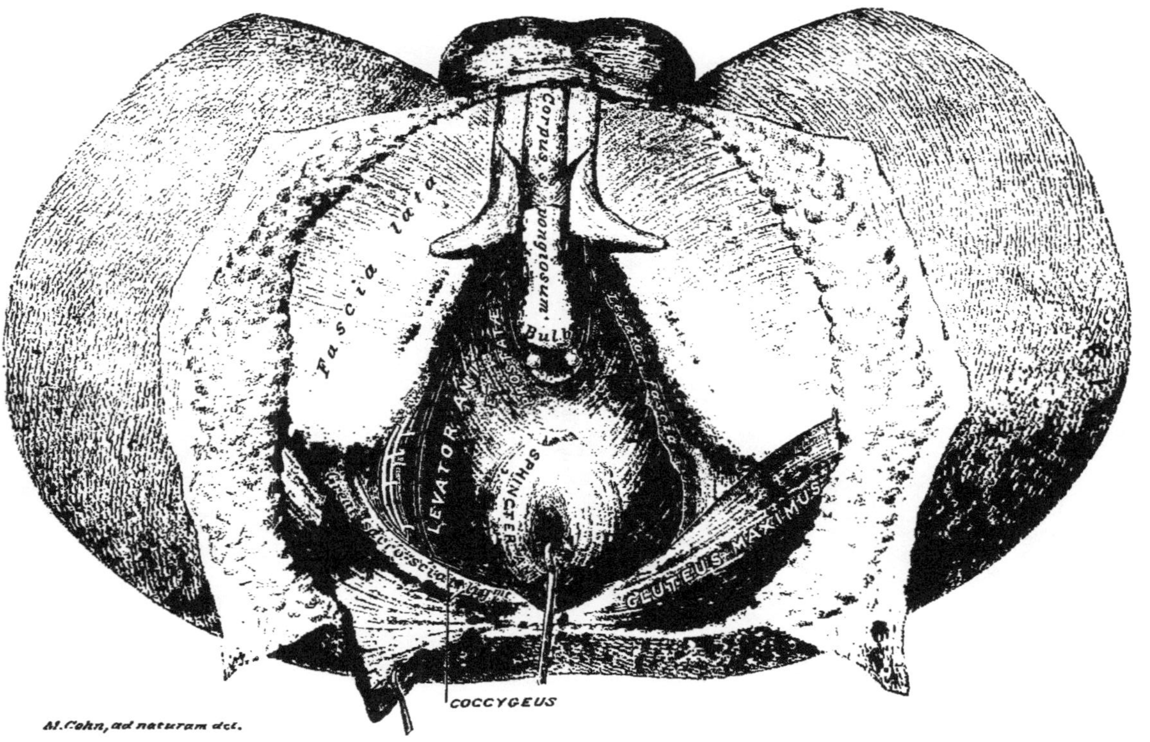
Corpus spongiosum
Bulb
Fascia lata
LEVATOR
SPHINCTER
GLUTEUS MAXIMUS
COCCYGEUS
M. Cohn, ad naturam del.

PLATE 10

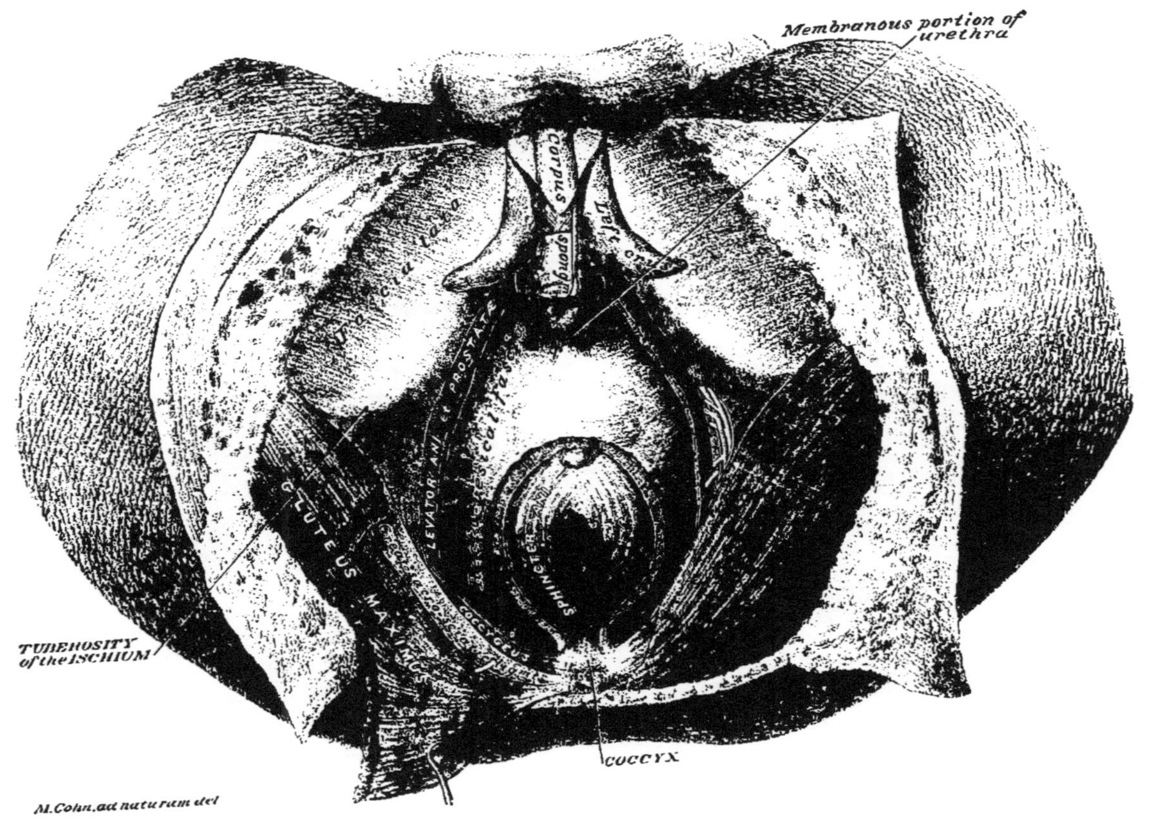

Membranous portion of urethra
Corpus
spong
SPHINCTER
LEVATOR ANI, et PROSTATE
COCCYGEUS
GLUTEUS MAXIMUS
TUBEROSITY of the ISCHIUM
COCCYX
M. Cohn, ad naturam del

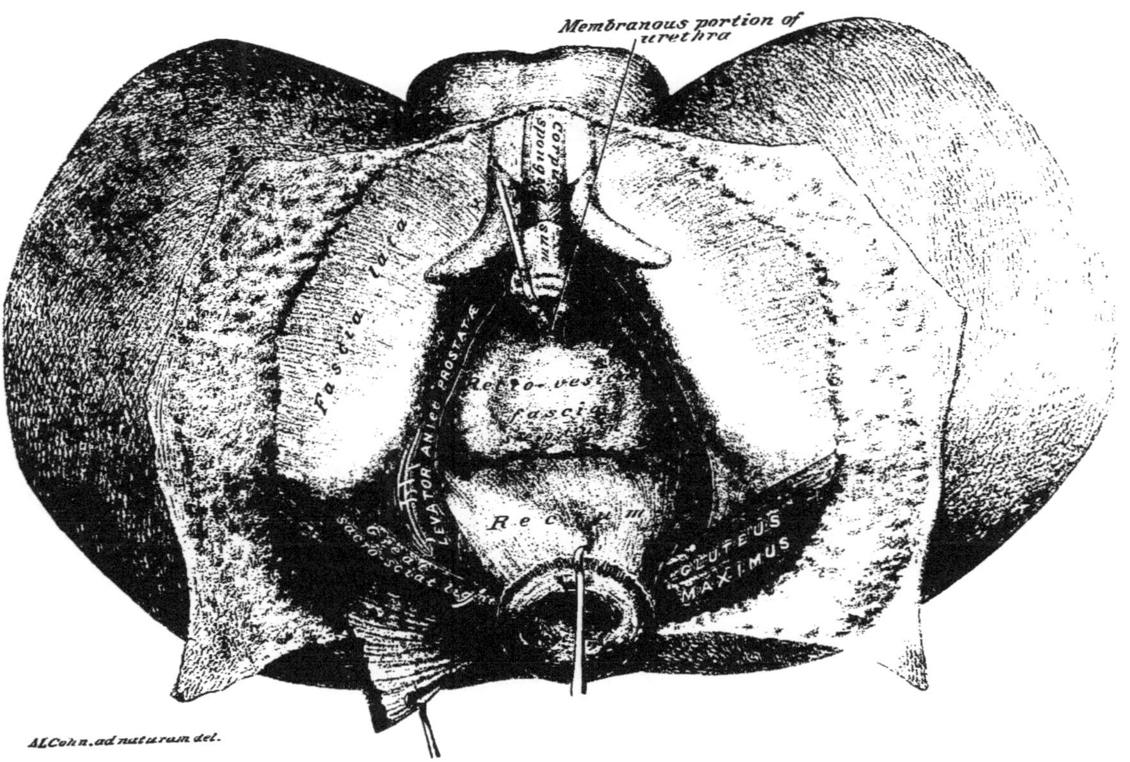
Membranous portion of urethra
Corpu
spongio sum
Fascia lata
LEVATOR ANI et PROSTATÆ
Recto-vesic
fascia
Rec m
Great sacro sciat ligt
GLUTEUS MAXIMUS
A.L.Cohn. ad naturam del.

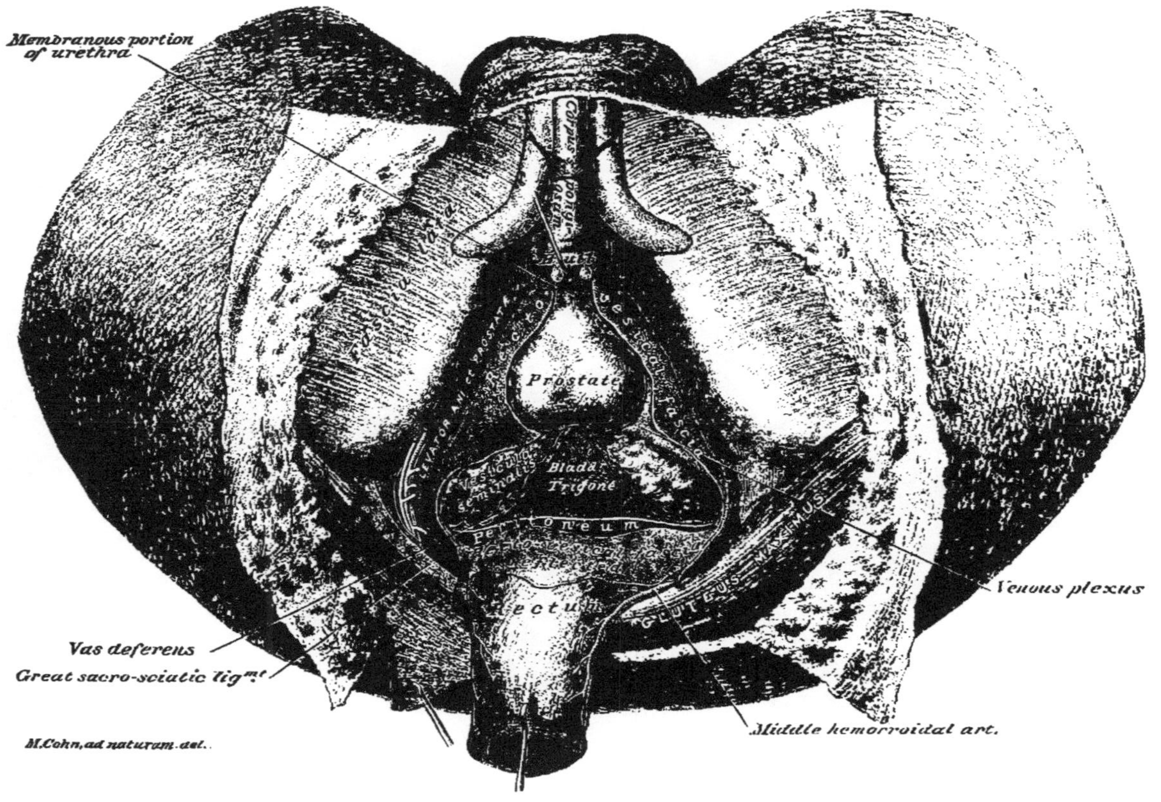
Membranous portion of urethra
Venous plexus
Vas deferens
Great sacro-sciatic lig^mt
Middle hemorrhoidal art.
Corpus spongiosum
Prostate
Trigone
M. Cohn, ad naturam del.

PLATE 13

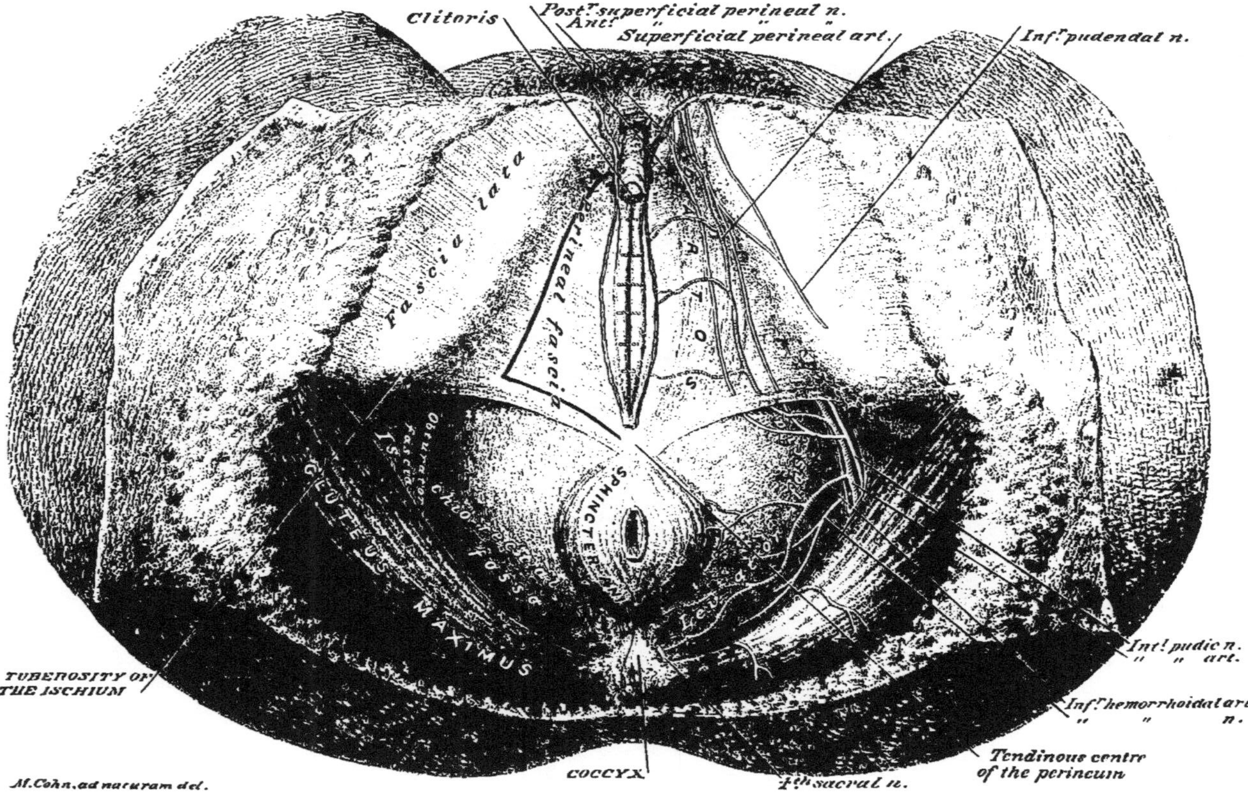
Clitoris
Post.r superficial perineal n.
Ant.r " "
Superficial perineal art.
Inf.r pudendal n.
fascia lata
perineal fascia
DARTOS
SPHINCTER ANI
Obturator fascia
Ischiorectal fossa
GLUTEUS MAXIMUS
TUBEROSITY OF THE ISCHIUM
Int.l pudic n.
" " art.
Inf.r hemorrhoidal art.
" " n.
Tendinous centre of the perineum
4th sacral n.
COCCYX
M. Cohn, ad naturam del.

PLATE 14

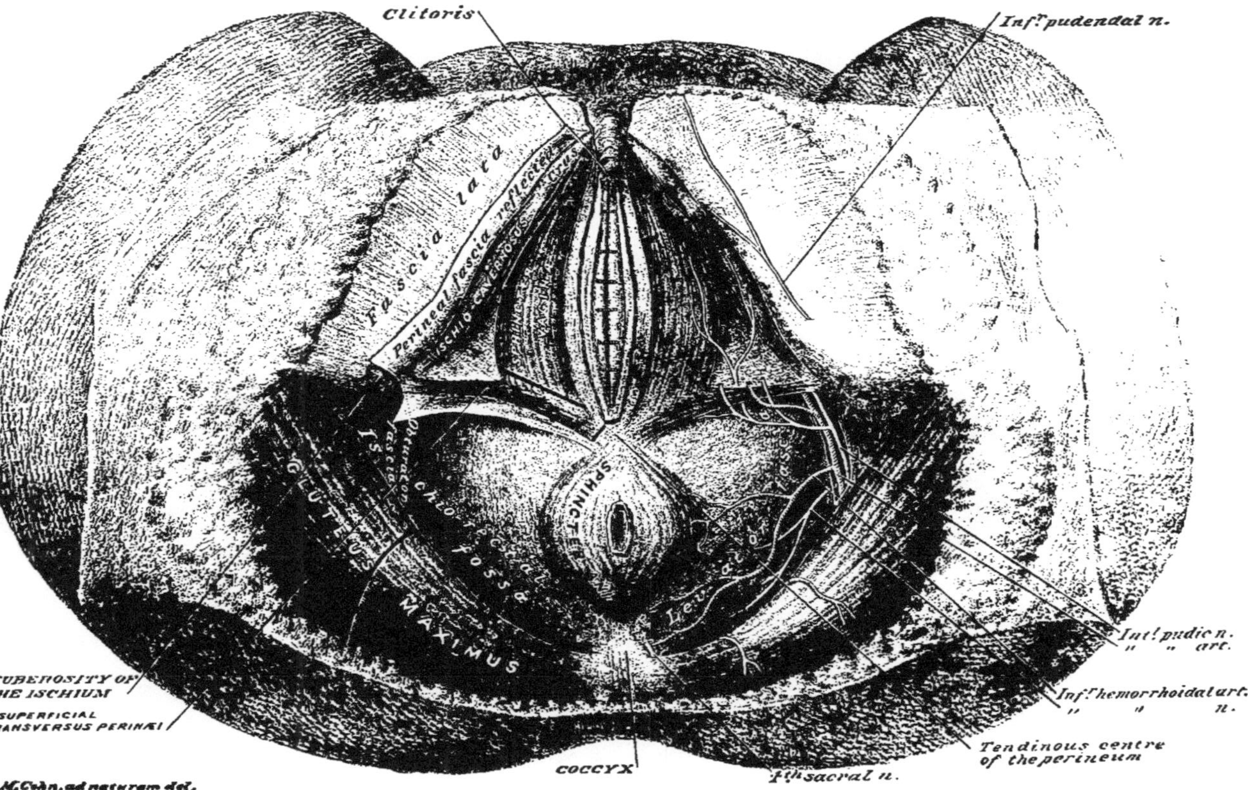
Clitoris
Infr. pudendal n.
Fascia lata
Perineal fascia reflected
Ischio cavernosus
Obturator fascia
Ischio rectal fossa
SPHINCTER
Levator ani
GLUTEUS MAXIMUS
Infr. pudic n.
" " art.
Infr. hemorrhoidal art.
" " n.
Tendinous centre of the perineum
5th sacral n.
COCCYX
TUBEROSITY OF THE ISCHIUM
SUPERFICIAL TRANSVERSUS PERINÆI
M. Cohn, ad naturam del.

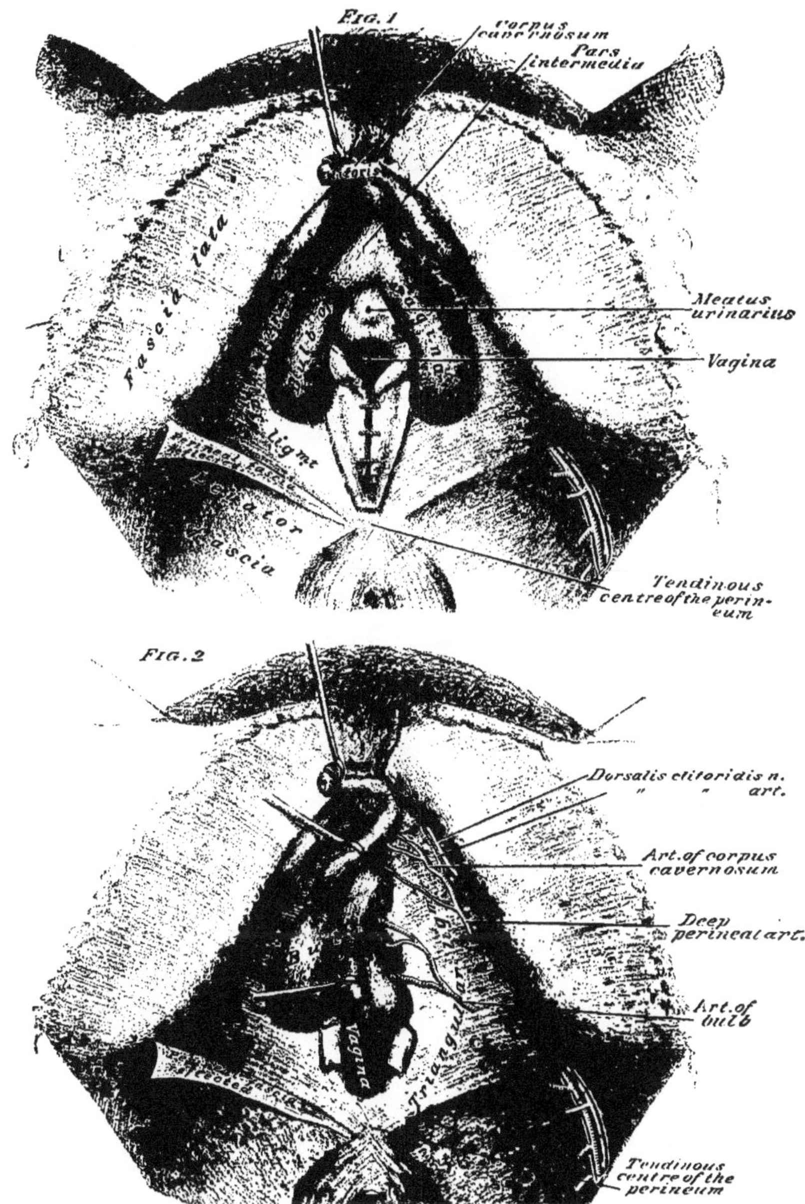
FIG. 1
Corpus cavernosum
Pars intermedia
Meatus urinarius
Vagina
Tendinous centre of the perineum
Fascia lata
Levator fascia
FIG. 2
Dorsalis clitoridis n.
" " art.
Art. of corpus cavernosum
Deep perineal art.
Art. of bulb
Vagina
Tendinous centre of the perineum

Cohn, ad naturam del.

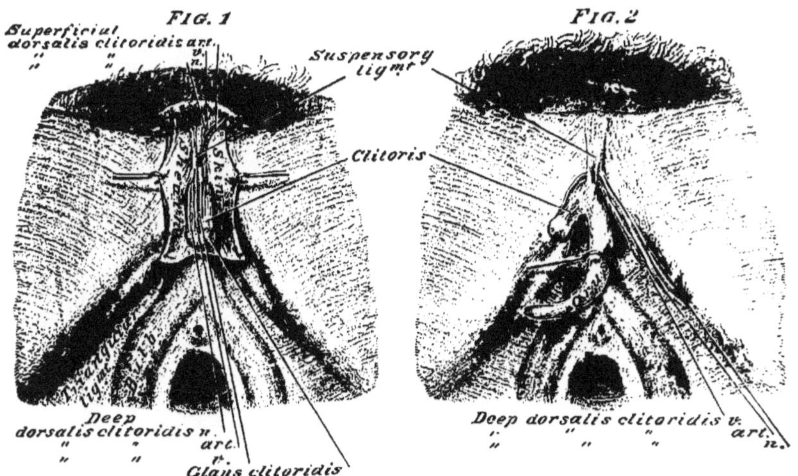
FIG. 1
Superficial dorsalis clitoridis art.
" " v.
" " n.
Suspensory ligmt
Skin
Clitoris
Triangular ligmt
Bulb
Deep dorsalis clitoridis n.
" " art.
" " v.
Glans clitoridis
FIG. 2
Deep dorsalis clitoridis v.
" " " art.
" " " n.

FIG. 3

Deep dors. clitoridis n.
" " art.
" " v.
Fascia lata
Meatus urinarius
Constrictor vaginæ
Vagina
Art. of bulb
Probe
Art. to gland
Vulvo-vaginal gland, and duct
Levator fascia
Tendinous centre of the perineum
DEEP TRANSVERSUS PERINÆI
Triangular ligmt, reflected
Perineal fascia reflected

M. Cohn, ad naturam del.

PLATE 17

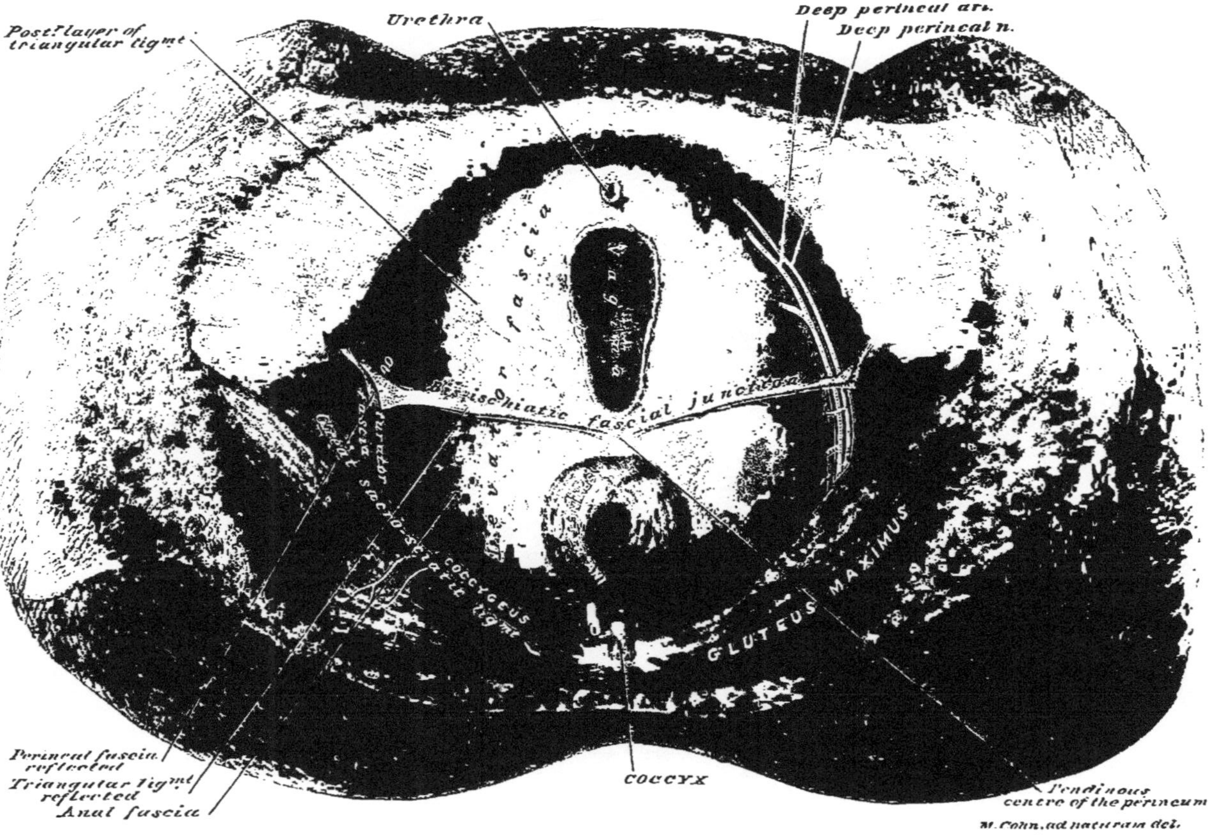

Post. layer of triangular ligt.
Urethra
Deep perineal art.
Deep perineal n.
Vagina
fascia
fascial junction
Obturator fascia
COCCYGEUS
GLUTEUS MAXIMUS
Perineal fascia reflected
Triangular ligt. reflected
Anal fascia
COCCYX
Tendinous centre of the perineum
M. Cohn, ad naturam del.

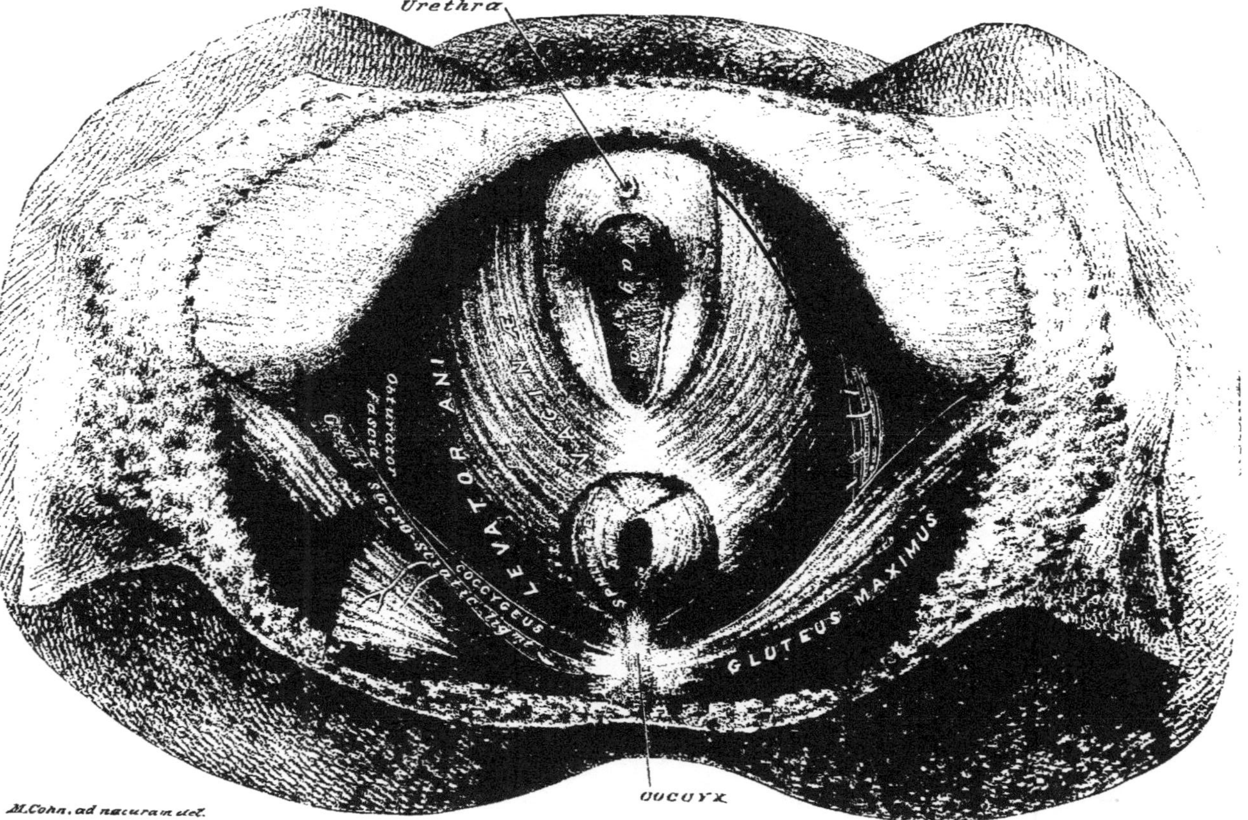
Urethra
Vagina
VAGINAE
LEVATOR ANI
Obturator Fascia
Great sacro-sciatic ligt
COCCYGEUS
SPHINCTER ANI
GLUTEUS MAXIMUS
Coccyx
M.Cohn. ad naturam del.

PLATE 19

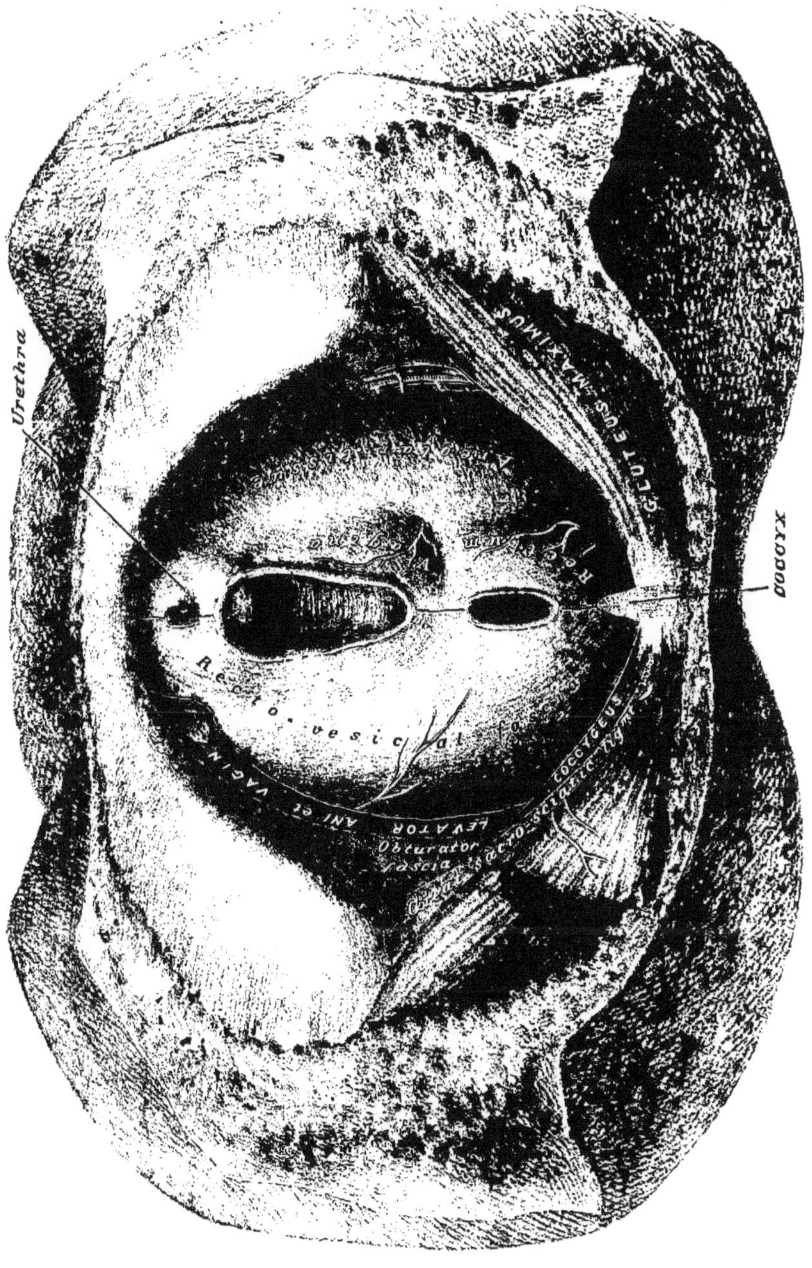
Urethra
GLUTEUS MAXIMUS
COCCYX
LEVATOR ANI et
COCCYGEUS
Obturator

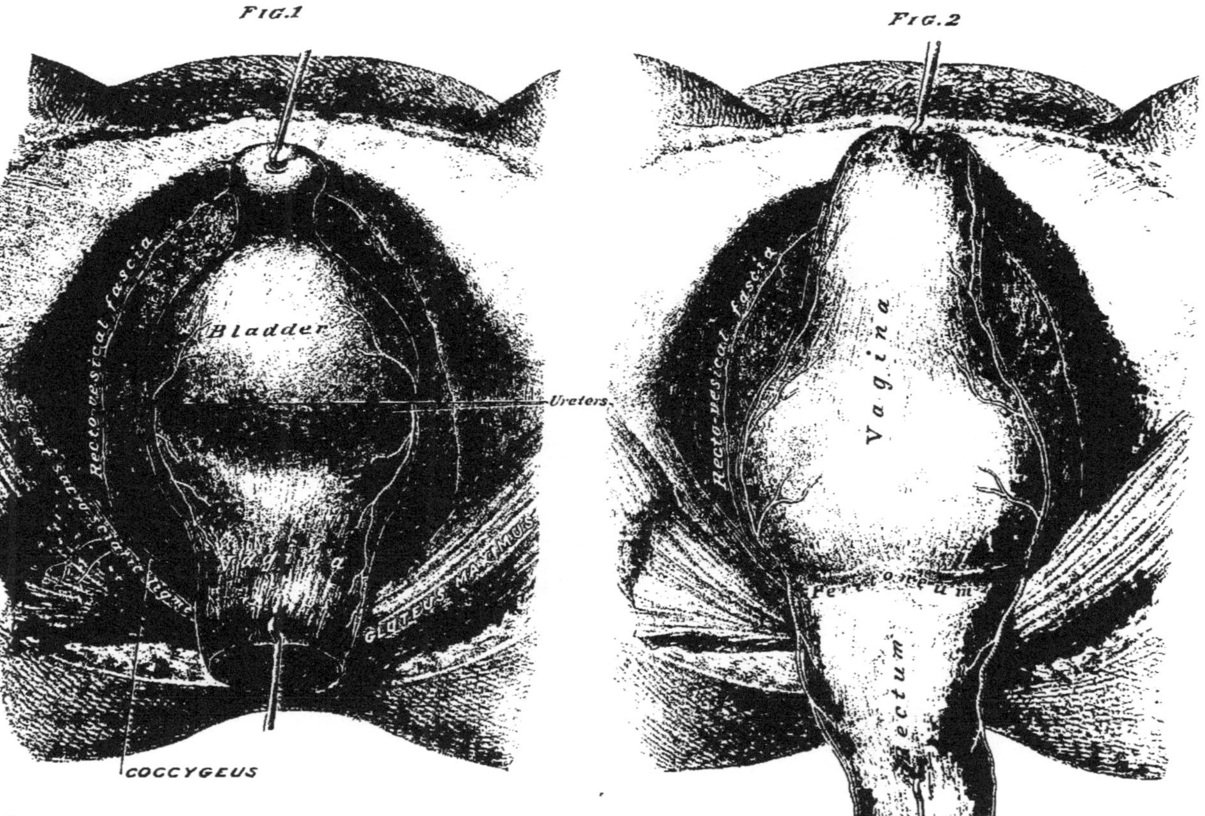
FIG.1
Recto-vesical fascia
Bladder
Ureters
GLUTEUS MAXIMUS
COCCYGEUS
FIG.2
Recto-vesical fascia
Vagina

M. Cohn, ad naturam del.

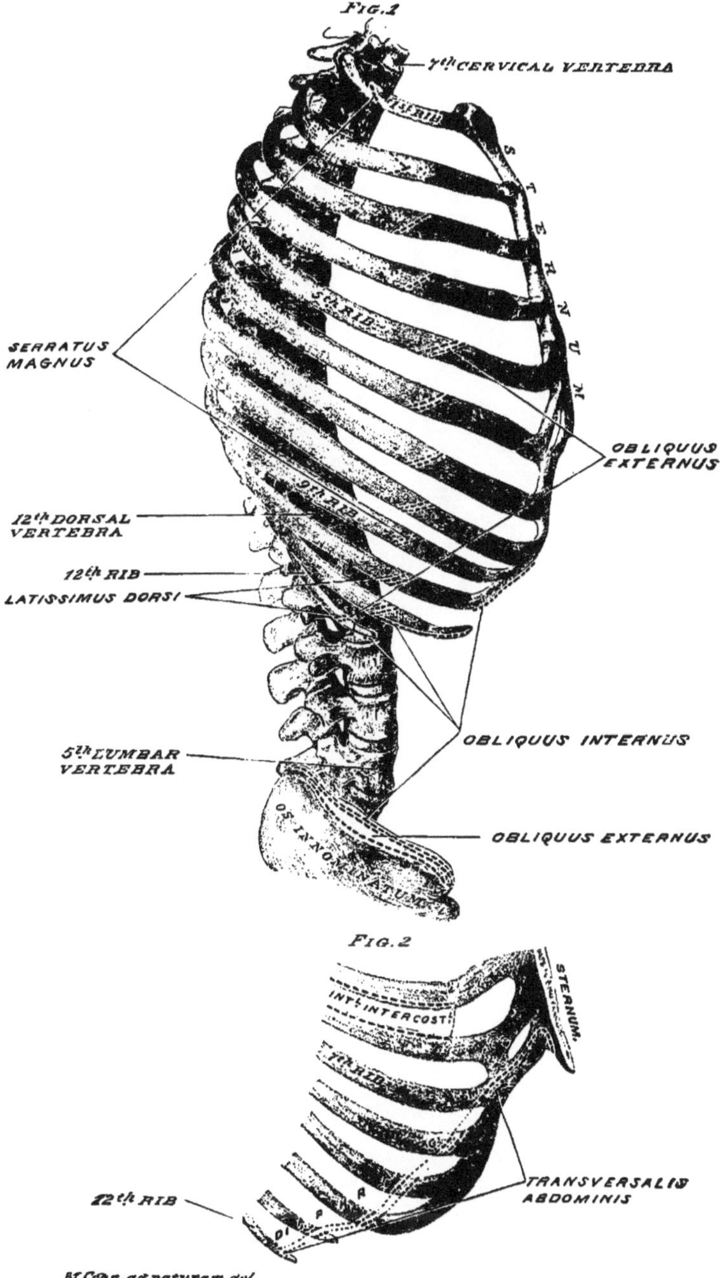
FIG. 1
7th CERVICAL VERTEBRA
1st RIB
STERNUM
SERRATUS MAGNUS
5th RIB
OBLIQUUS EXTERNUS
12th DORSAL VERTEBRA
9th RIB
12th RIB
LATISSIMUS DORSI
OBLIQUUS INTERNUS
5th LUMBAR VERTEBRA
OS INNOMINATUM
OBLIQUUS EXTERNUS
FIG. 2
STERNUM.
INT'L INTERCOST.
7th RIB
TRANSVERSALIS ABDOMINIS
12th RIB
DI
A
R

M. Cohn, ad naturam del.

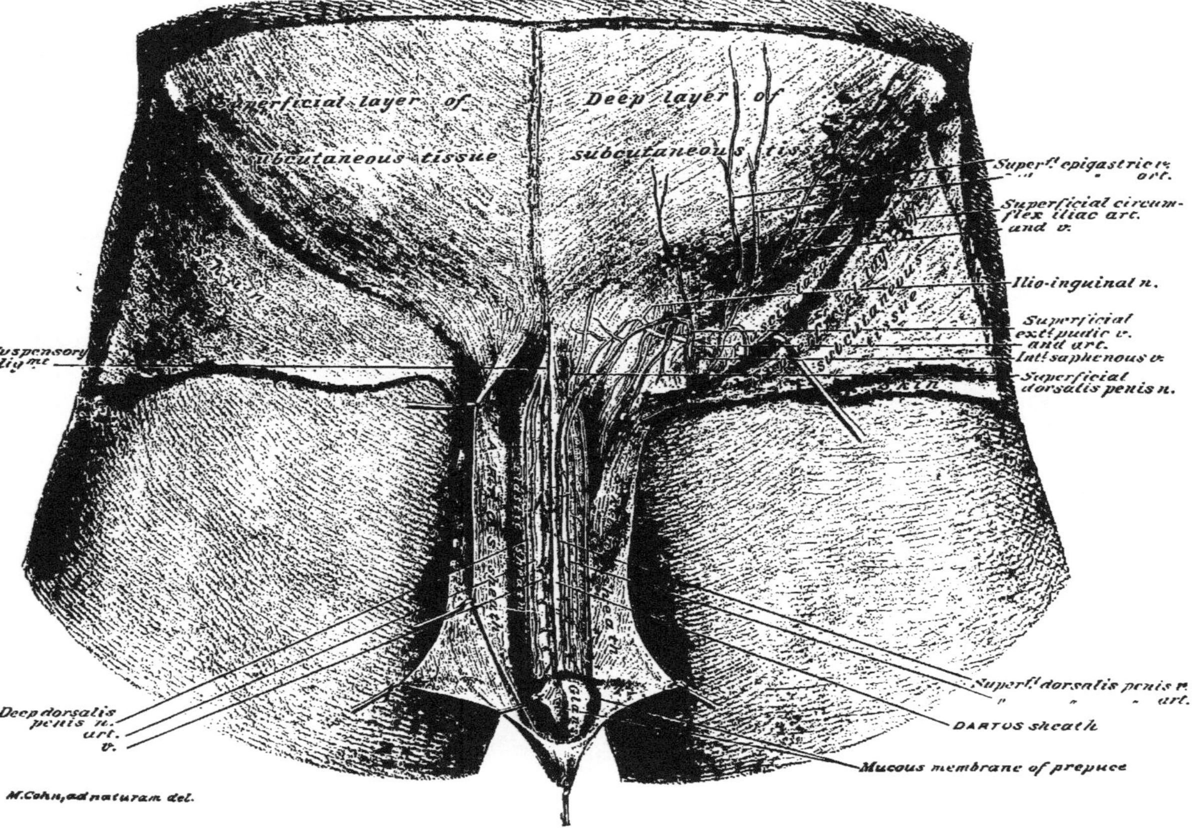
Superficial layer of
subcutaneous tissue
Deep layer of
subcutaneous tissue
Skin
Suspensory ligmt
Superfl epigastric v.
" " art.
Superficial circumflex iliac art. and v.
Ilio-inguinal n.
Superficial extl pudic v. and art.
Intl saphenous v.
Superficial dorsalis penis n.
Elastic sheath
Glans
Superfl dorsalis penis n.
" " " art.
DARTOS sheath
Mucous membrane of prepuce
Deep dorsalis penis n.
art.
v.
M.Cohn, ad naturam del.

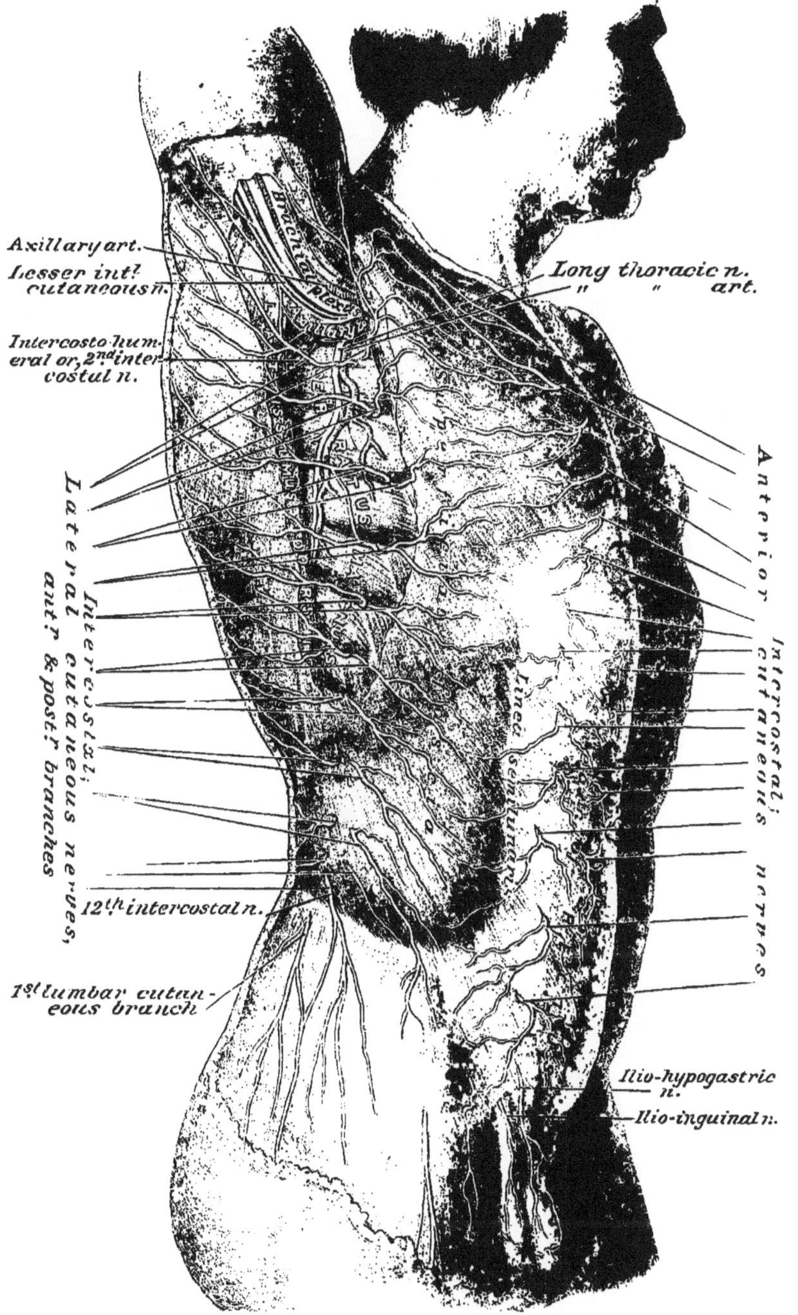
Axillary art.
Lesser intl. cutaneous n.
Intercosto-humeral or, 2nd intercostal n.
Long thoracic n.
" " art.
Lateral cutaneous nerves, ant.r & post.r branches
Intercostal;
Anterior intercostal; cutaneous nerves
Linea semilunaris
12th intercostal n.
1st lumbar cutaneous branch
Ilio-hypogastric n.
Ilio-inguinal n.

PLATE 24

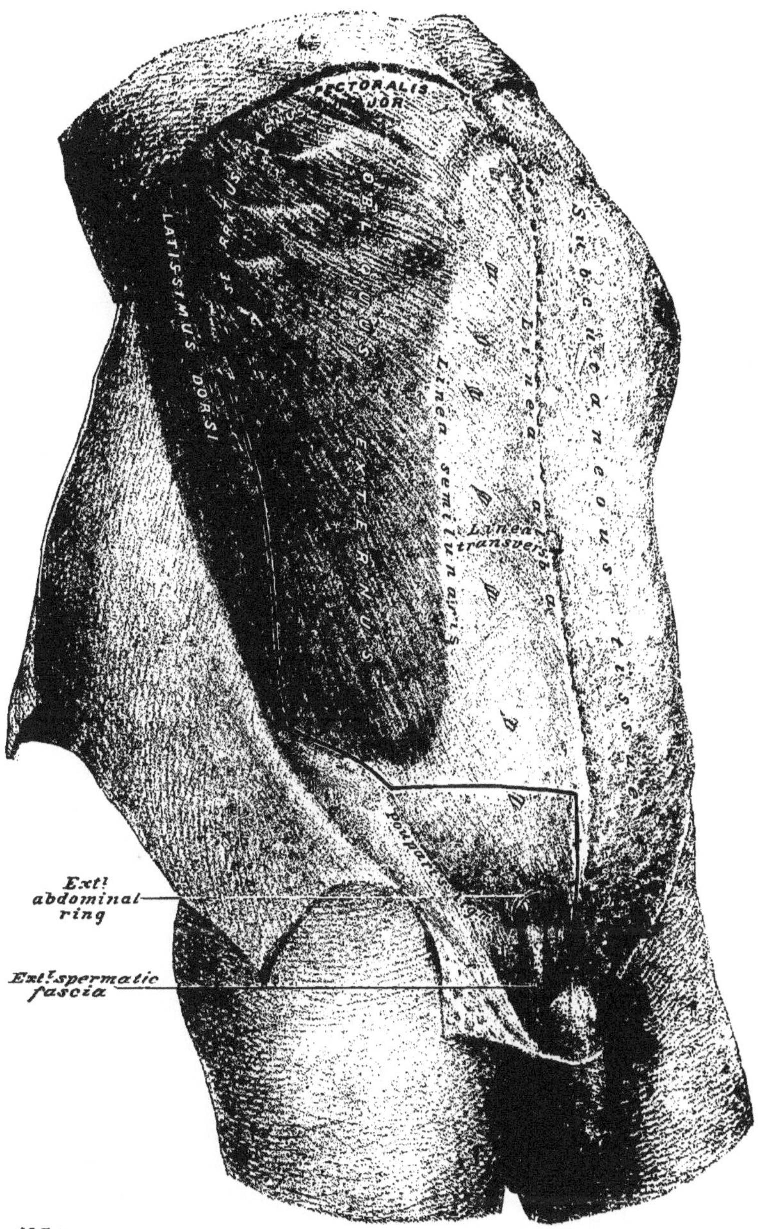

PECTORALIS
JOR
LATISSIMUS DORSI
OBLIQUUS EXTERNUS
Linea semilunaris
Linea alba
Subcutaneous tissue
Linea transversa
Ext! abdominal ring
Ext! spermatic fascia

M Cohn, ad naturam del.

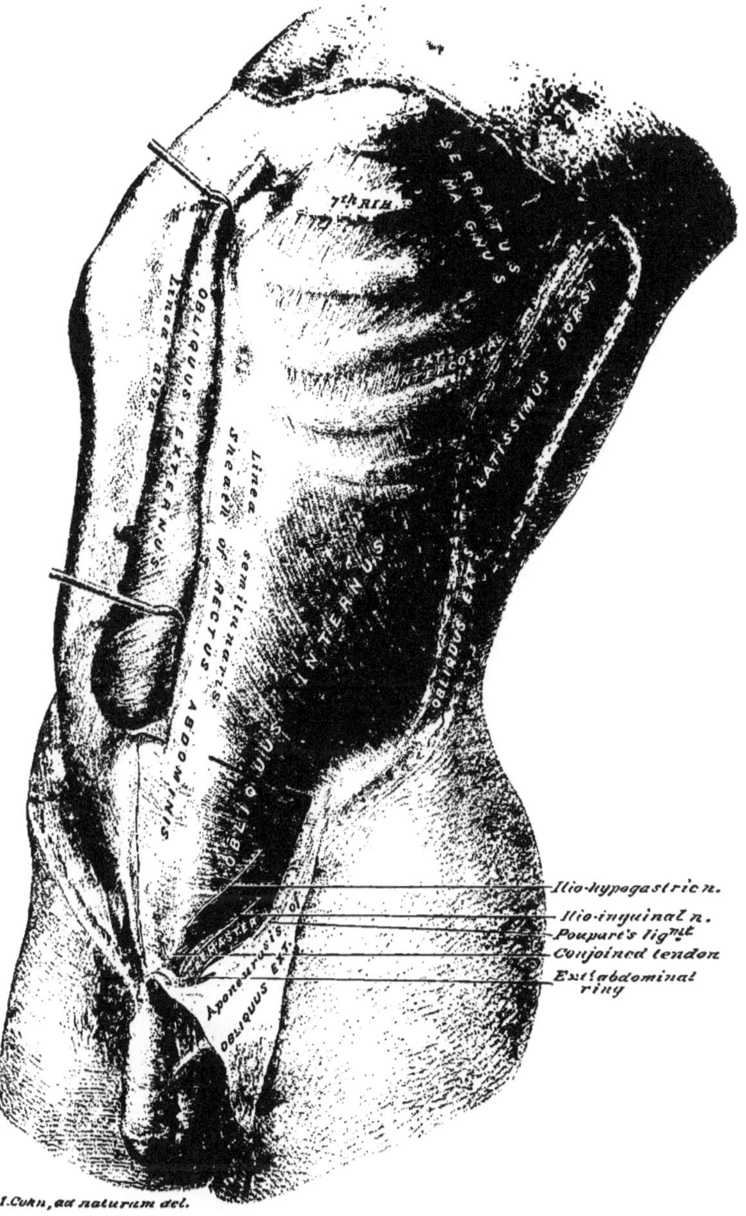
7th RIB
SERRATUS MAGNUS
EXT. INTERCOSTAL
LATISSIMUS DORSI
OBLIQUUS EXTERNUS
Linea alba
Linea semilunaris
Sheath of RECTUS ABDOMINIS
OBLIQUUS INTERNUS
OBLIQUUS EXT.
CREMASTER
Aponeurosis OBLIQUUS EXT.
Ilio-hypogastric n.
Ilio-inguinal n.
Poupart's ligmt.
Conjoined tendon
Ext. abdominal ring

M. Cohn, ad naturam del.

PLATE 26

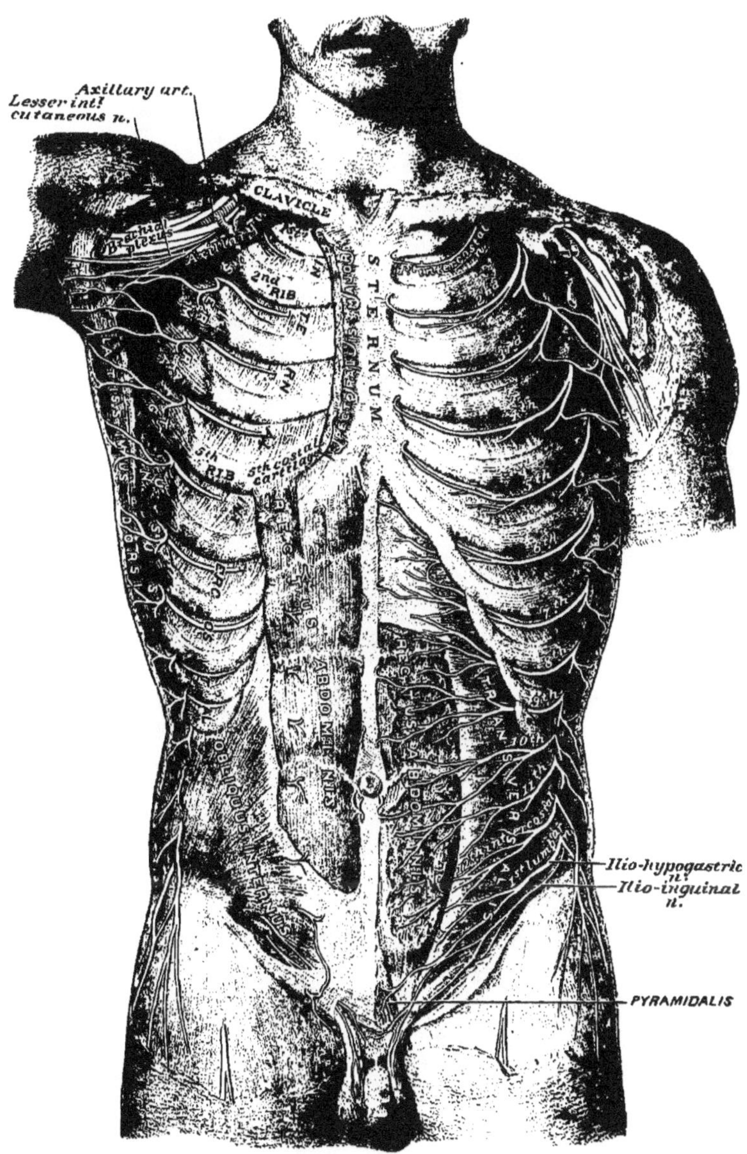

Axillary art.
Lesser int! cutaneous n.
CLAVICLE
Brachial plexus
2nd RIB
STERNUM
5th RIB
5th costal cartilage
Ilio-hypogastric n.
Ilio-inguinal n.
PYRAMIDALIS

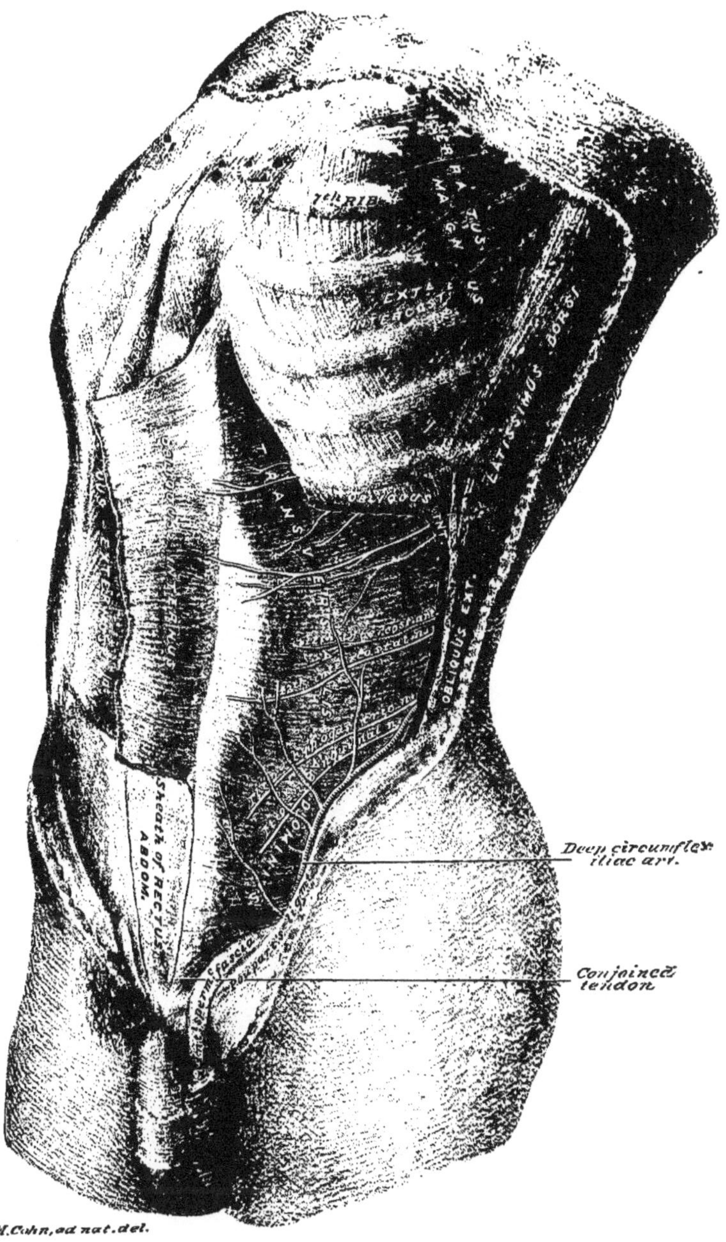
7th RIB
LATISSIMUS DORSI
OBLIQUUS EXT.
TRANSV
Sheath of RECTUS ABDOM.
Deep circumflex iliac art.
Conjoined tendon
M.Cohn, ad nat. del.

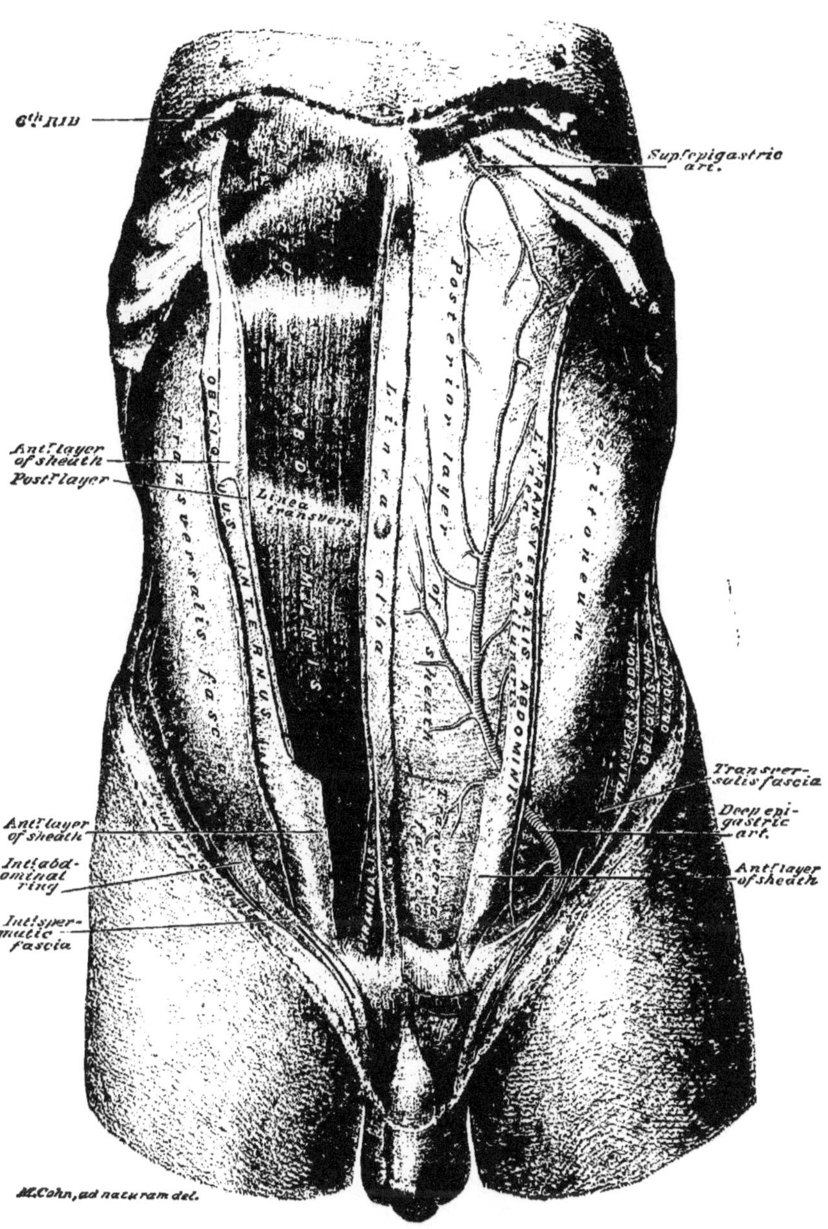
6th RIB
Sup. epigastric art.
Ant. layer of sheath
Post. layer
Linea transvers.
Transversalis fascia
Deep epigastric art.
Ant. layer of sheath
Ant. layer of sheath
Int. abdominal ring
Int. spermatic fascia
Posterior layer of sheath
Linea alba
Linea semilunaris
TRANSVERSALIS ABDOMINIS
Peritoneum
PYRAMIDALIS
M. Cohn, ad naturam del.

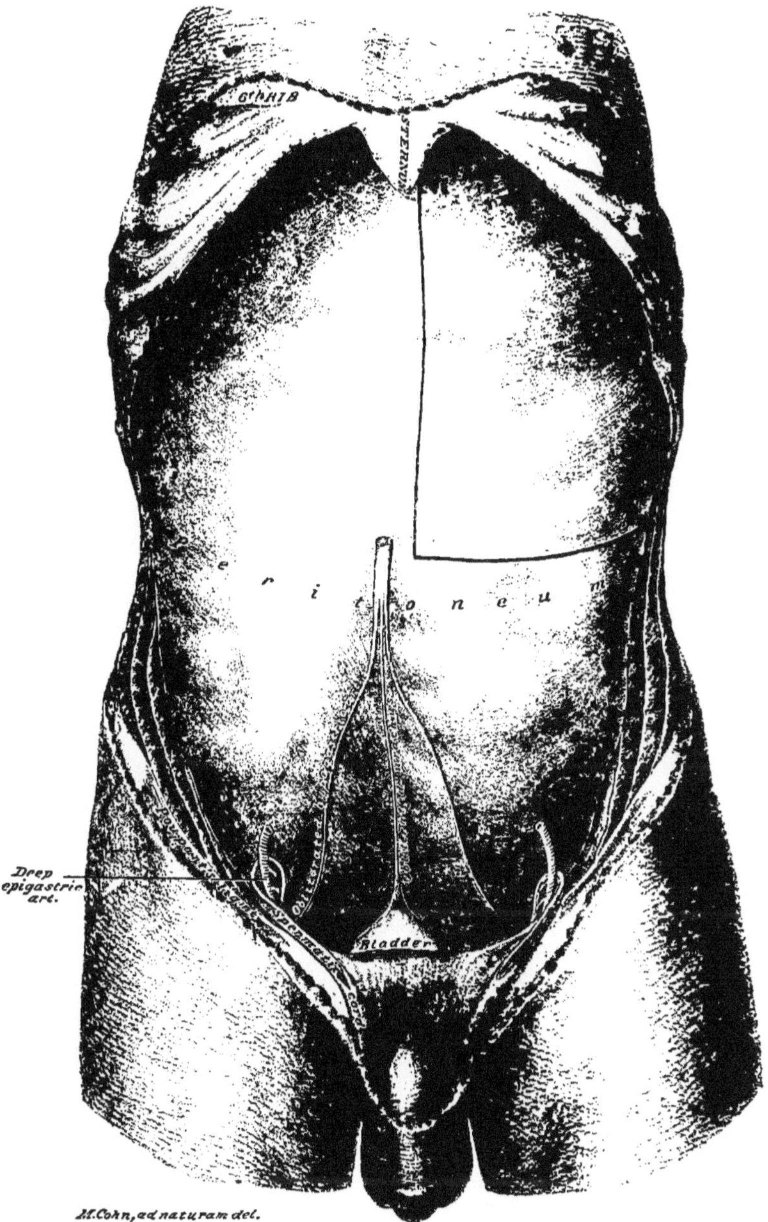
6th RIB
Deep
epigastric
art.
Bladder
M.Cohn, ad naturam del.

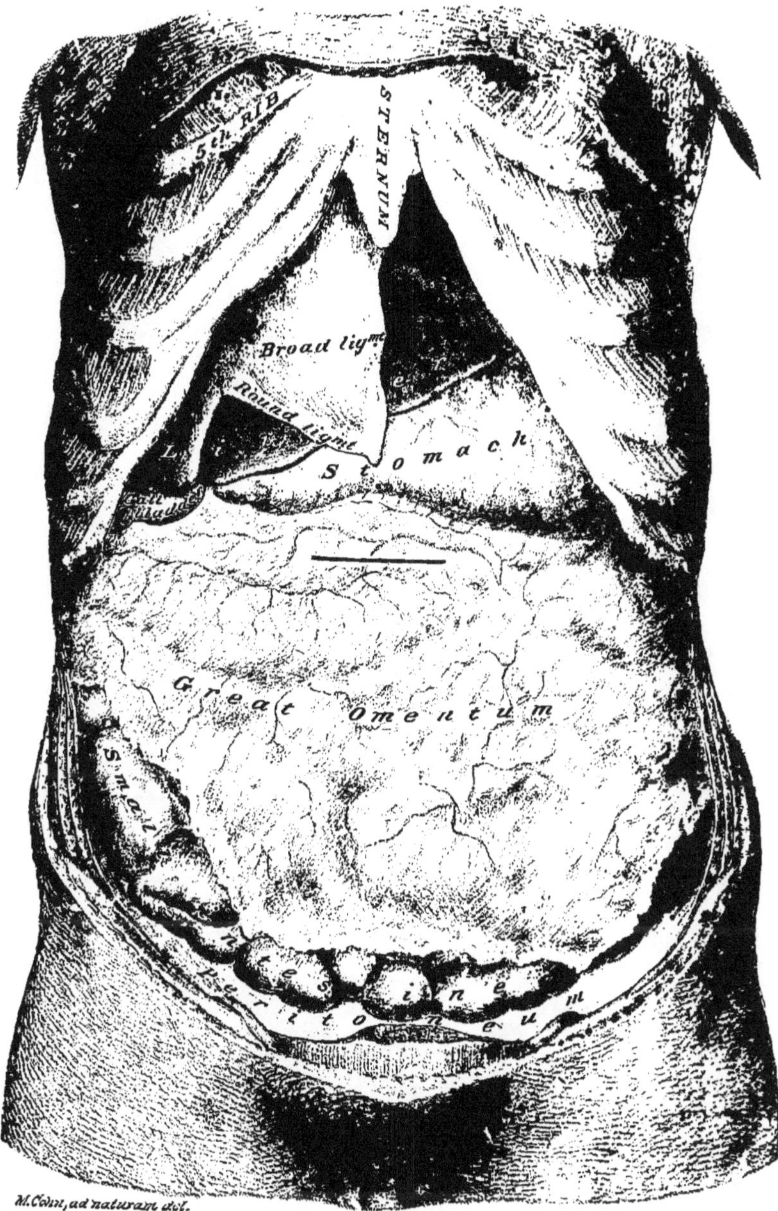
5th RIB
STERNUM
Broad ligmt
Round ligmt
Stomach
Great Omentum
Peritoneum

M. Cohn, ad naturam del.

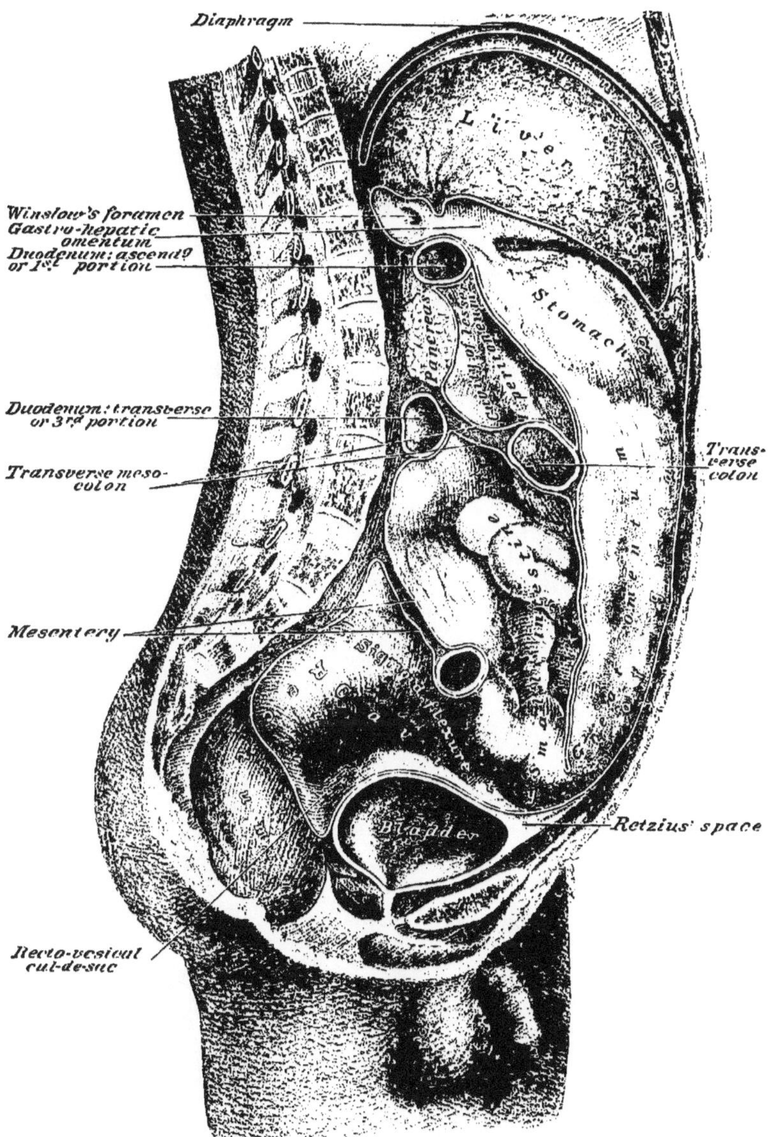
Diaphragm
Liver
Winslow's foramen
Gastro-hepatic omentum
Duodenum: ascend'g or 1st portion
Stomach
Pancreas
Duodenum: transverse or 3rd portion
Transverse meso-colon
Trans-verse colon
Mesentery
Sigmoid flexure
Bladder
Retzius' space
Recto-vesical cul-de-sac

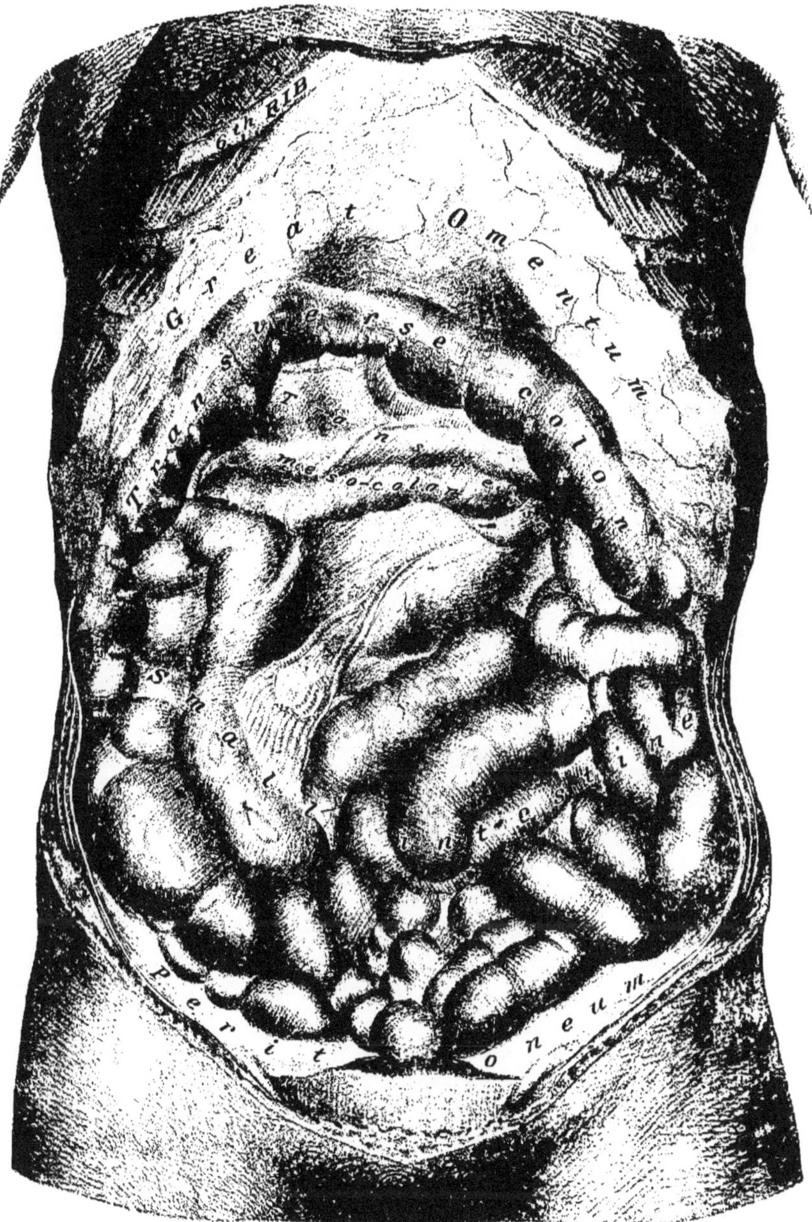
6th RIB
Great Omentum
Transverse colon
Transverse meso-colon
Small intestine
Peritoneum

M. Cohn, ad naturam del.

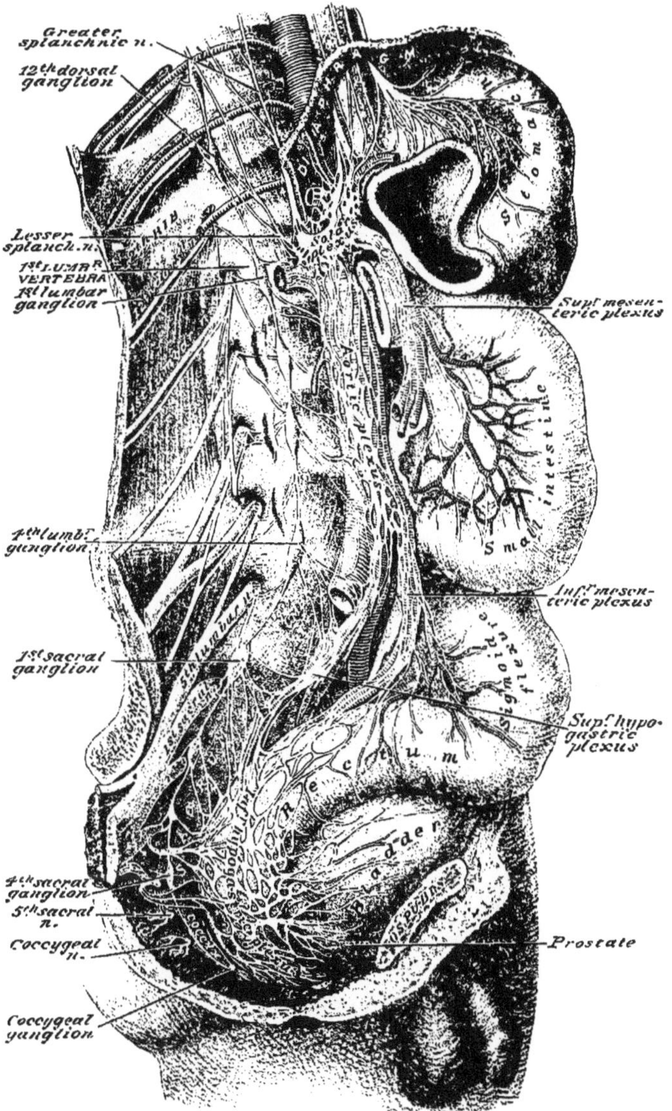
Greater splanchnic n.
12th dorsal ganglion
Lesser splanch. n.
1st LUMBR VERTEBRA
1st lumbar ganglion
4th lumbr ganglion
1st sacral ganglion
4th sacral ganglion
5th sacral n.
Coccygeal n.
Coccygeal ganglion
Stomach
Supr mesenteric plexus
Small intestine
Infr mesenteric plexus
Sigmoid flexure
Supr hypogastric plexus
Rectum
Bladder
Prostate

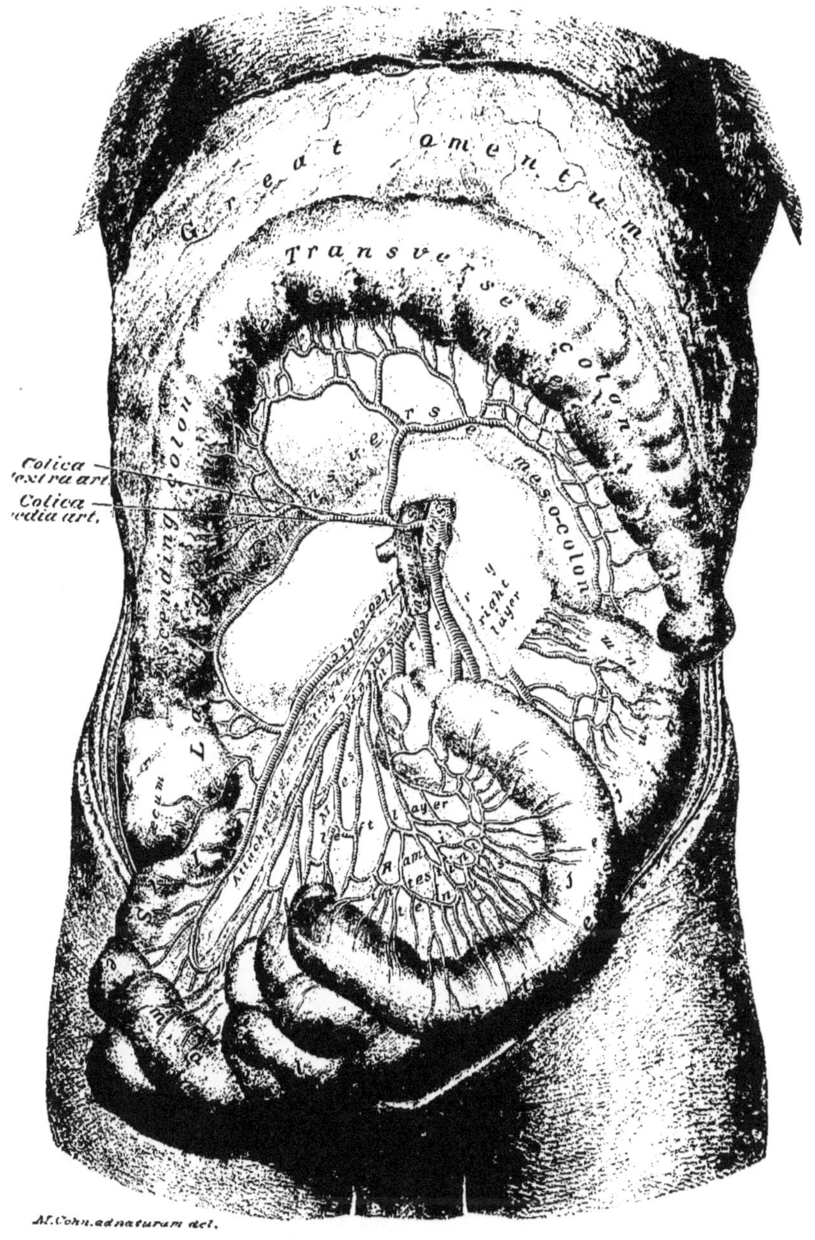
Great omentum
Transverse colon
Transverse meso-colon
Colica dextra art.
Colica media art.
Ascending colon
right layer
left layer
Attachment of mesentery
Rami intestini
Caecum
Jejunum
Ileum

M. Cohn. ad naturam del.

PLATE 35

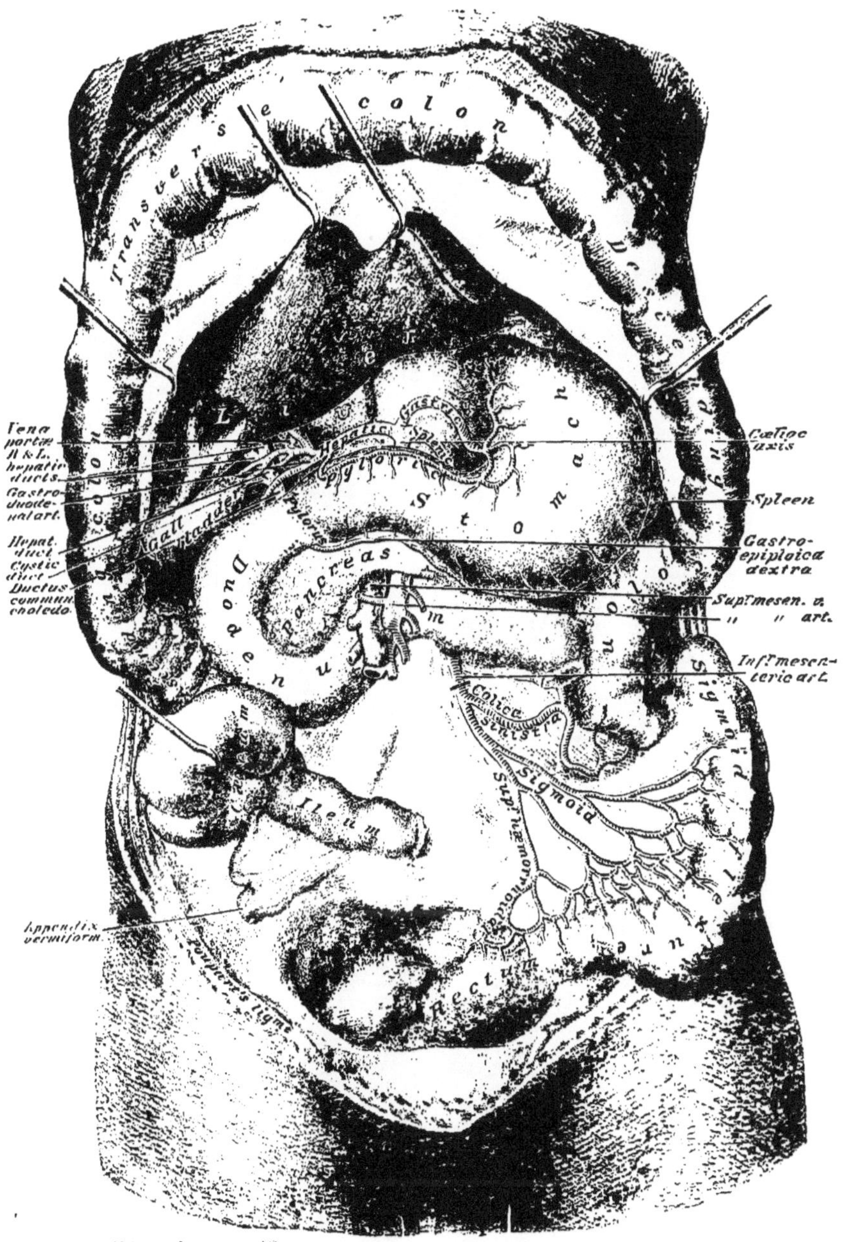
Transverse colon
Liver
Stomach
Gall bladder
Pancreas
Duodenum
Hepatic
Gastric
Splenic
Pyloric
Ileum
Colica sinistra
Sigmoid
Rectum
Vena portæ
R & L. hepatic ducts.
Gastro-duodenal art.
Hepat. duct
Cystic duct
Ductus commun choledo.
Cœliac axis
Spleen
Gastro-epiploica dextra
Supr mesen. v.
" " art.
Infr mesenteric art.
Appendix vermiform.

M. Cohn, ad naturam del.

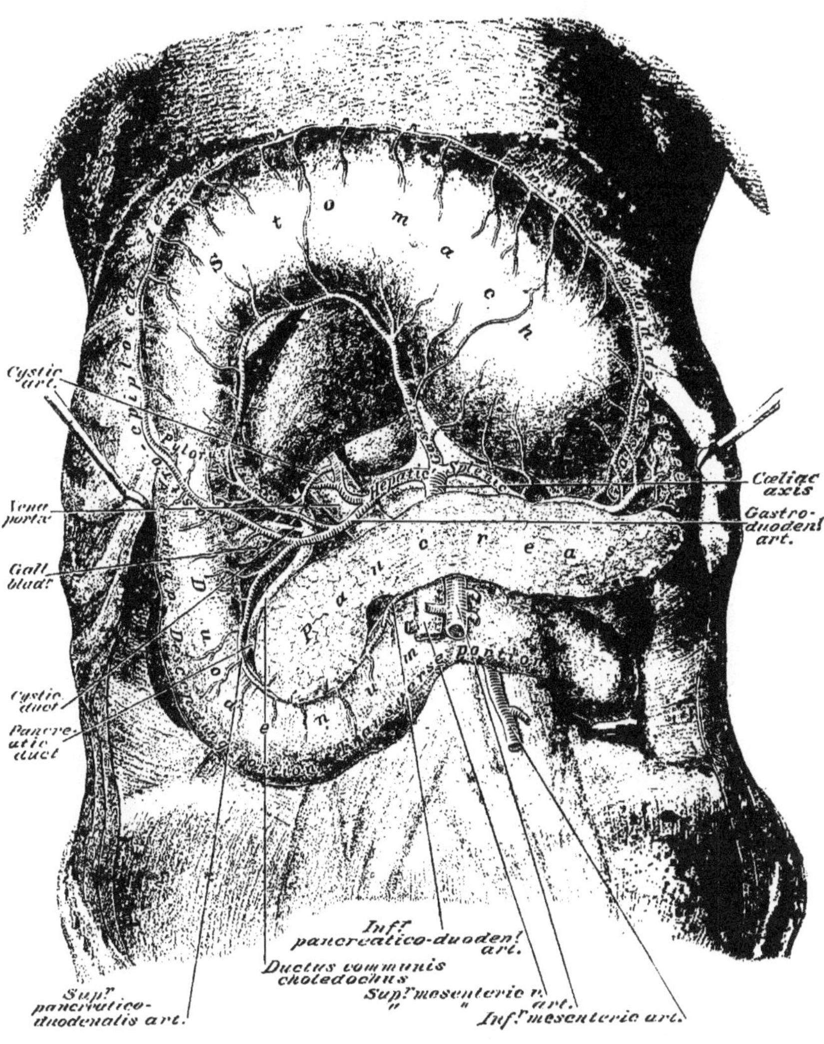

Stomach
Pancreas
Duodenum
Pylorus
Hepatic
Splenic
Cystic art.
Vena portæ
Gall bladr
Cystic duct
Pancreatic duct
Cœliac axis
Gastro-duodenl art.
Infr pancreatico-duodenl art.
Ductus communis choledochus
Supr mesenteric v.
" " art.
Infr mesenteric art.
Supr pancreatico-duodenalis art.

M. Cohn, ad naturam del.

FIG. 1

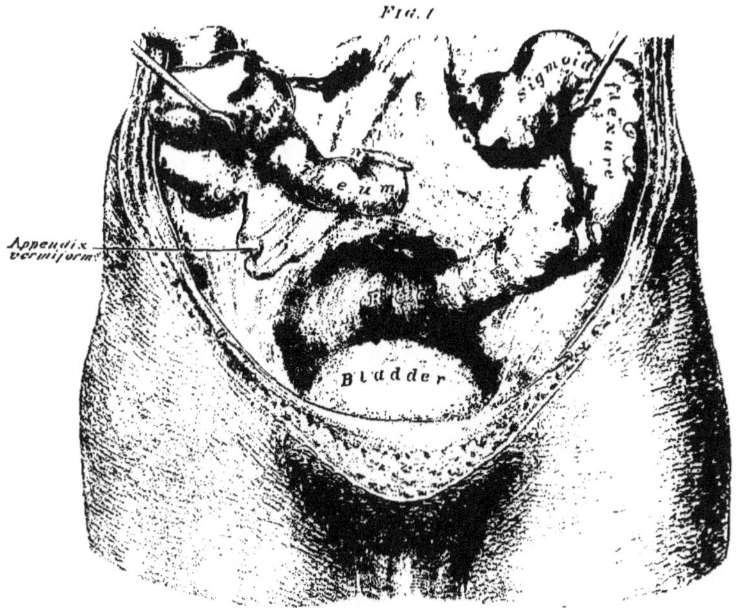
Sigmoid flexure
Ileum
Appendix vermiform.
Rectum
Bladder

FIG. 2

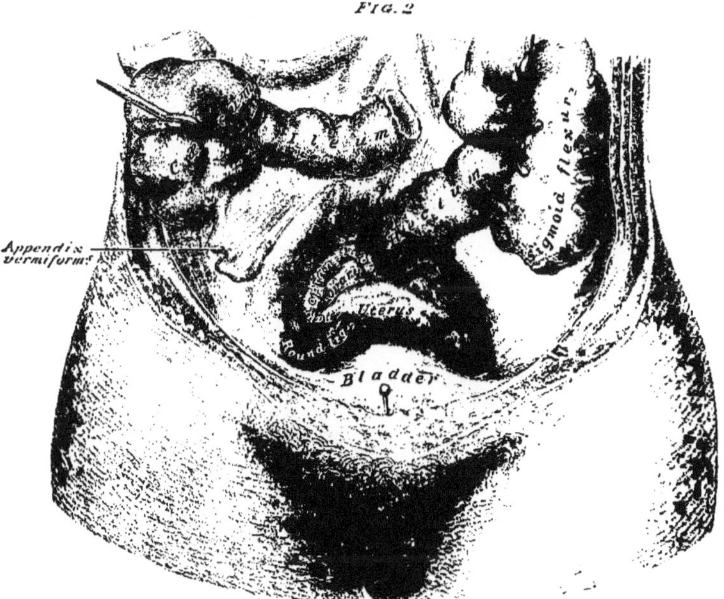
Ileum
Sigmoid flexure
Appendix vermiform.
Uterus
Round lig.
Bladder

M. Cohn, ad naturam del.

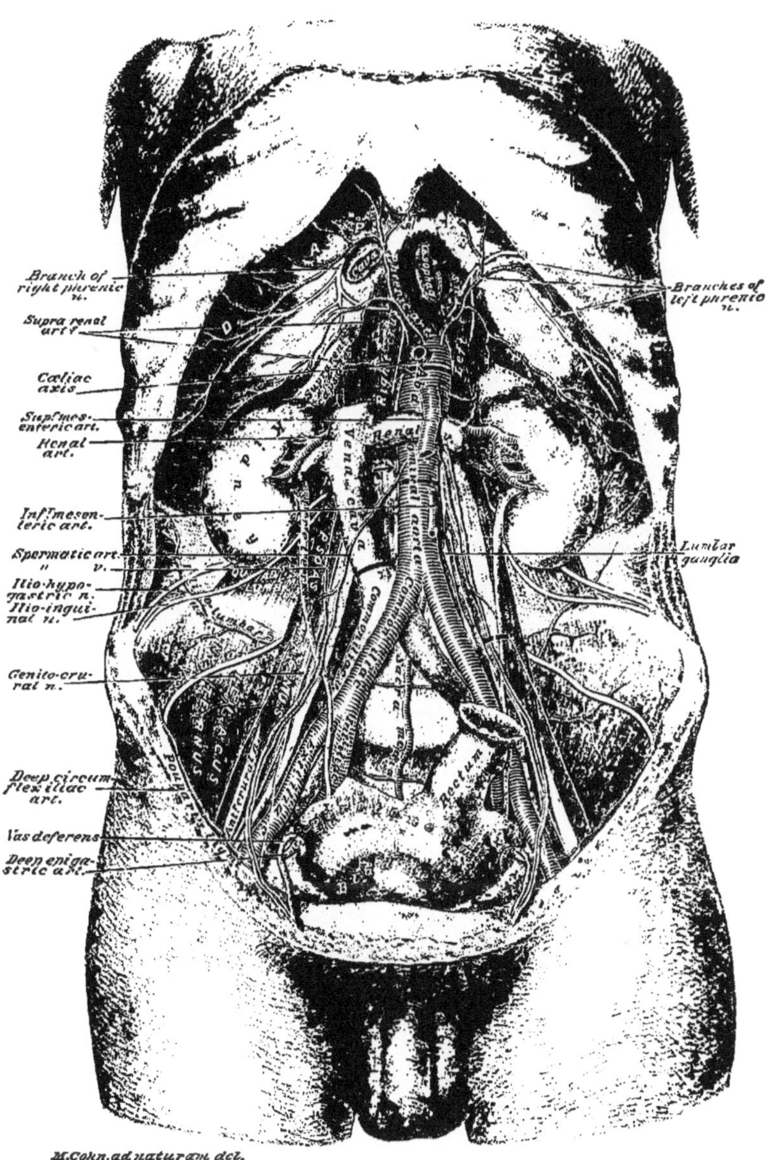
Branch of right phrenic n.
Supra renal art.
Cœliac axis
Supt mesenteric art.
Renal art.
Inf. mesenteric art.
Spermatic art.
" v.
Ilio-hypogastric n.
Ilio-inguinal n.
Genito-crural n.
Deep circumflex iliac art.
Vas deferens
Deep epigastric art.
Branches of left phrenic n.
Lumbar ganglia
Rectum

M. Cohn, ad naturam del.

College of Physicians & Surgeons,

New York City.

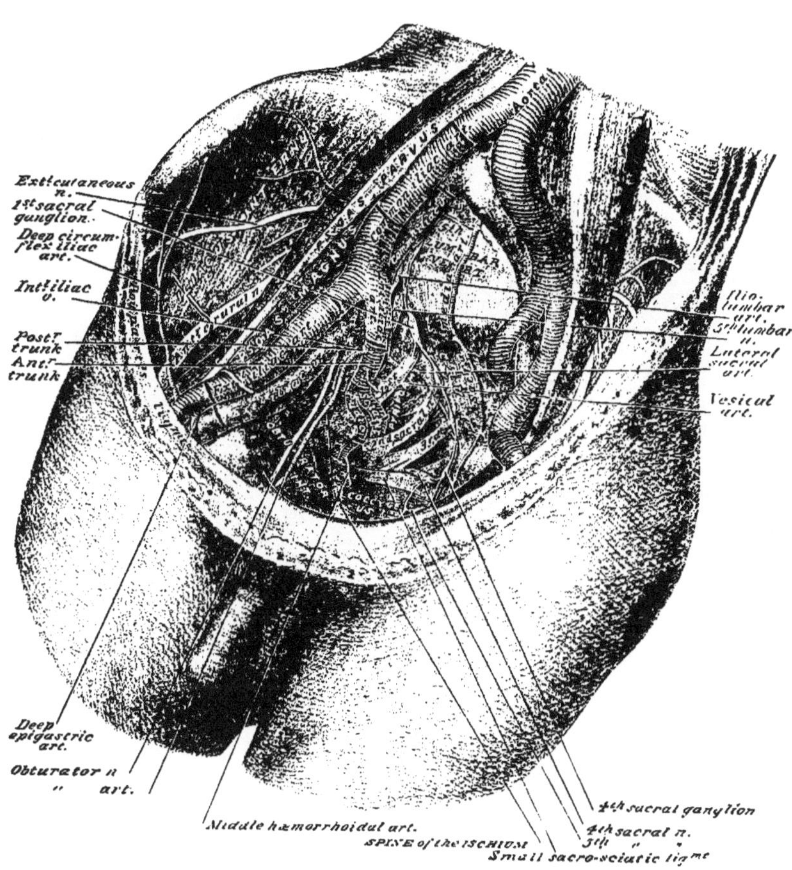
Ext! cutaneous n.
1st sacral ganglion.
Deep circum-flex iliac art.
Int! iliac v.
Post! trunk
Ant! trunk
Ilio-lumbar art.
5th lumbar n.
Lateral sacral art.
Vesical art.
Deep epigastric art.
Obturator n
" art.
Middle hæmorrhoidal art.
SPINE of the ISCHIUM
4th sacral ganglion
4th sacral n.
3th " "
Small sacro-sciatic ligmt
Aorta
PSOAS PARVUS
PSOAS MAGNUS
5th LUMBAR VERTEBRA
COCCYGEUS

M. Cohn. ad naturam del.

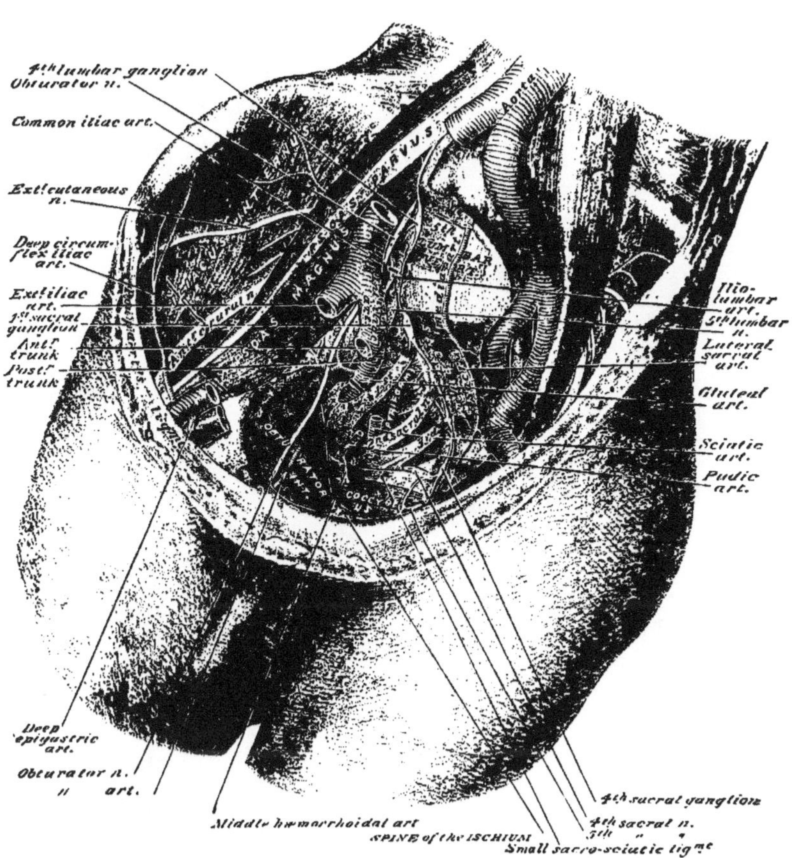
4th lumbar ganglion
Obturator n.
Common iliac art.
Ext! cutaneous n.
Deep circumflex iliac art.
Ext! iliac art.
1st sacral ganglion
Ant! trunk
Post! trunk
Aorta
PSOAS PARVUS
PSOAS MAGNUS
Ilio-lumbar art.
5th lumbar n.
Lateral sacral art.
Gluteal art.
Sciatic art.
Pudic art.
Deep epigastric art.
Obturator n.
" art.
Middle hæmorrhoidal art
SPINE of the ISCHIUM
Small sacro-sciatic ligmt
4th sacral ganglion
4th sacral n.
5th " "

M. Cohn. ad naturam del

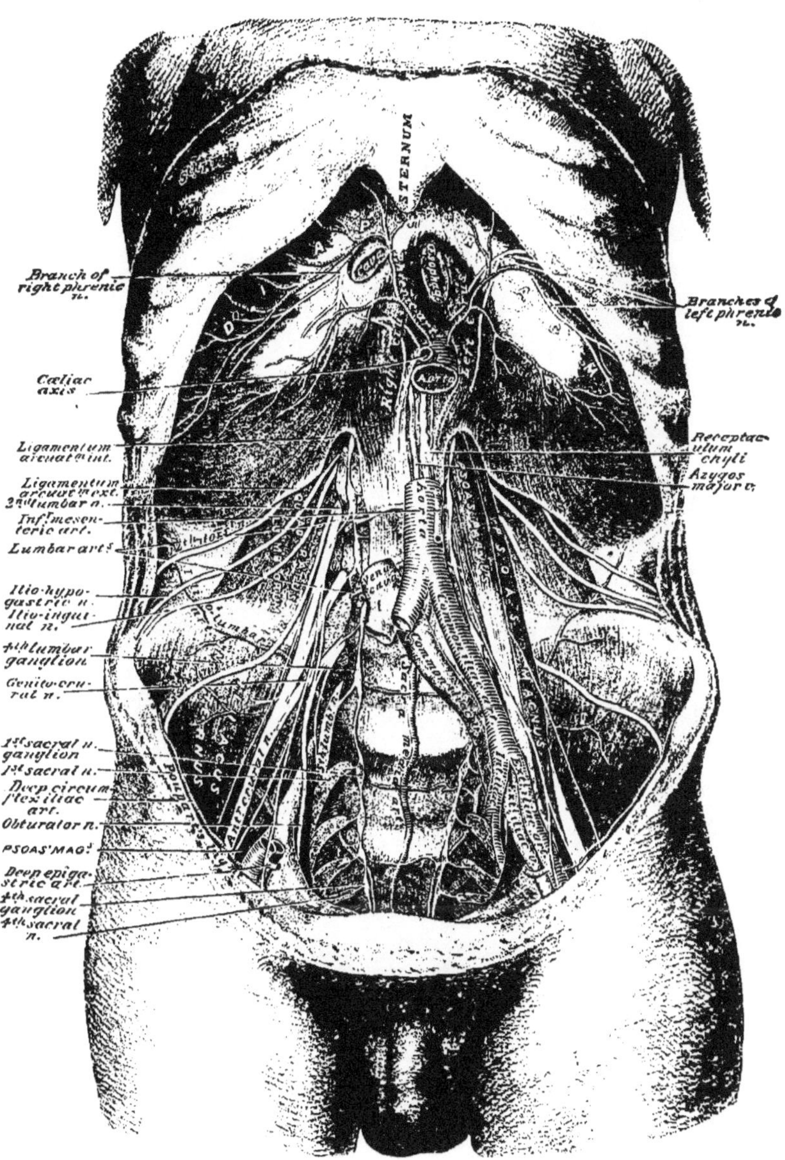
STERNUM
Branch of right phrenic n.
Branches of left phrenic n.
Cœliac axis
Aorta
Receptaculum chyli
Azygos major v.
Ligamentum arcuat.m int.
Ligamentum arcuat.m ext.
2nd lumbar n.
Inf.r mesenteric art.
Lumbar art.s
Ilio-hypogastric n.
Ilio-inguinal n.
4th lumbar ganglion
Genito-crural n.
1st sacral n. ganglion
1st sacral n.
Deep circumflex iliac art.
Obturator n.
PSOAS MAG.s
Deep epigastric art.
4th sacral ganglion
4th sacral n.

M. Cohn, ad naturam del.

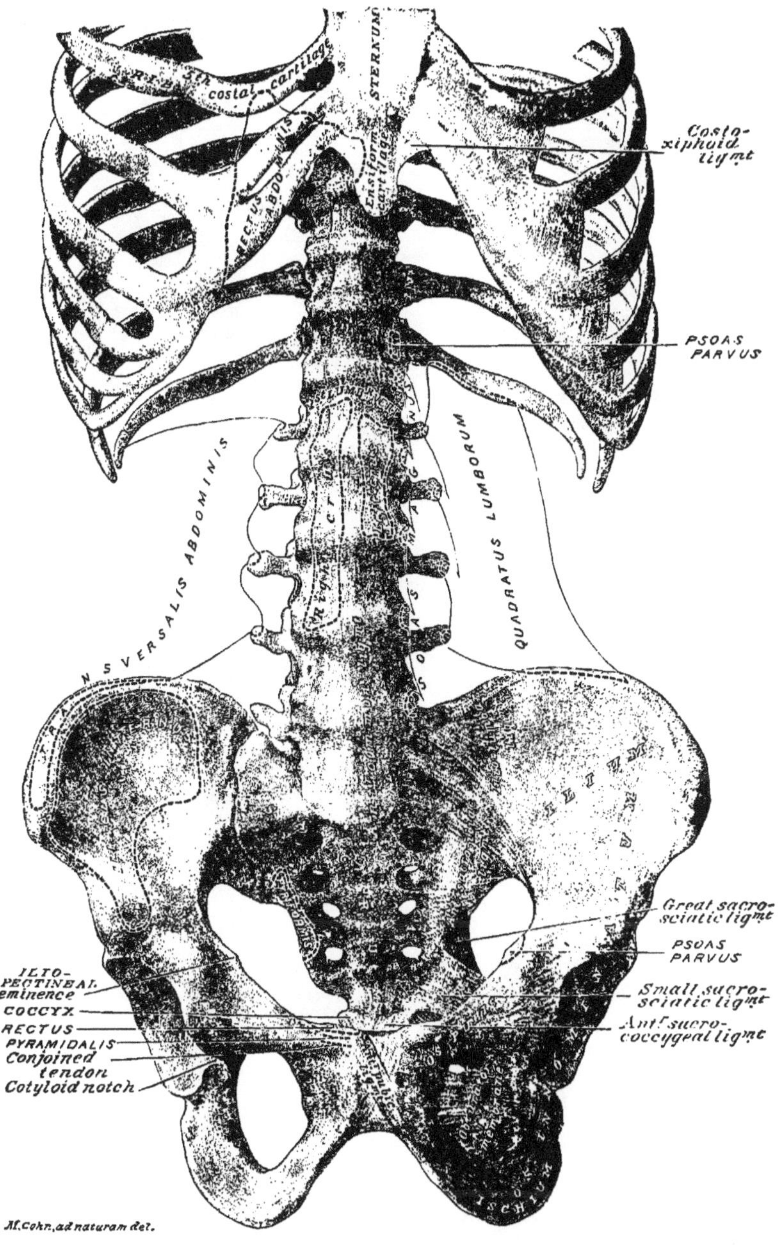
Costo-xiphoid ligmt
PSOAS PARVUS
QUADRATUS LUMBORUM
TRANSVERSALIS ABDOMINIS
RECTUS ABDOMINIS
STERNUM
Ensiform cartilage
5th costal cartilage
Great sacro-sciatic ligmt
PSOAS PARVUS
Small sacro-sciatic ligmt
Antr sacro-coccygeal ligmt
ILIO-PECTINEAL eminence
COCCYX
RECTUS
PYRAMIDALIS
Conjoined tendon
Cotyloid notch
ISCHIUM

M. Cohn, ad naturam del.

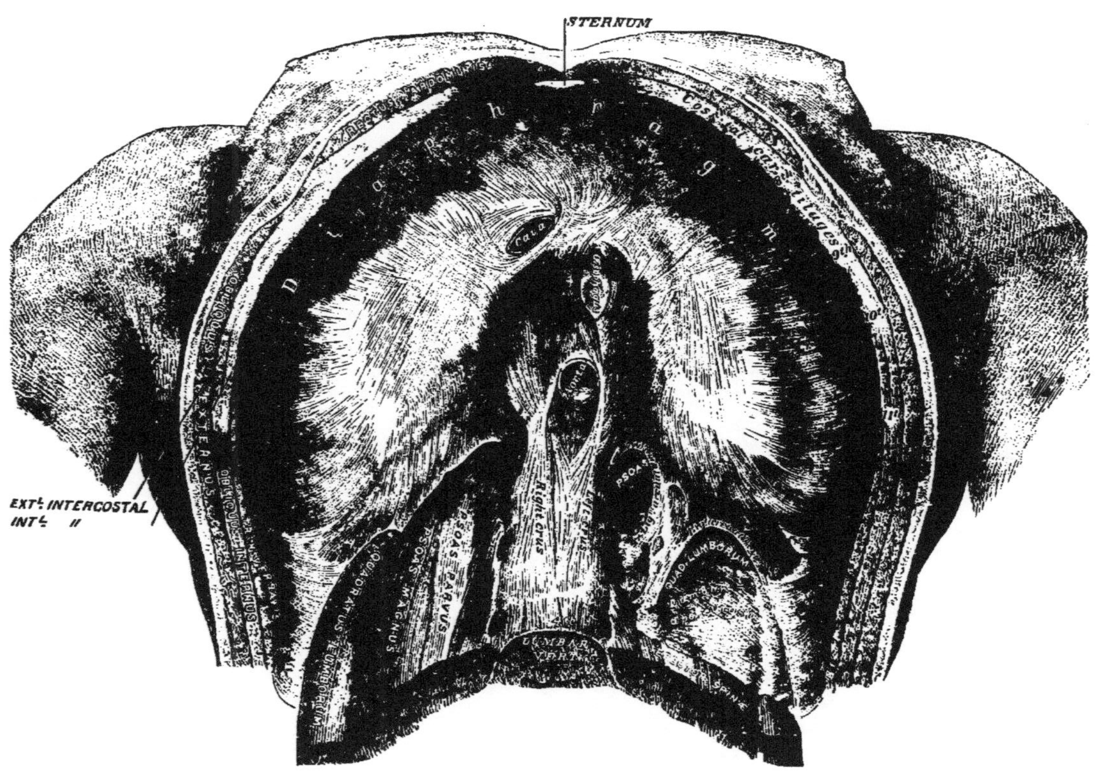
STERNUM
Diaphragm
Costal cartilages
Cava
Œsophagus
Aorta
Left crus
Right crus
PSOAS
PSOAS PARVUS
PSOAS MAGNUS
QUADRATUS LUMBORUM
QUAD. LUMBORUM
LUMBAR VERT.
OBLIQUUS INTERNUS
OBLIQUUS EXTERNUS
EXT'L INTERCOSTAL
INT'L "

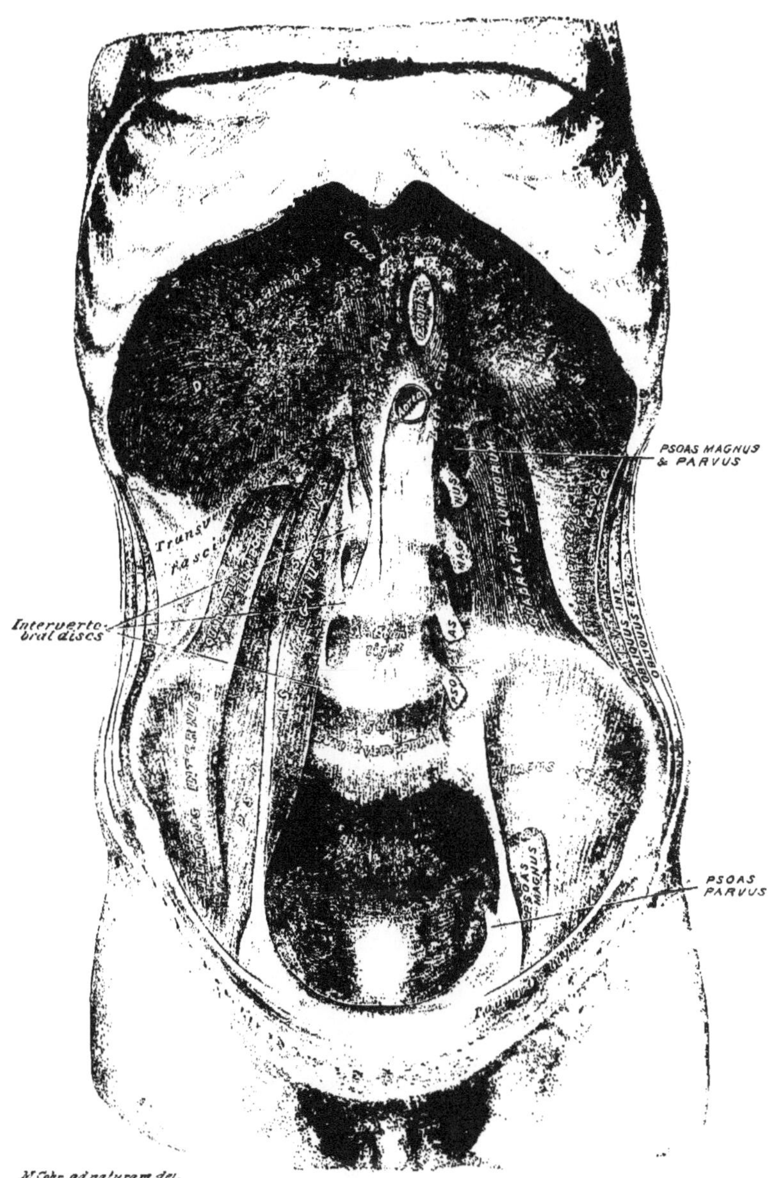
PSOAS MAGNUS
& PARVUS
Intervertebral discs
PSOAS
PARVUS
PSOAS MAGNUS
QUADRATUS LUMBORUM
OBLIQUUS INT.
OBLIQUUS EXT.
Aorta
Cava

M. Cohn, ad naturam del.

FIG. 1

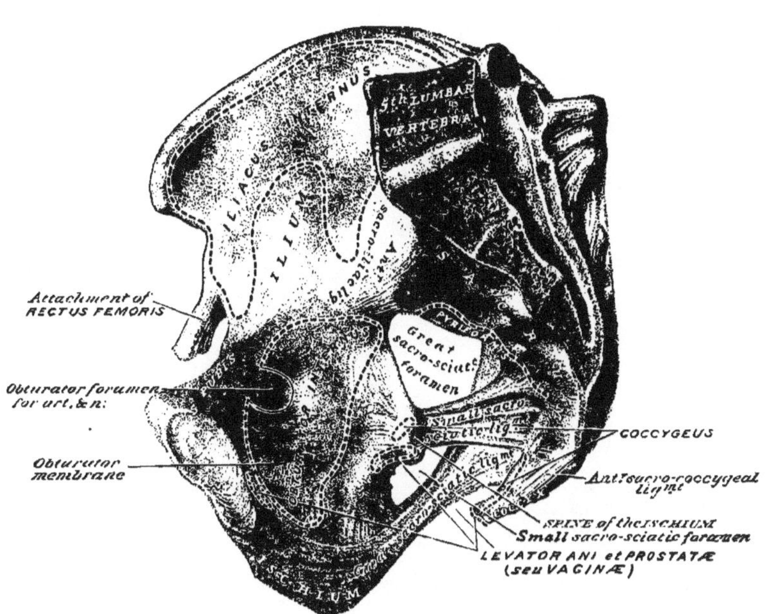

Attachment of RECTUS FEMORIS
Obturator foramen for art. & n.
Obturator membrane
5th LUMBAR VERTEBRA
ILIACUS INTERNUS
ILIUM
Antr sacro-iliac ligt
Great sacro-sciatic foramen
Small sacro-sciatic ligt
Great sacro-sciatic ligmt
COCCYGEUS
Antr sacro-coccygeal ligmt
SPINE of the ISCHIUM
Small sacro-sciatic foramen
LEVATOR ANI et PROSTATÆ (seu VAGINÆ)
ISCHIUM

FIG. 2

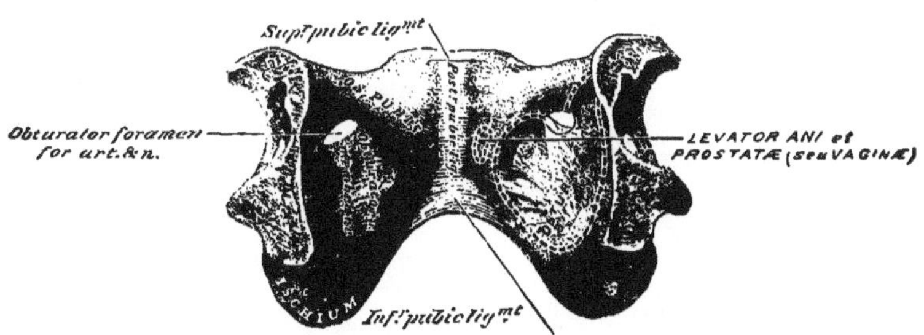

Supr pubic ligmt
Obturator foramen for art. & n.
LEVATOR ANI et PROSTATÆ (seu VAGINÆ)
ISCHIUM
Infr pubic ligmt

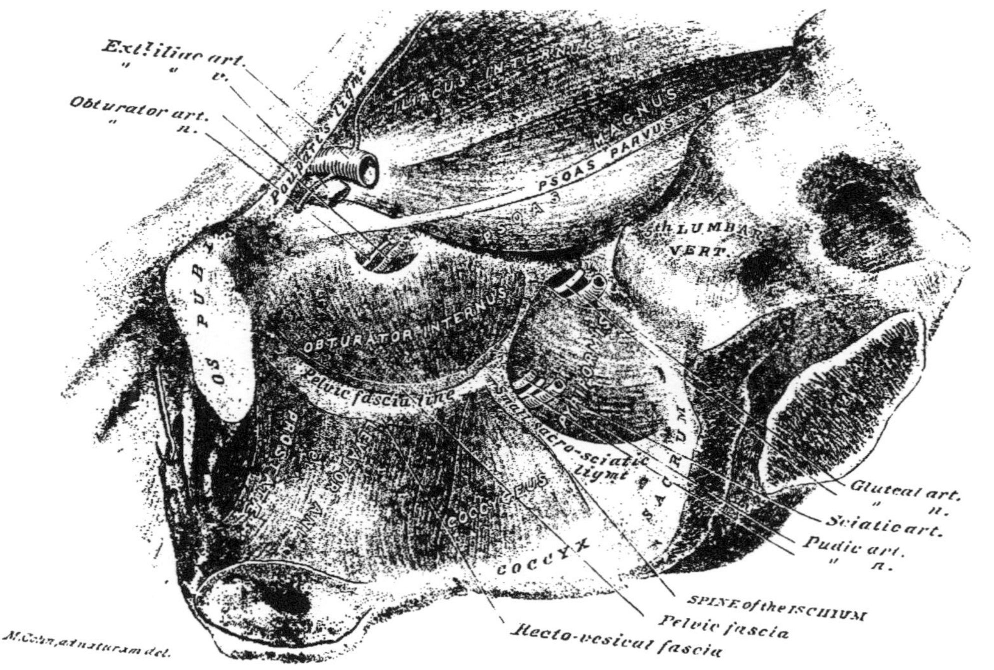
Ext. iliac art.
" " v.
Obturator art.
" n.
Poupart's ligmt
ILIACUS INTERNUS
PSOAS MAGNUS
PSOAS PARVUS
PSOAS
5th LUMBAR VERT.
OS PUBIS
OBTURATOR INTERNUS
Pelvic fascia line
Small sacro-sciatic ligmt
PYRIFORMIS
SACRUM
PROSTATE
LEVATOR ANI
COCCYGEUS
COCCYX
Gluteal art.
" n.
Sciatic art.
Pudic art.
" n.
SPINE of the ISCHIUM
Pelvic fascia
Recto-vesical fascia
M. Cohn, ad naturam del.

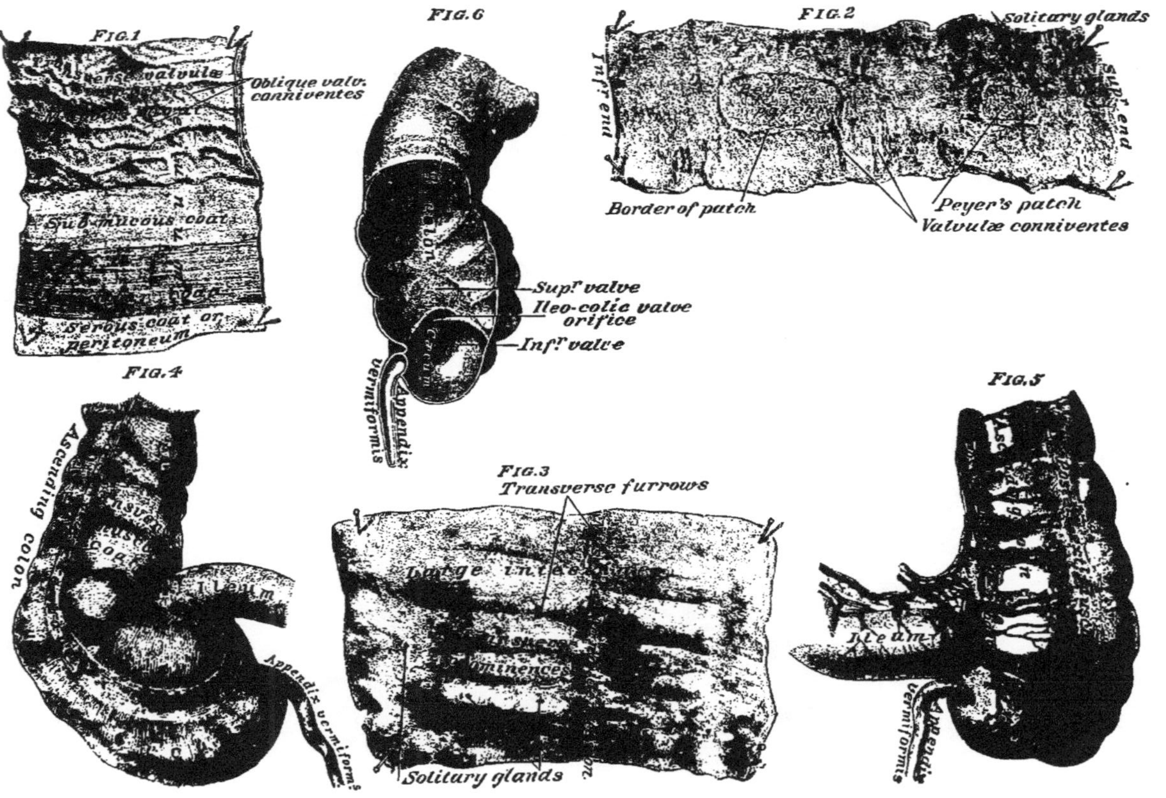
FIG.1
Oblique valv.
conniventes
Submucous coat
Serous coat or
peritoneum
FIG.6
Supr. valve
Ileo-colic valve
orifice
Infr. valve
Appendix
vermiformis
FIG.2
Solitary glands
Infr. end
Supr. end
Border of patch
Peyer's patch
Valvulæ conniventes
FIG.4
Ascending colon
Ileum
Appendix vermiformis
FIG.3
Transverse furrows
Large intestine
Prominences
Solitary glands
FIG.5
Ileum
Appendix
vermiformis

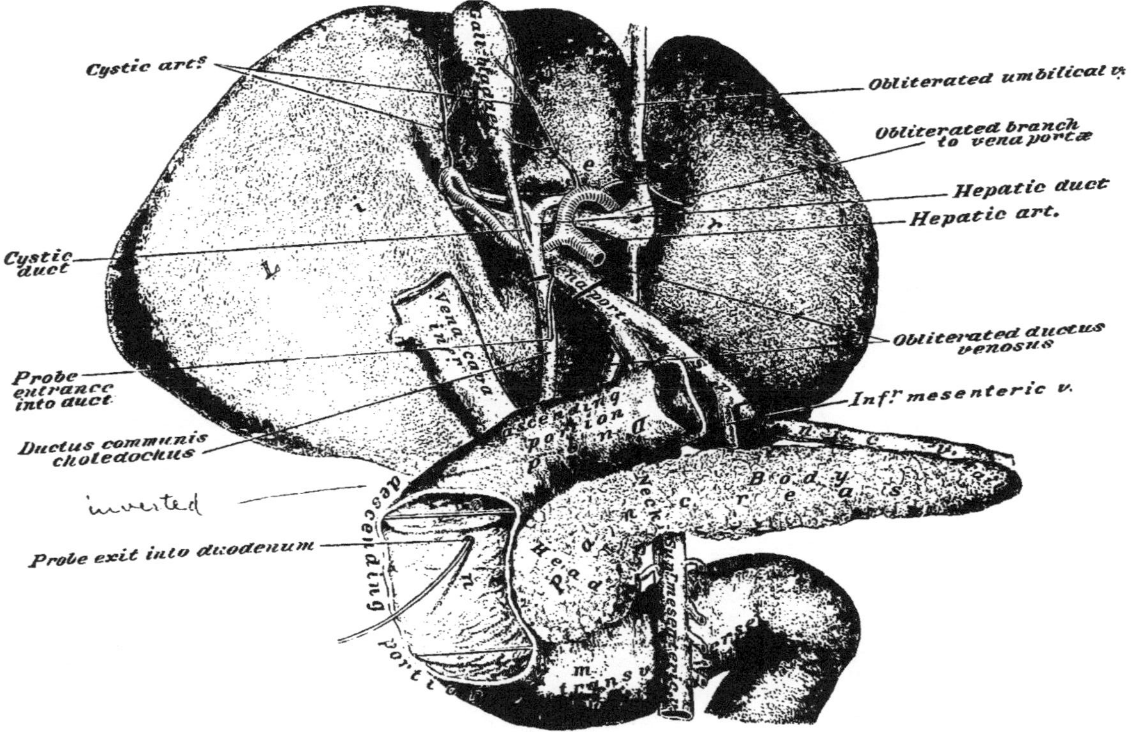
Cystic arts.
Obliterated umbilical v.
Obliterated branch to vena portæ
Hepatic duct
Hepatic art.
Cystic duct
Obliterated ductus venosus
Probe entrance into duct
Inf.r mesenteric v.
Ductus communis choledochus
inverted
Probe exit into duodenum
Gall-bladder
Vena cava infr.
Vena porta
ascending portion
descending portion
Body
Neck
Head
Sup. mesenteric v.

FIG. 1

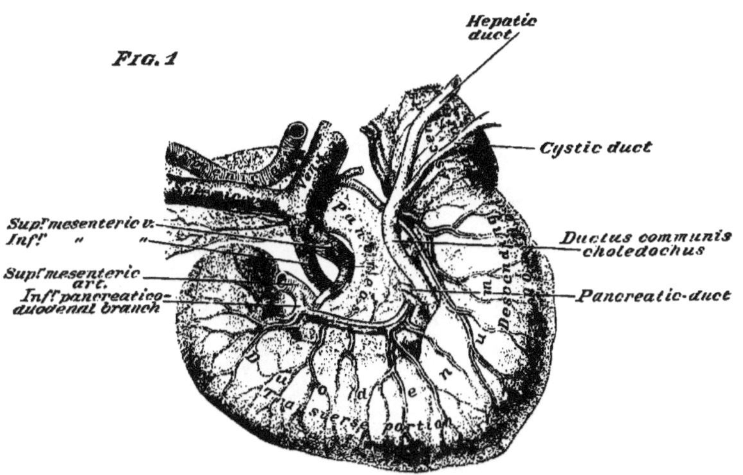
Hepatic duct
Cystic duct
Supr mesenteric v.
Infr " "
Ductus communis choledochus
Supr mesenteric art.
Infr pancreatico-duodenal branch
Pancreatic-duct
Descending
Transverse portion

FIG. 2

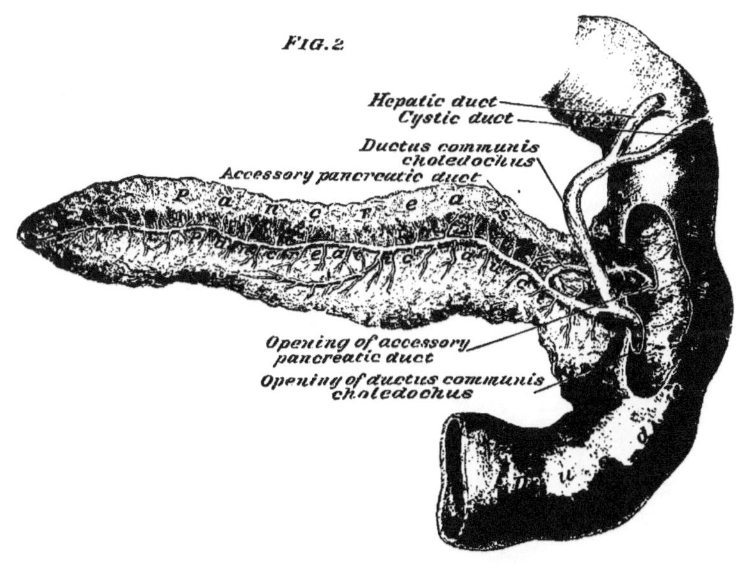
Hepatic duct
Cystic duct
Ductus communis choledochus
Accessory pancreatic duct
Pancreas
Pancreatic duct
Opening of accessory pancreatic duct
Opening of ductus communis choledochus

FIG. 1

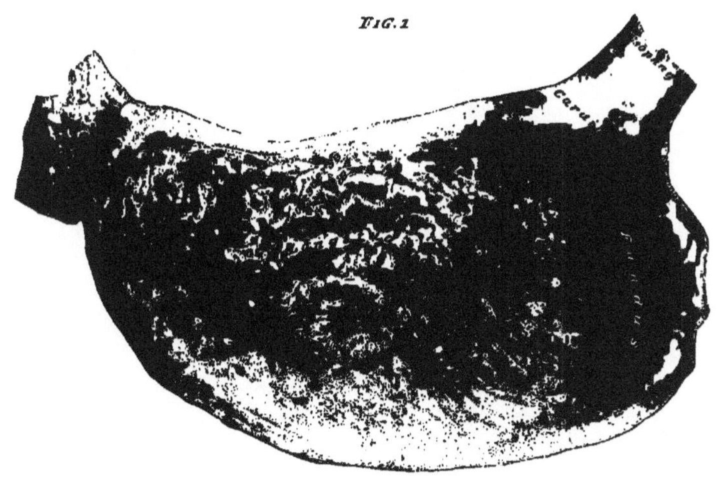

FIG. 2

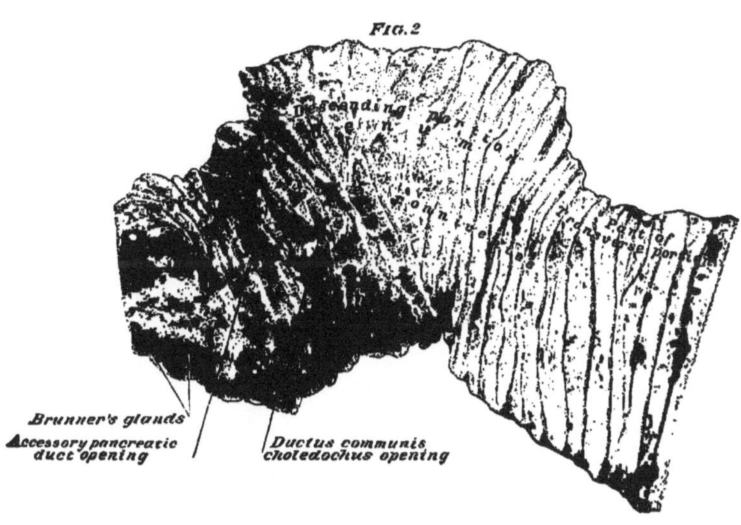
Brunner's glands
Accessory pancreatic duct opening
Ductus communis choledochus opening

PLATE 51

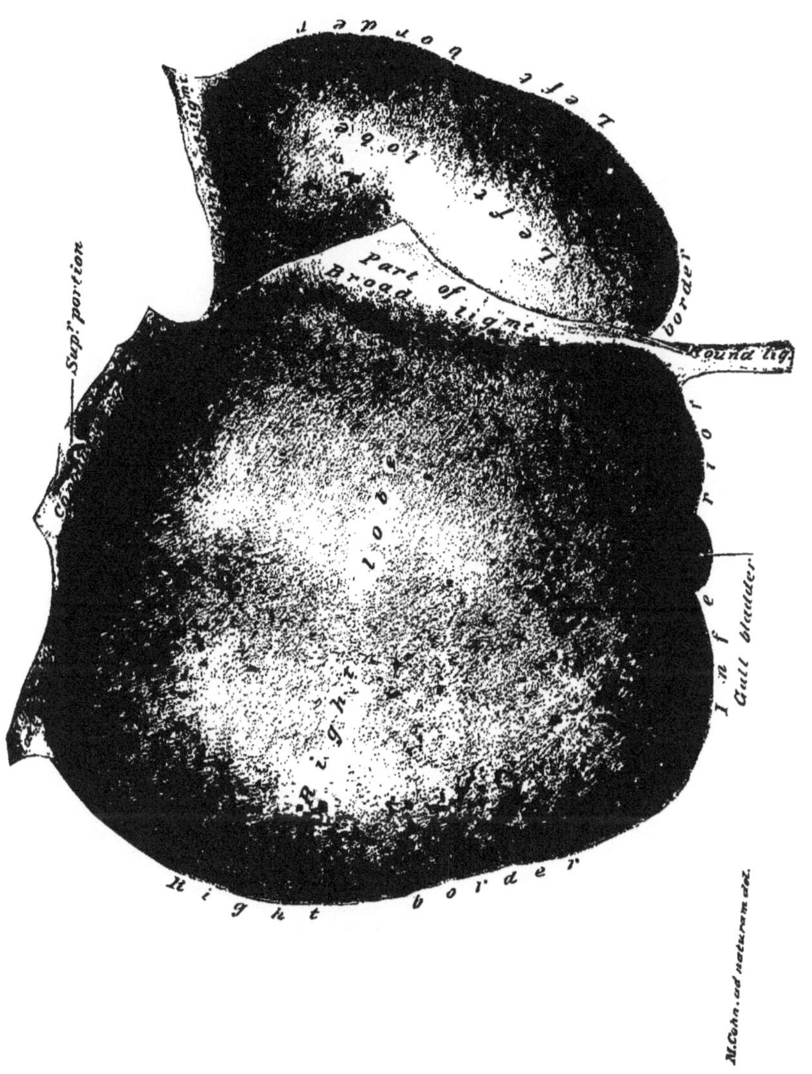
Left border
Left lobe
Part of Broad ligmt
Round lig.
Sup.r portion
Right lobe
Inferior border
Gall bladder
Right border
M.Cohn. ad naturam del.

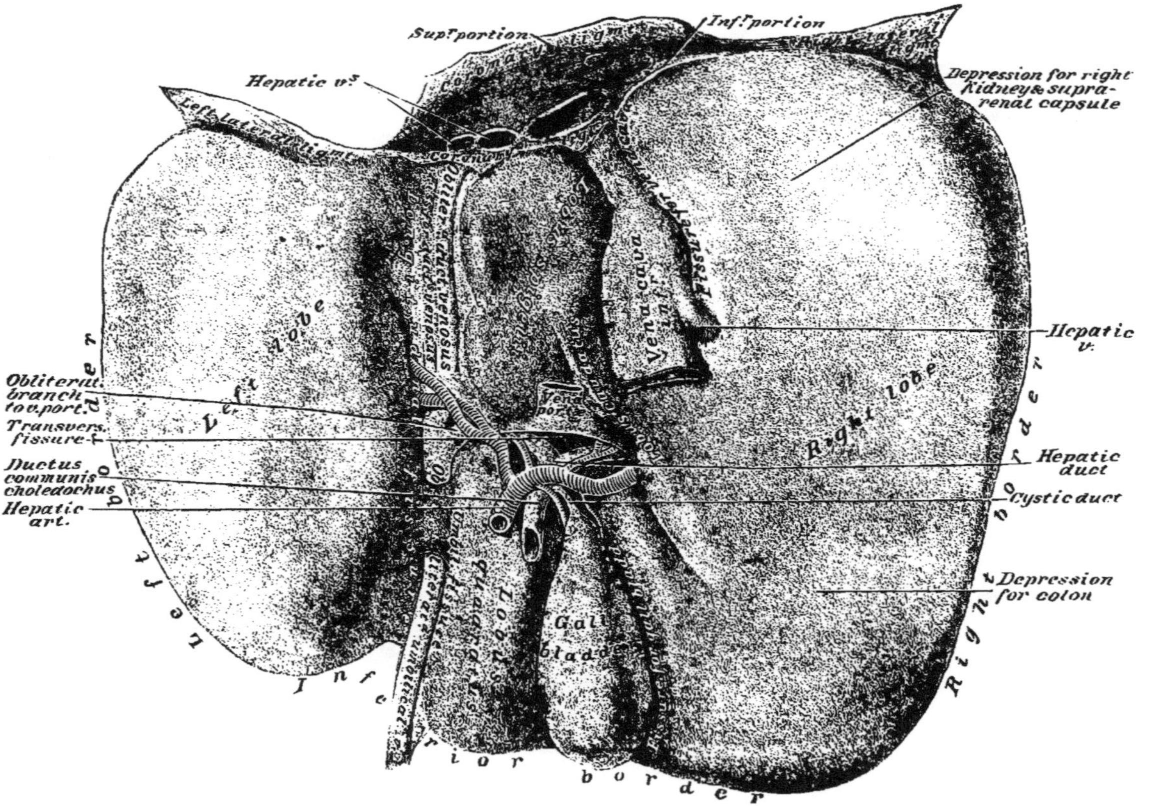
Sup.r portion
Inf.r portion
Hepatic v.s
Left lateral ligament
Depression for right kidney & supra-renal capsule
Hepatic v.
Obliterat. branch to v. port.
Transvers. fissure
Ductus communis choledochus
Hepatic art.
Hepatic duct
Cystic duct
Depression for colon
Left lobe
Right lobe
Vena cava inf.
Gall bladd.
Lobus quadratus
Left border
Inferior border
Right border

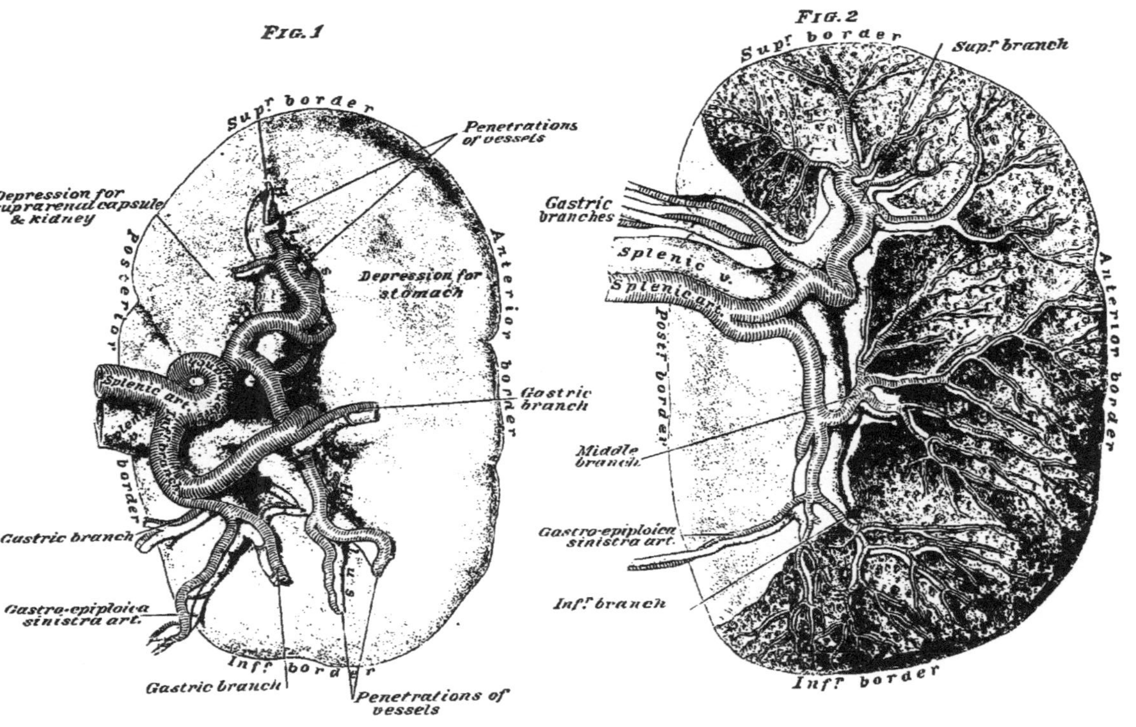
FIG.1
Sup.r border
Penetrations of vessels
Depression for suprarenal capsule & kidney
Posterior border
Depression for stomach
Anterior border
Splenic art.
Gastric branch
Gastric branch
Gastra-epiploica sinistra art.
Inf.r border
Gastric branch
Penetrations of vessels
FIG.2
Sup.r border
Sup.r branch
Gastric branches
Splenic v.
Splenic art.
Post.r border
Anterior border
Middle branch
Gastro-epiploica sinistra art.
Inf.r branch
Inf.r border

FIG. 1

FIG. 3

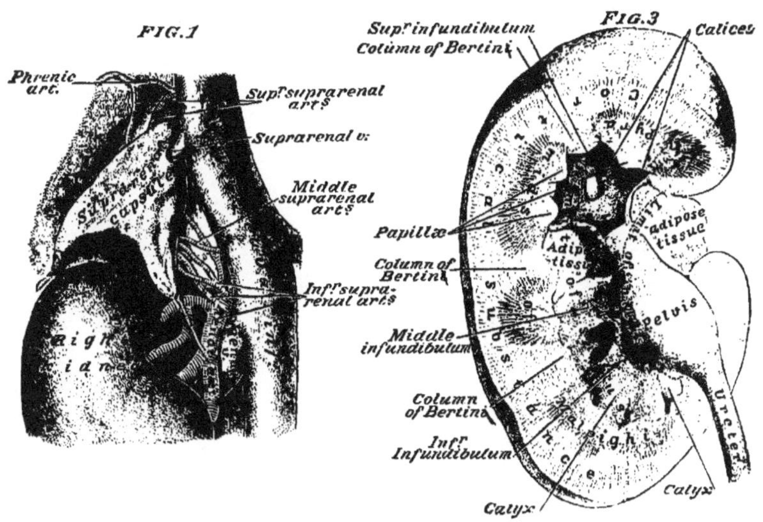
Phrenic art.
Supr. suprarenal arts.
Suprarenal v.
Middle suprarenal arts.
Infr. suprarenal arts.
Suprarenal capsule
Right kidney
Supr. infundibulum
Column of Bertini
Calices
Papillæ
Column of Bertini
Middle infundibulum
Column of Bertini
Infr. Infundibulum
Calyx
Calyx
Adipose tissue
adipose tissue
pelvis
Ureter

FIG. 2

FIG. 4

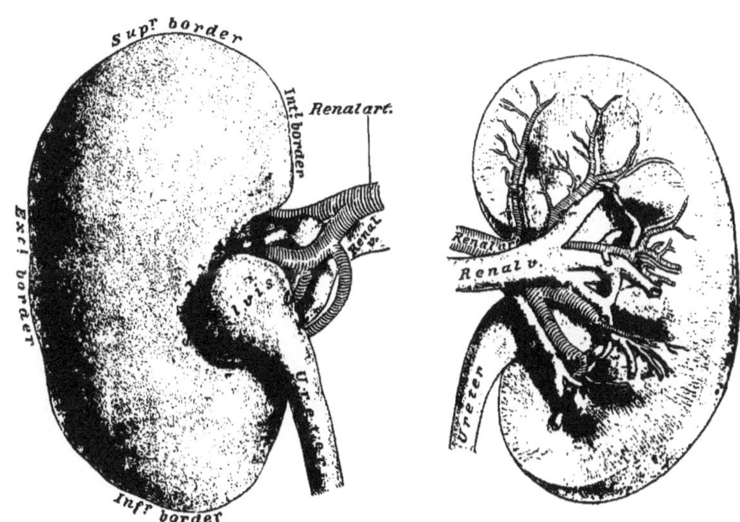
Supr. border
Intl. border
Renal art.
Renal v.
Extl. border
Pelvis
Ureter
Infr. border
Renal a.
Renal v.
Ureter

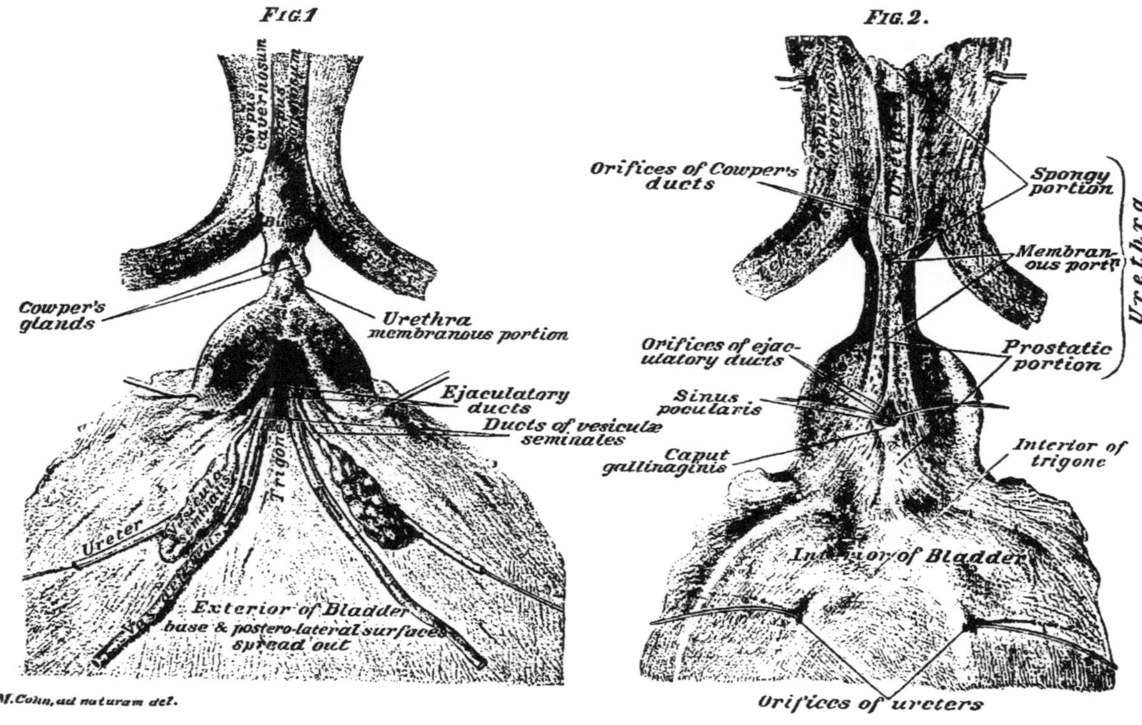

FIG. 1
Corpus cavernosum
Cowper's glands
Urethra
membranous portion
Ejaculatory ducts
Ducts of vesiculæ seminales
Trigone
Ureter
Vesicula seminalis
Exterior of Bladder
base & postero-lateral surfaces
spread out
FIG. 2.
Orifices of Cowper's ducts
Spongy portion
Membranous portn
Urethra
Prostatic portion
Orifices of ejaculatory ducts
Sinus pocularis
Caput gallinaginis
Interior of trigone
Interior of Bladder
Orifices of ureters

M. Cohn, ad naturam del.

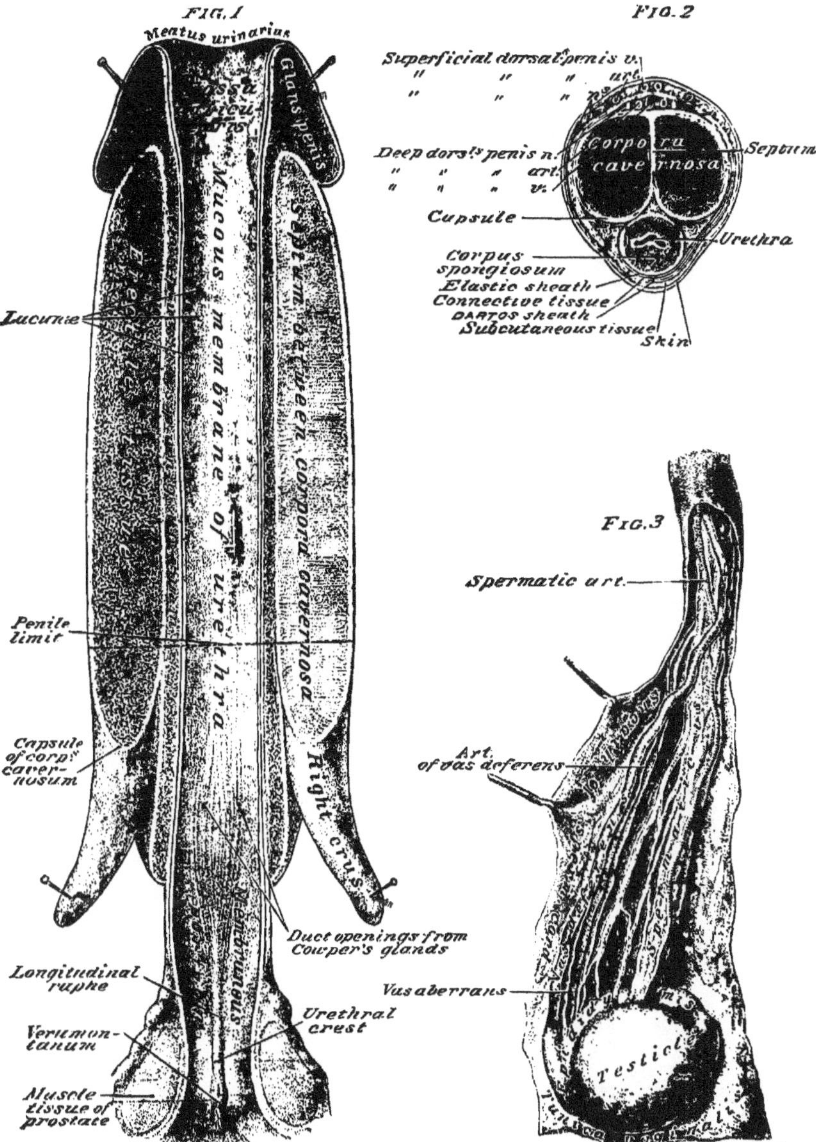
FIG. 1
Meatus urinarius
Glans penis
Fossa navicularis
Mucous membrane of urethra
Septum between corpora cavernosa
Erectile tissue
Lacunae
Penile limit
Capsule of corp. cavernosum
Right crus
Membranous
Duct openings from Cowper's glands
Longitudinal raphe
Urethral crest
Verumontanum
Muscle tissue of prostate
FIG. 2
Superficial dorsal penis v.
" " " art.
" " " n.
Deep dors.ls penis n.
" " " art.
" " " v.
Corpora cavernosa
Septum
Cupsule
Urethra
Corpus spongiosum
Elastic sheath
Connective tissue
Dartos sheath
Subcutaneous tissue
Skin
FIG. 3
Spermatic art.
Art. of vas deferens
Vas aberrans
Testicle
Tunica vaginalis
M.C ad.nat.del.

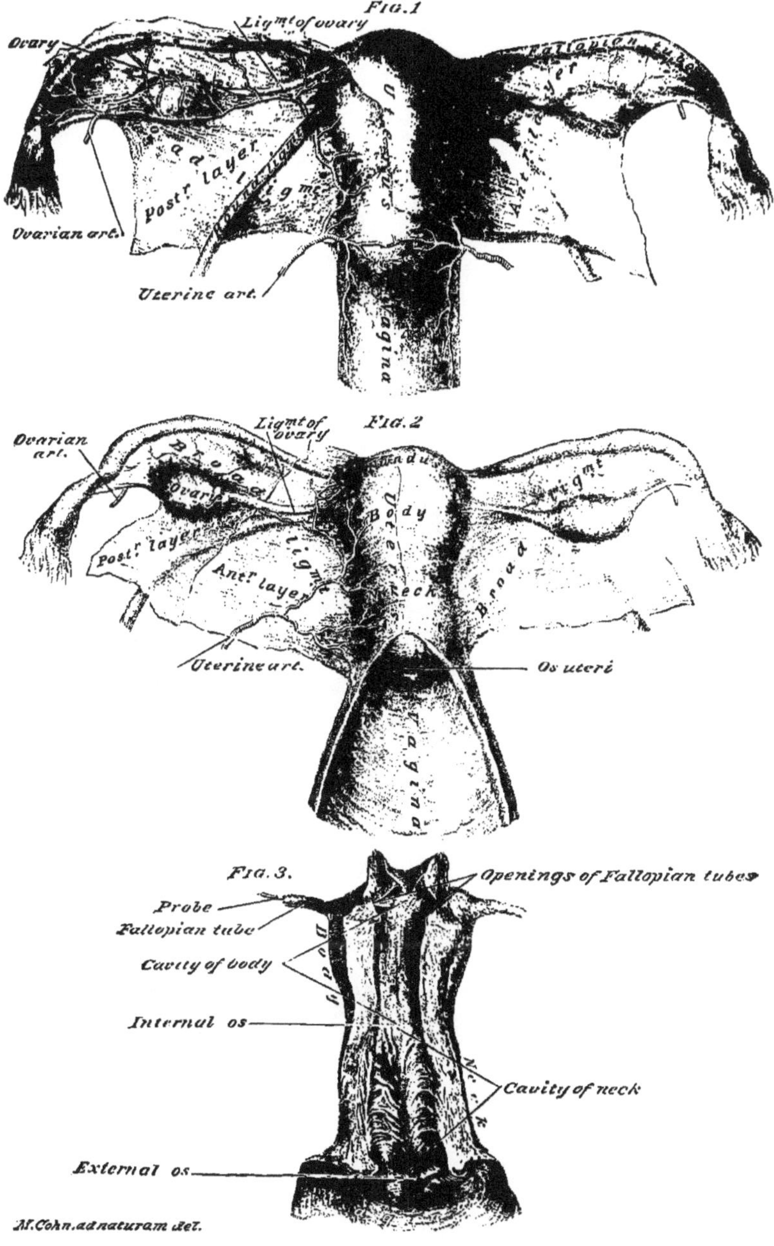
Fig. 1
Ligmt of ovary
Ovary
Fallopian tube
Uterus
Post.r layer
Round ligmt
Broad ligmt
Ant.r layer
Ovarian art.
Uterine art.
Vagina
Fig. 2
Ovarian art.
Ligmt of ovary
Broad
Ovary
Fundus
Body
Uterus
Neck
Post.r layer
Ant.r layer
Broad ligmt
Uterine art.
Os uteri
Vagina
Fig. 3.
Openings of Fallopian tubes
Probe
Fallopian tube
Body
Cavity of body
Internal os
Neck
Cavity of neck
External os

M. Cohn. ad naturam del.

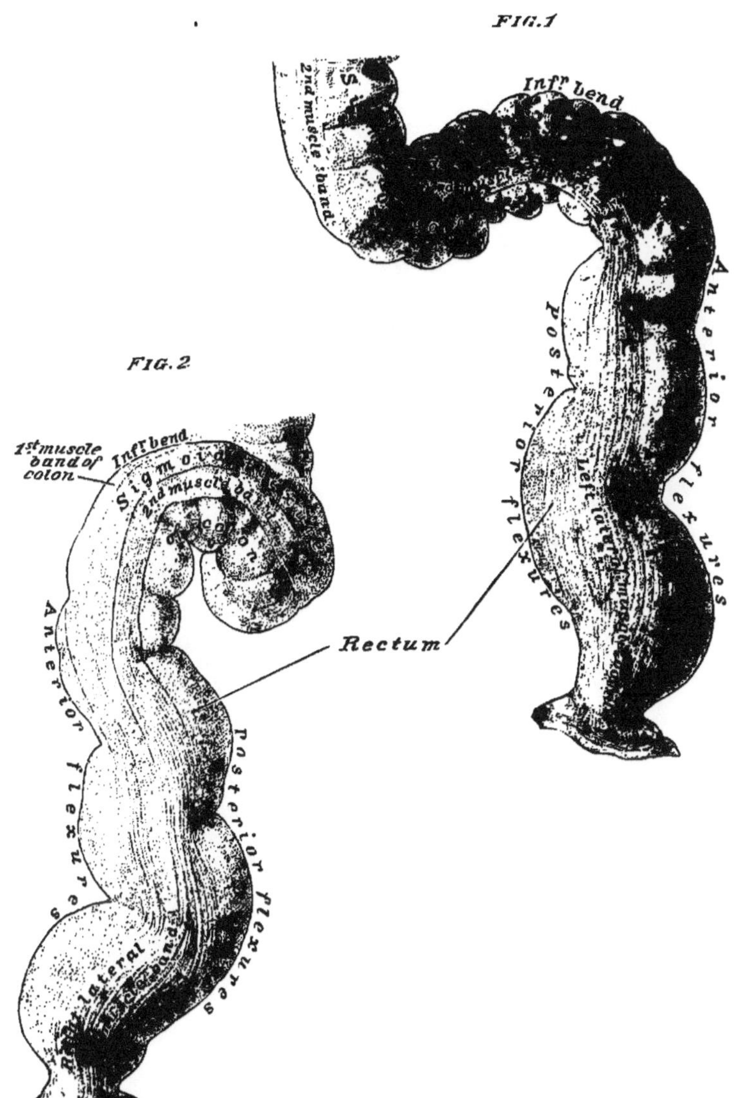
FIG. 1
Infr. bend
2nd muscle band
Anterior flexures
Posterior flexures
Left lateral muscle band
FIG. 2
1st muscle band of colon
Infr. bend
Sigmoid
2nd muscle band
Anterior flexures
Posterior flexures
Right lateral muscle band
Rectum

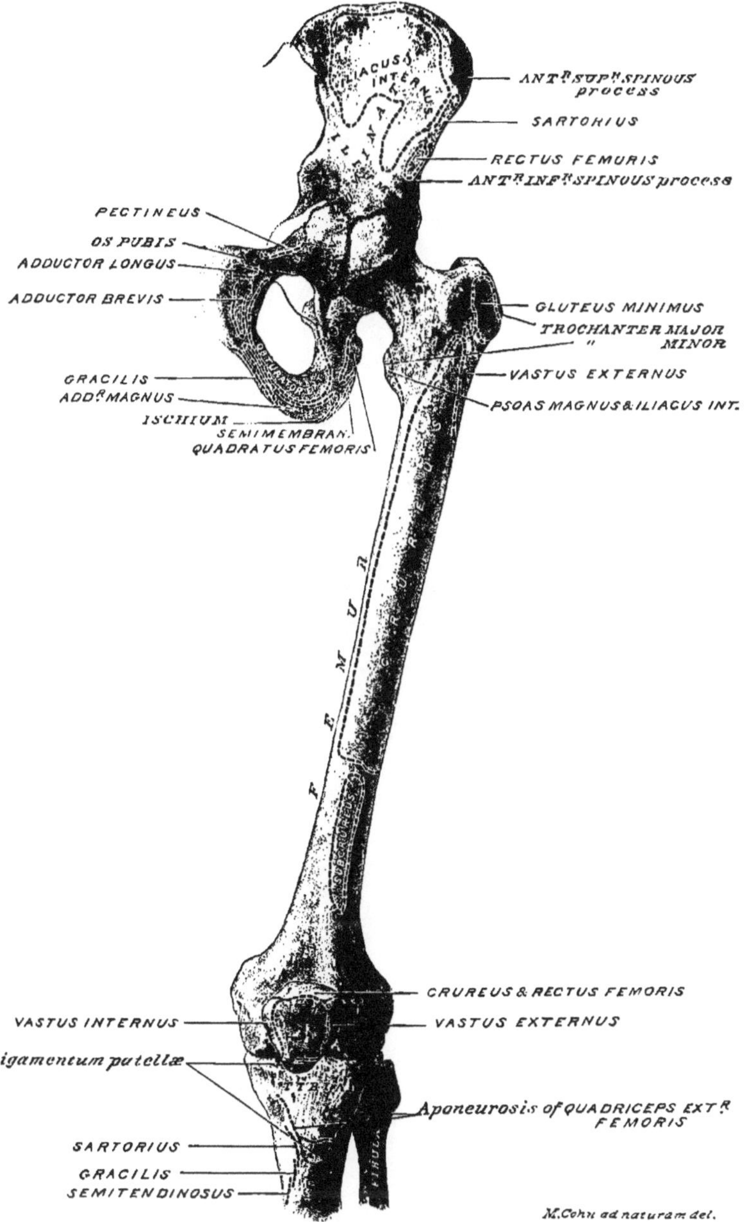
ILIACUS INTERNUS
ANTR. SUPR. SPINOUS process
SARTORIUS
RECTUS FEMORIS
ANTR. INFR. SPINOUS process
PECTINEUS
OS PUBIS
ADDUCTOR LONGUS
ADDUCTOR BREVIS
GLUTEUS MINIMUS
TROCHANTER MAJOR
" MINOR
VASTUS EXTERNUS
GRACILIS
ADDR. MAGNUS
PSOAS MAGNUS & ILIACUS INT.
ISCHIUM
SEMIMEMBRAN.
QUADRATUS FEMORIS
FEMUR
CRUREUS
CRUREUS & RECTUS FEMORIS
VASTUS INTERNUS
VASTUS EXTERNUS
Ligamentum patellæ
Aponeurosis of QUADRICEPS EXTR. FEMORIS
SARTORIUS
GRACILIS
SEMITENDINOSUS
FIBULA
M. Cohn ad naturam del.

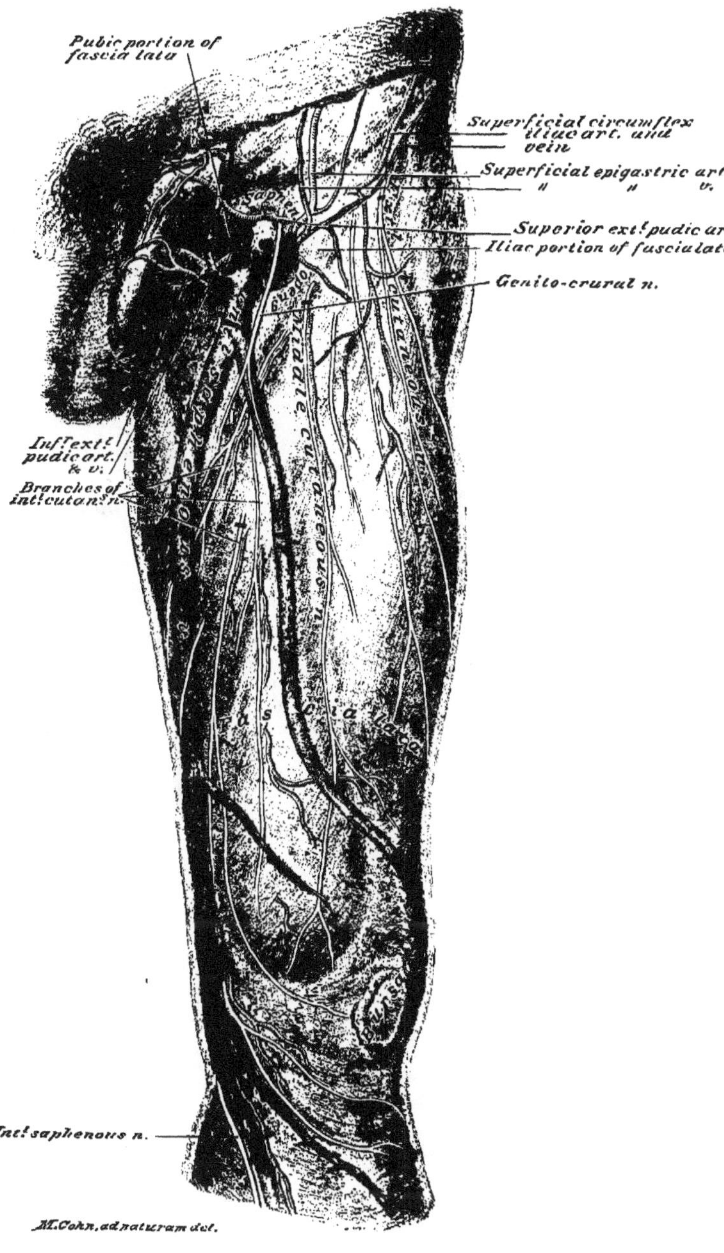

Pubic portion of fascia lata
Superficial circumflex iliac art. and vein
Superficial epigastric art.
" " v.
Superior ext.l pudic art.
Iliac portion of fascia lata
Genito-crural n.
Saphen
Middle cutaneous n.
fascia lata
Inf. ext.l pudic art. & v.
Branches of int.l cutan.s n.
Int.l saphenous n.
M. Cohn, ad naturam del.

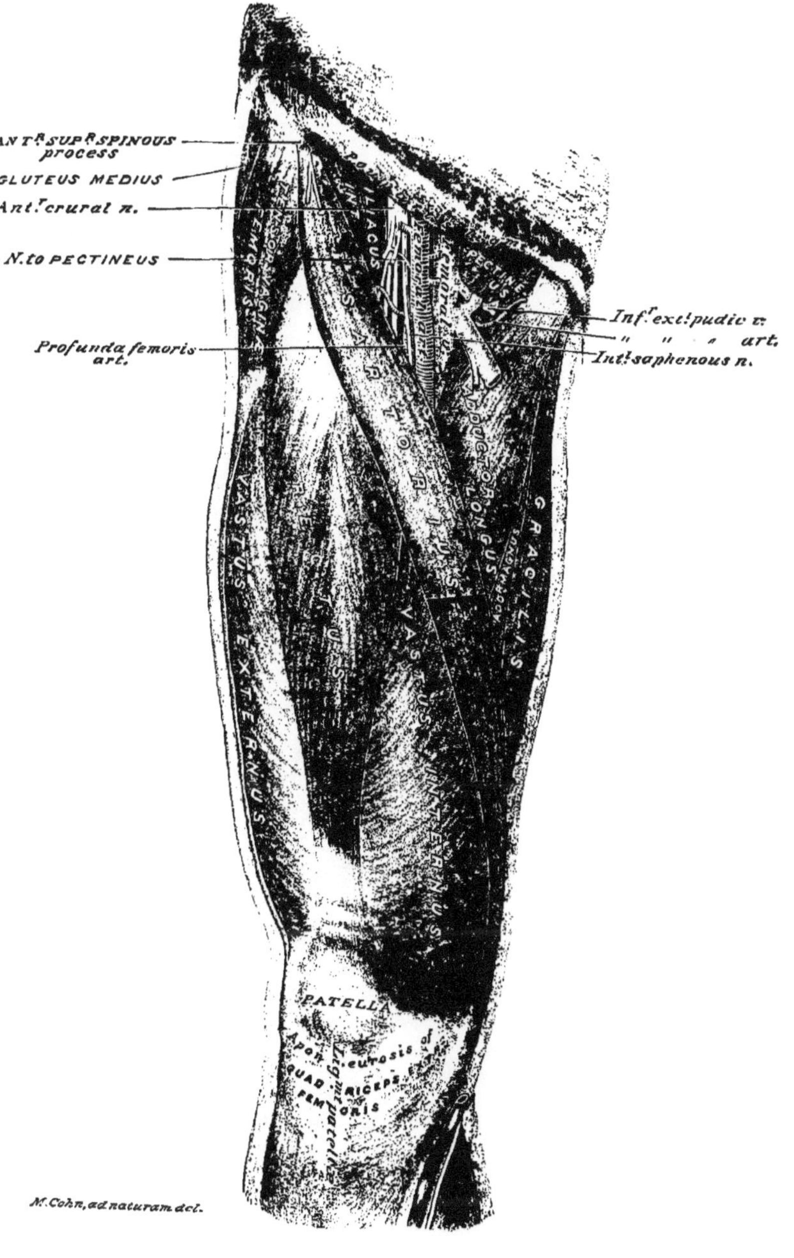
ANTR. SUPR. SPINOUS process
GLUTEUS MEDIUS
Ant.r crural n.
N. to PECTINEUS
Profunda femoris art.
Infr ext. pudic v.
" " " art.
Int. saphenous n.
ILIACUS
PECTINEUS
SARTORIUS
ADDUCTOR LONGUS
GRACILIS
VASTUS EXTERNUS
VASTUS INTERNUS
PATELLA
Aponeurosis of QUADRICEPS EXTENSOR FEMORIS
Ligm. patellæ

M. Cohn, ad naturam del.

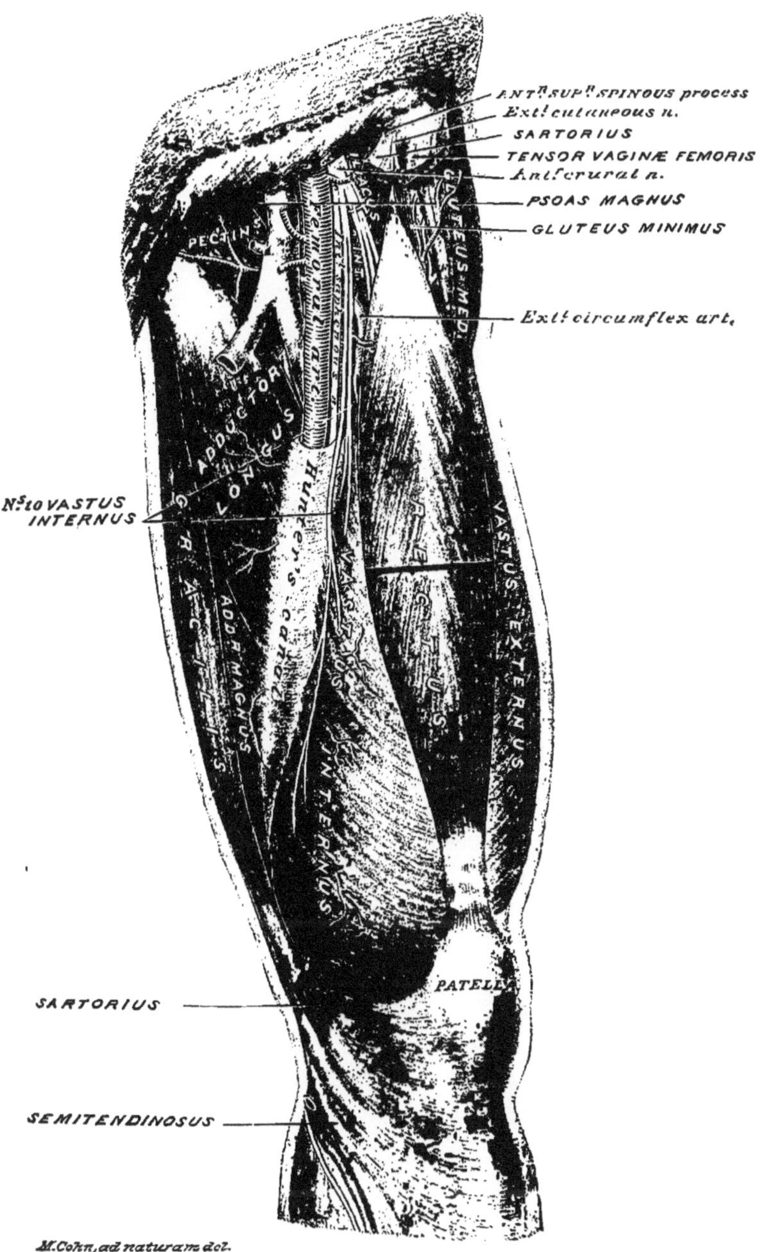
ANT^R SUP^R SPINOUS process
Ext! cutaneous n.
SARTORIUS
TENSOR VAGINÆ FEMORIS
Ant! crural n.
PSOAS MAGNUS
GLUTEUS MINIMUS
Ext! circumflex art.
N^S to VASTUS INTERNUS
SARTORIUS
SEMITENDINOSUS
PECTINS
ADDUCTOR LONGUS
femoral art.
Hunter's canal
GRACILIS
ADDR MAGNUS
VASTUS INTERNUS
RECTUS
VASTUS EXTERNUS
GLUTEUS MED.
PATELL
ILIACUS

M.Cohn, ad naturam del.

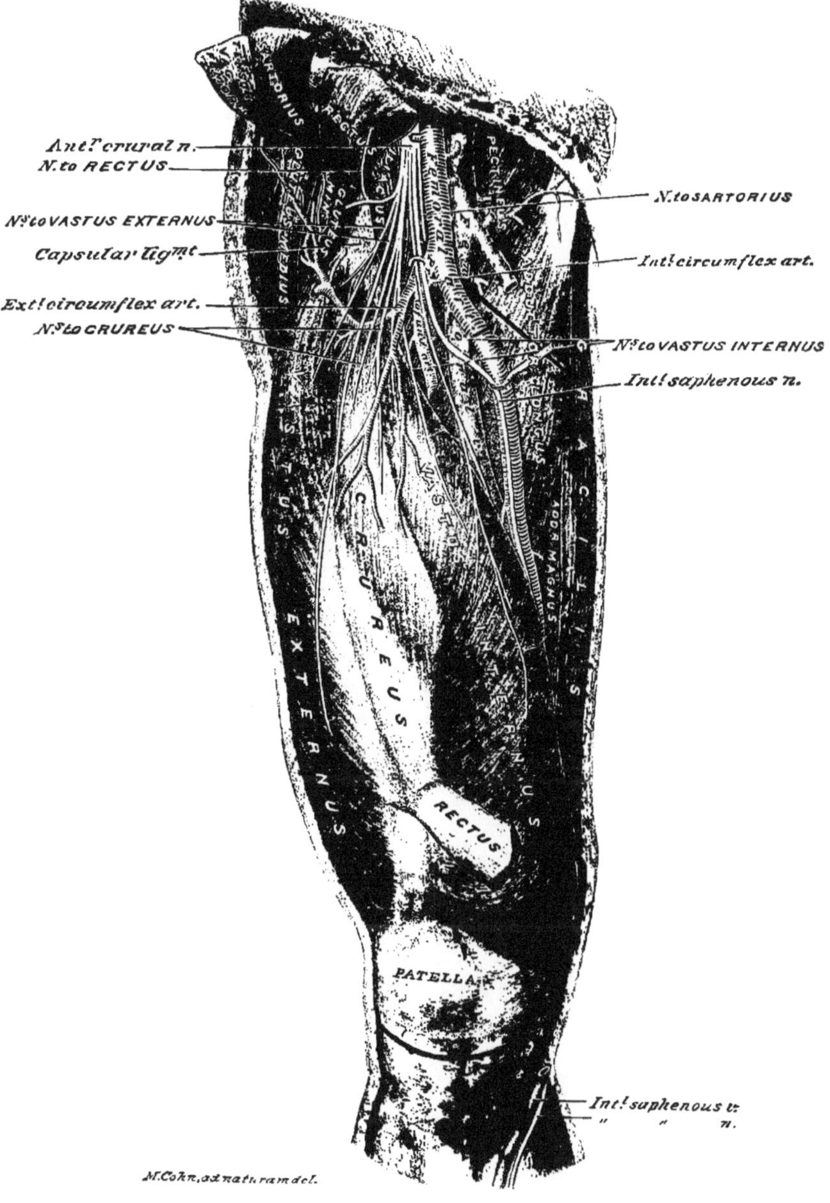
Antr crural n.
N. to RECTUS
Ns to VASTUS EXTERNUS
Capsular ligmt
Extl circumflex art.
Ns to CRUREUS
N. to SARTORIUS
Intl circumflex art.
Ns to VASTUS INTERNUS
Intl saphenous n.
PECTINEUS
CRUREUS
VASTUS EXTERNUS
ADD. MAGNUS
RECTUS
PATELLA
Intl saphenous v.
" " n.
M. Cohn, ad naturam del.

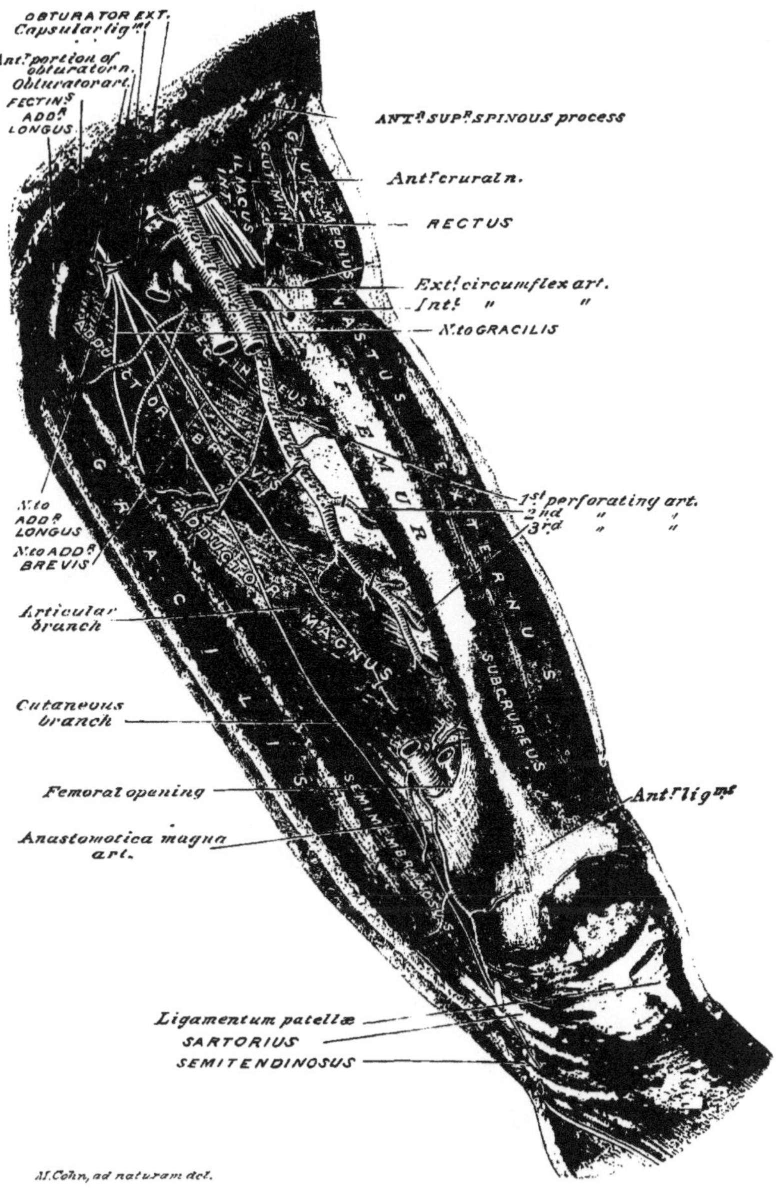
OBTURATOR EXT.
Capsular ligmt
Antr portion of obturator n.
Obturator art.
PECTINS
ADDR LONGUS
ANTR SUPR SPINOUS process
Antr crural n.
RECTUS
Extl circumflex art.
Intl " "
N. to GRACILIS
1st perforating art.
2nd " "
3rd " "
N. to ADDR LONGUS
N. to ADDR BREVIS
Articular branch
Cutaneous branch
Femoral opening
Anastomotica magna art.
Antr ligmt
Ligamentum patellæ
SARTORIUS
SEMITENDINOSUS
GLUT MIN
GLUT MEDIUS
ILIACUS INT.
ADDUCTOR
PECTIN
ADDUCTOR BREVIS
ADDUCTOR MAGNUS
GRACILIS
VASTUS EXTERNUS
FEMUR
SUBCRUREUS
SEMIMEMBRANOSUS
M. Cohn, ad naturam del.

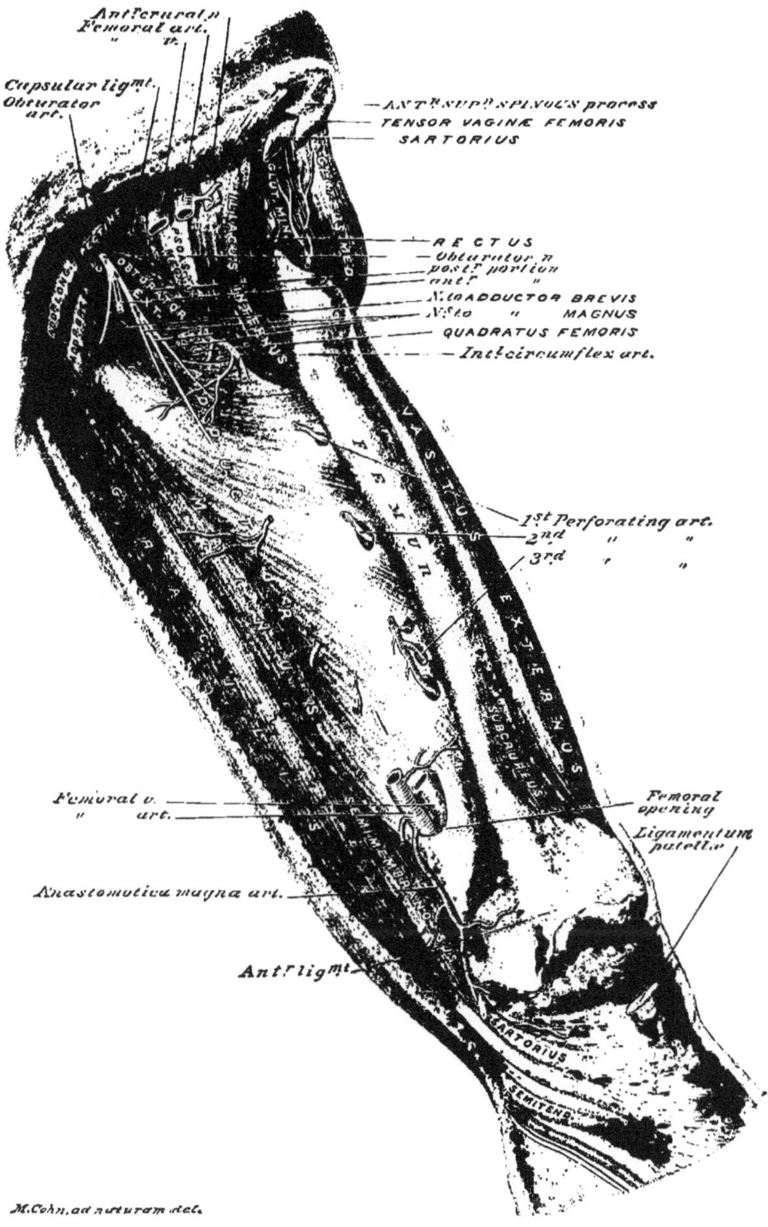
Antr crural n.
Femoral art.
" v.
Capsular ligmt.
Obturator art.
ANTR SUPR SPINOUS process
TENSOR VAGINÆ FEMORIS
SARTORIUS
RECTUS
Obturator n
postr portion
antr "
N. to ADDUCTOR BREVIS
N. to " MAGNUS
QUADRATUS FEMORIS
Intl circumflex art.
VASTUS EXTERNUS
FEMUR
1st Perforating art.
2nd " "
3rd " "
SUBCRUREUS
Femoral v.
" art.
Femoral opening
Ligamentum patellæ
Anastomotica magna art.
Antr ligmt.
SARTORIUS
SEMITEND.
GRACILIS
SEMIMEMBRANOSUS
GLUT. MIN.
GLUT. MED.
ILIACUS
PSOAS
PECTINEUS
OBTURATOR EXT.
ADDUCTOR BREVIS
ADDUCTOR LONGUS
ADDUCTOR MAGNUS
M. Cohn, ad naturam del.

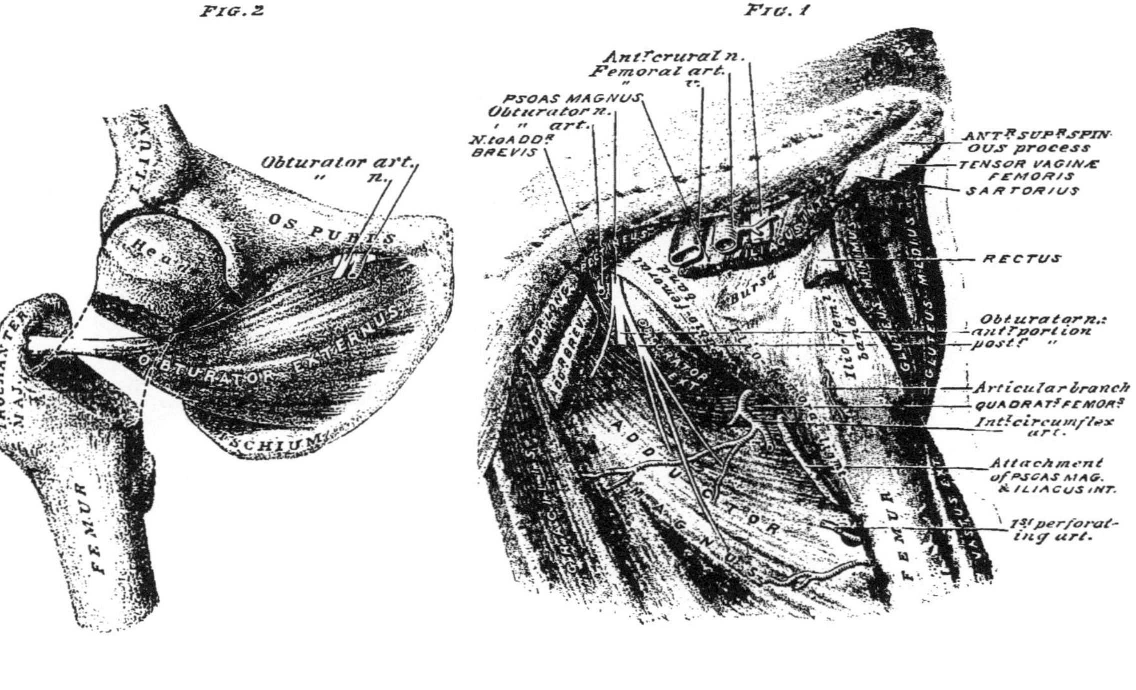
FIG. 1
ANT. SUP. SPIN. OUS process
TENSOR VAGINÆ FEMORIS
SARTORIUS
RECTUS
Obturator n.: ant. portion
post. "
Articular branch
QUADRAT. FEMOR.
Int. circumflex art.
Attachment of PSOAS MAG. & ILIACUS INT.
1st perforating art.
Ant. crural n.
Femoral art.
" v.
PSOAS MAGNUS
Obturator n.
" " art.
N. to ADDR. BREVIS
GLUTEUS MEDIUS
GLUTEUS MINIMUS
VASTUS EXT.
FEMUR
Ilio-femoral band
Ilio-femoral band
Bursa
OBTURATOR EXT.
ADDUCTOR MAGNUS
ADD. BREV.
GRACILIS
FIG. 2
Obturator art.
" n.
OS PUBIS
OBTURATOR EXTERNUS
ISCHIUM
ILIUM
Head
FEMUR
TROCHANTER MAJ.

M. Cohn, ad. nat. del.

PLATE 67

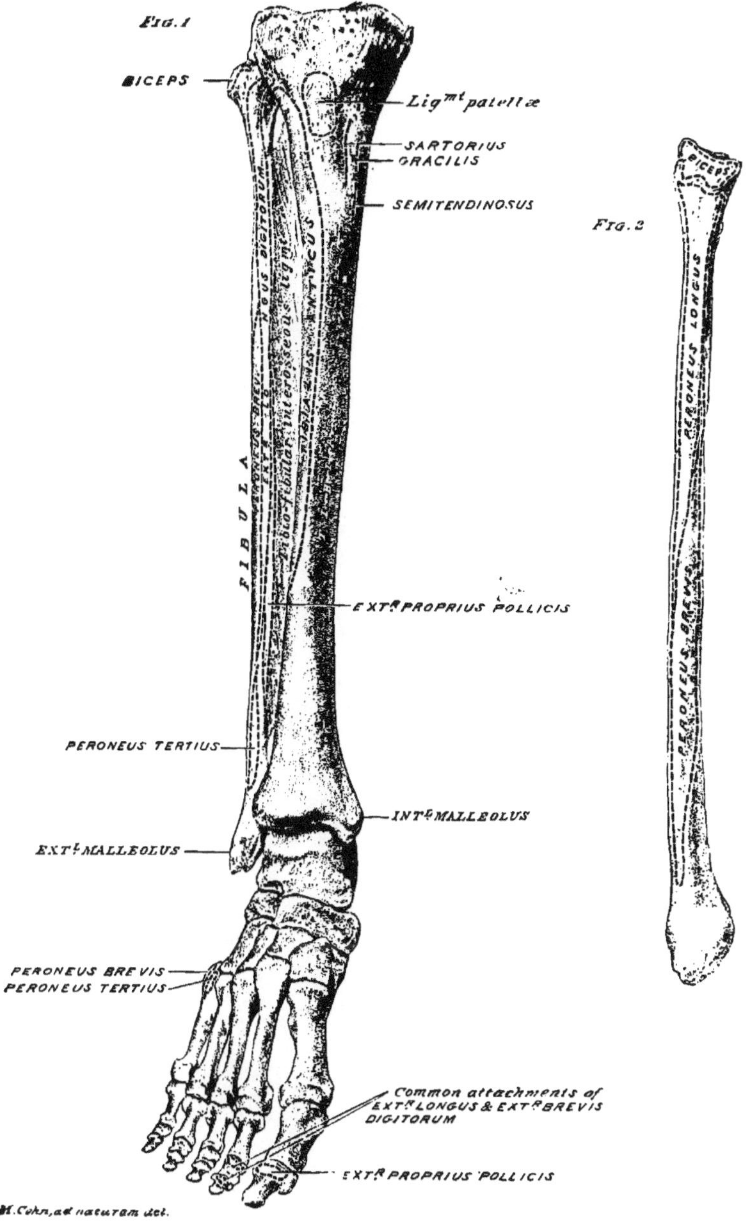

FIG. 1
BICEPS
Ligmt patellæ
SARTORIUS
GRACILIS
SEMITENDINOSUS
FIBULA
EXTR PROPRIUS POLLICIS
PERONEUS TERTIUS
INTL MALLEOLUS
EXTL MALLEOLUS
PERONEUS BREVIS
PERONEUS TERTIUS
Common attachments of
EXTR LONGUS & EXTR BREVIS
DIGITORUM
EXTR PROPRIUS POLLICIS
FIG. 2
BICEPS
PERONEUS LONGUS
PERONEUS BREVIS
M. Cohn, ad naturam del.

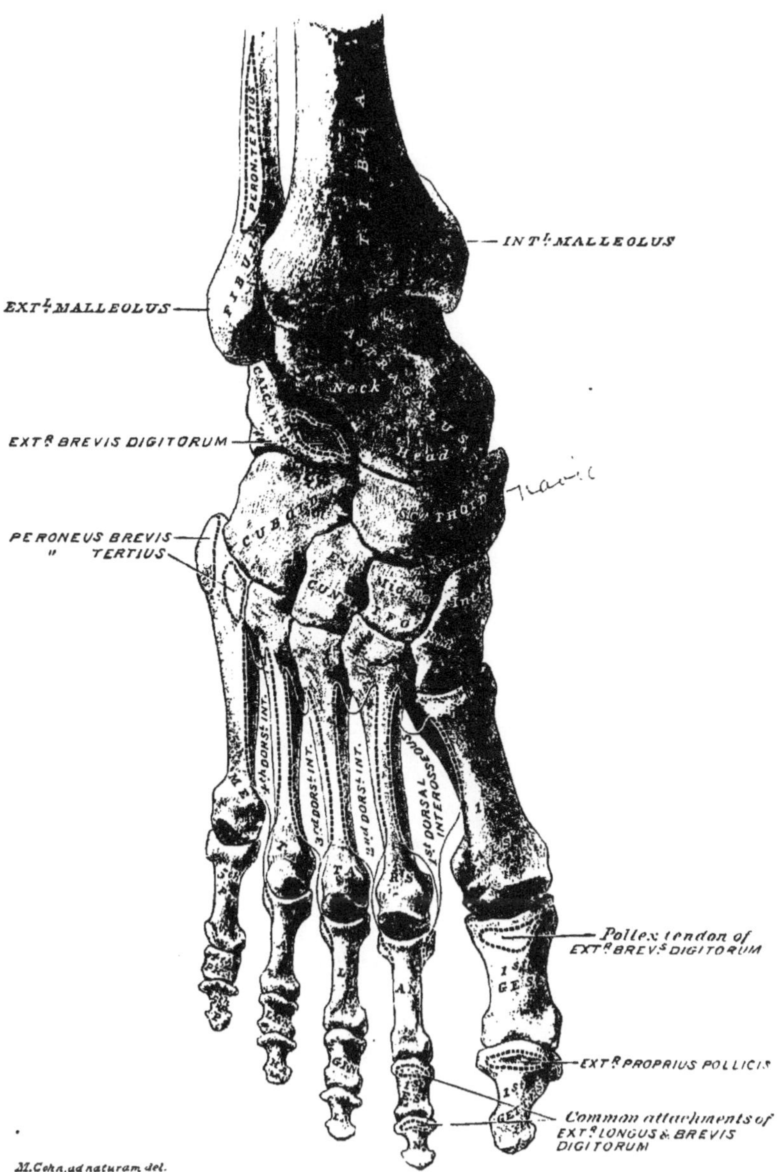
INT^L MALLEOLUS
EXT^L MALLEOLUS
EXT^R BREVIS DIGITORUM
PERONEUS BREVIS
" TERTIUS
Pollex tendon of
EXT^R BREV^S DIGITORUM
EXT^R PROPRIUS POLLICIS
Common attachments of
EXT^S LONGUS & BREVIS
DIGITORUM
Neck
Head
CUBOID
SCAPHOID
CALCANEUS
PERONEUS TERTIUS
FIBULA
TIBIA
4th DORS^L INT.
3rd DORS^L INT.
2nd DORS^L INT.
1st DORSAL
INTEROSSEOUS

M. Cohn, ad naturam del.

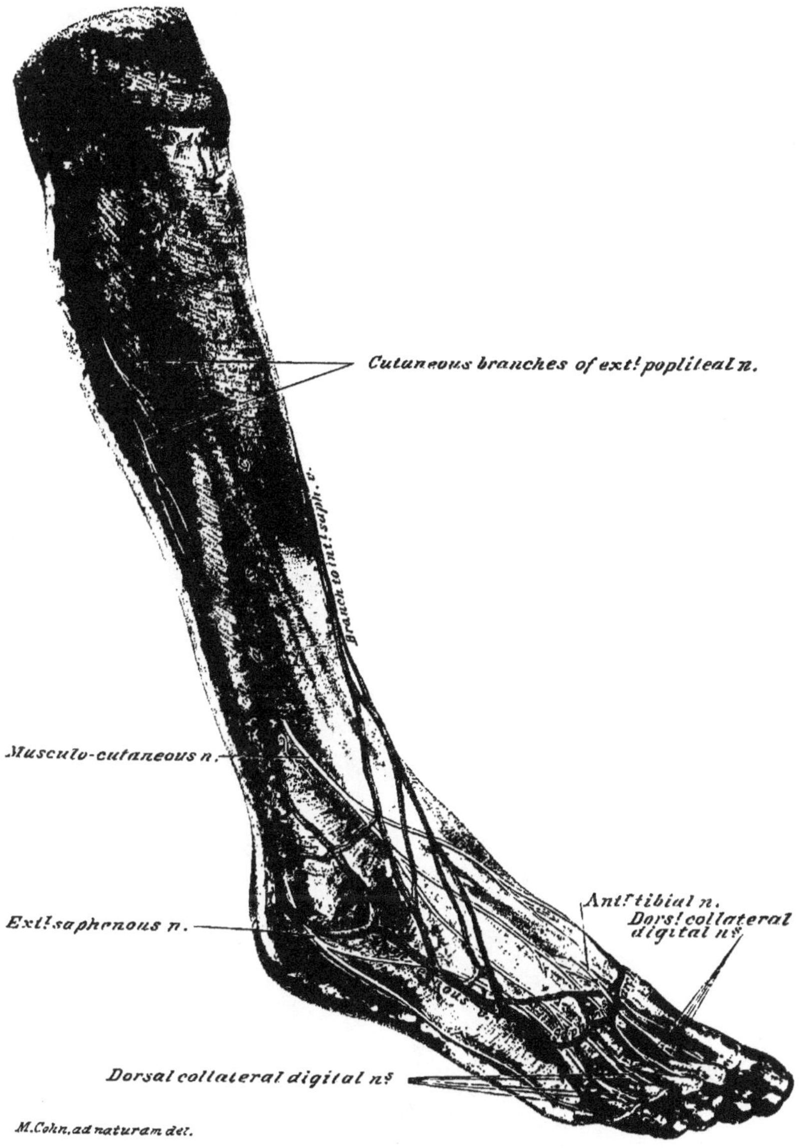
Cutaneous branches of ext! popliteal n.
Branch to int! saph. v.
Musculo-cutaneous n.
Ant! tibial n.
Dors! collateral digital n^s
Ext! saphenous n.
Dorsal collateral digital n^s
M. Cohn, ad naturam del.

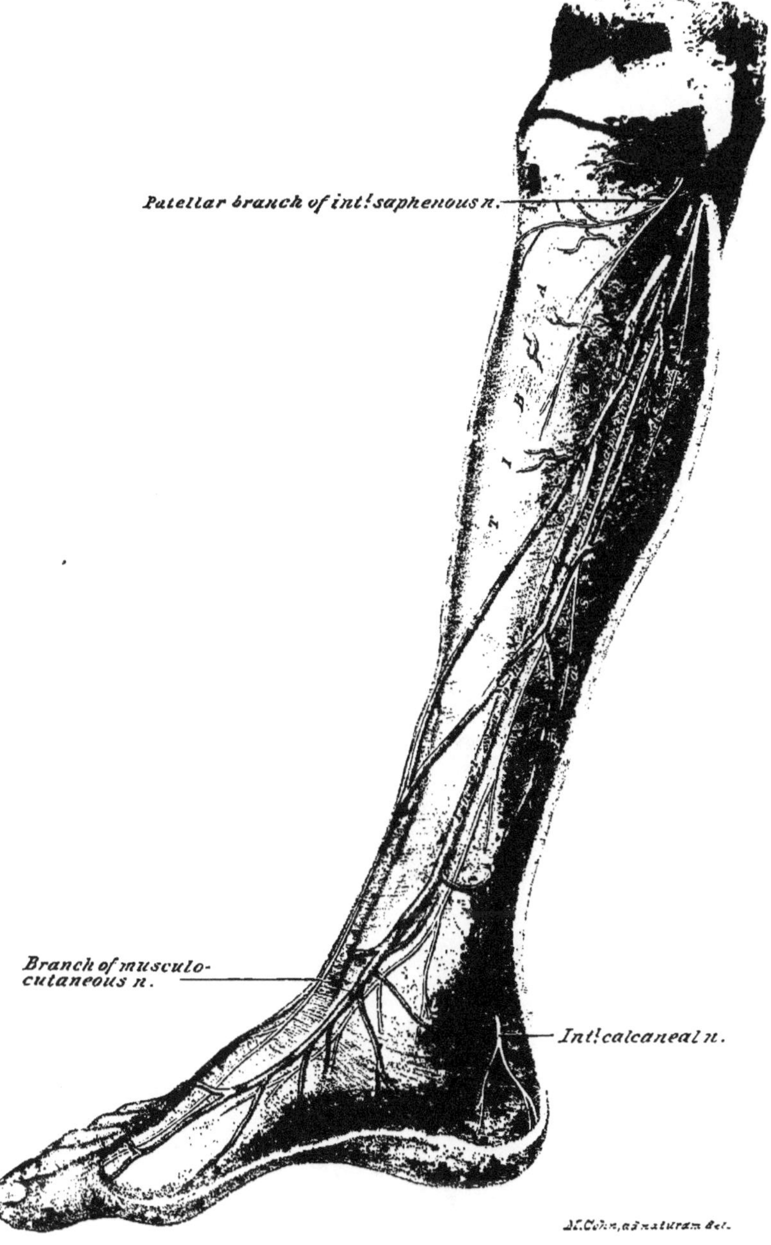
Patellar branch of int! saphenous n.
Branch of musculo-cutaneous n.
Int! calcaneal n.
M. Cohn, ad naturam del.

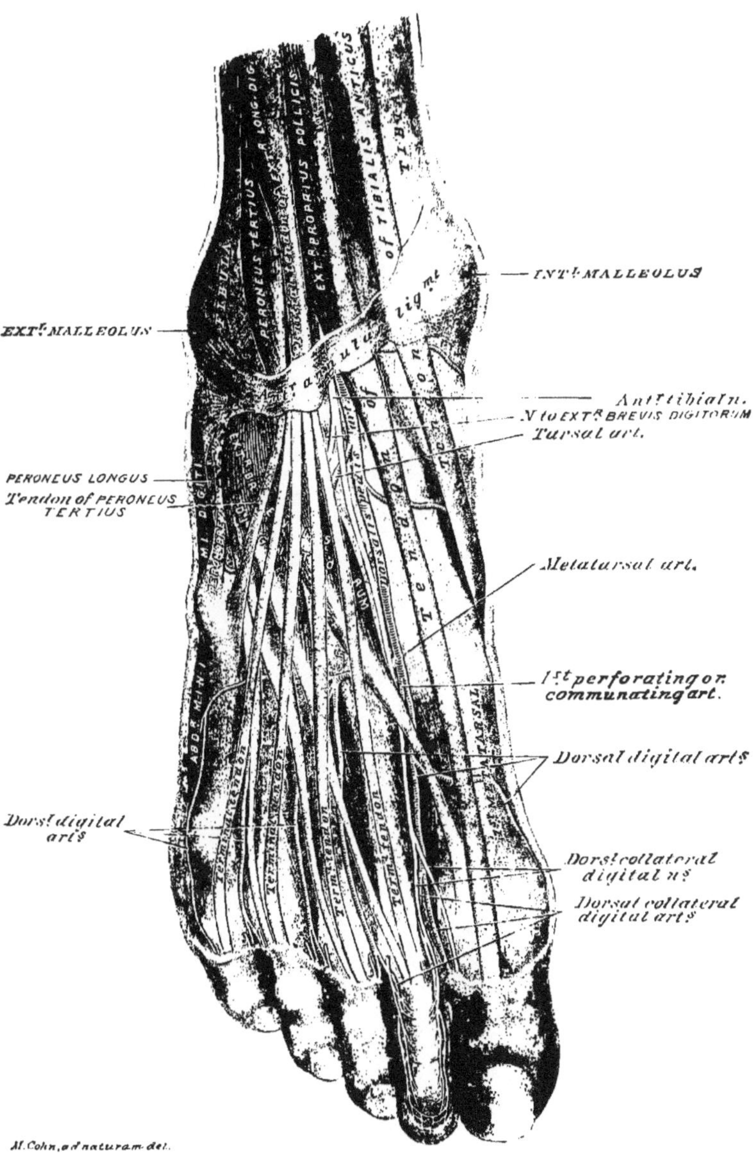
INTᴸ MALLEOLUS
EXTᴸ MALLEOLUS
Antᴿ tibial n.
N to EXTᴿ BREVIS DIGITORUM
Tarsal art.
PERONEUS LONGUS
Tendon of PERONEUS TERTIUS
Metatarsal art.
1st perforating or communating art.
Dorsal digital artˢ
Dorsˡ digital artˢ
Dorsˡ collateral digital nˢ
Dorsal collateral digital artˢ
FIBULA
PERONEUS TERTIUS
EXT. PROPRIUS POLLICIS
of TIBIALIS ANTICUS
TIBIA
Ant annular ligᵐᵗ
Tendon
Term. tendon
M. Cohn, ad naturam del.

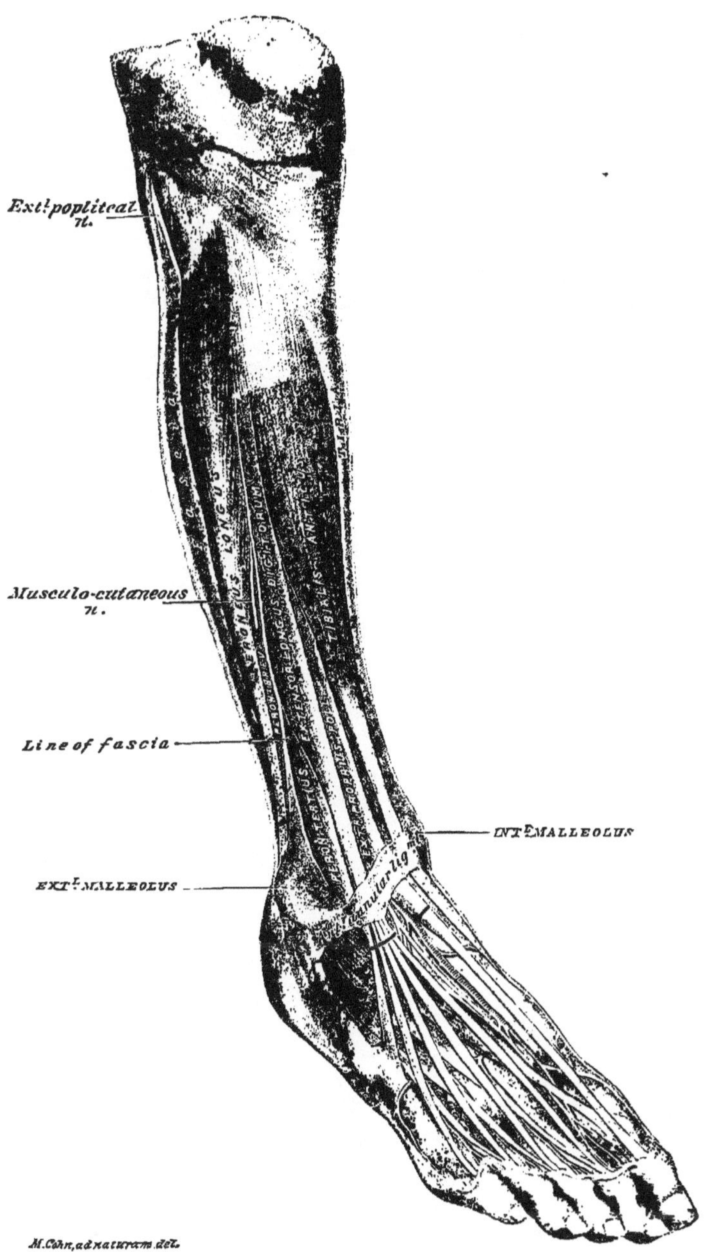
Extl popliteal n.
Musculo-cutaneous n.
Line of fascia
INTL MALLEOLUS
EXTL MALLEOLUS
Annular lig.

M. Cohn, ad naturam del.

PLATE 73

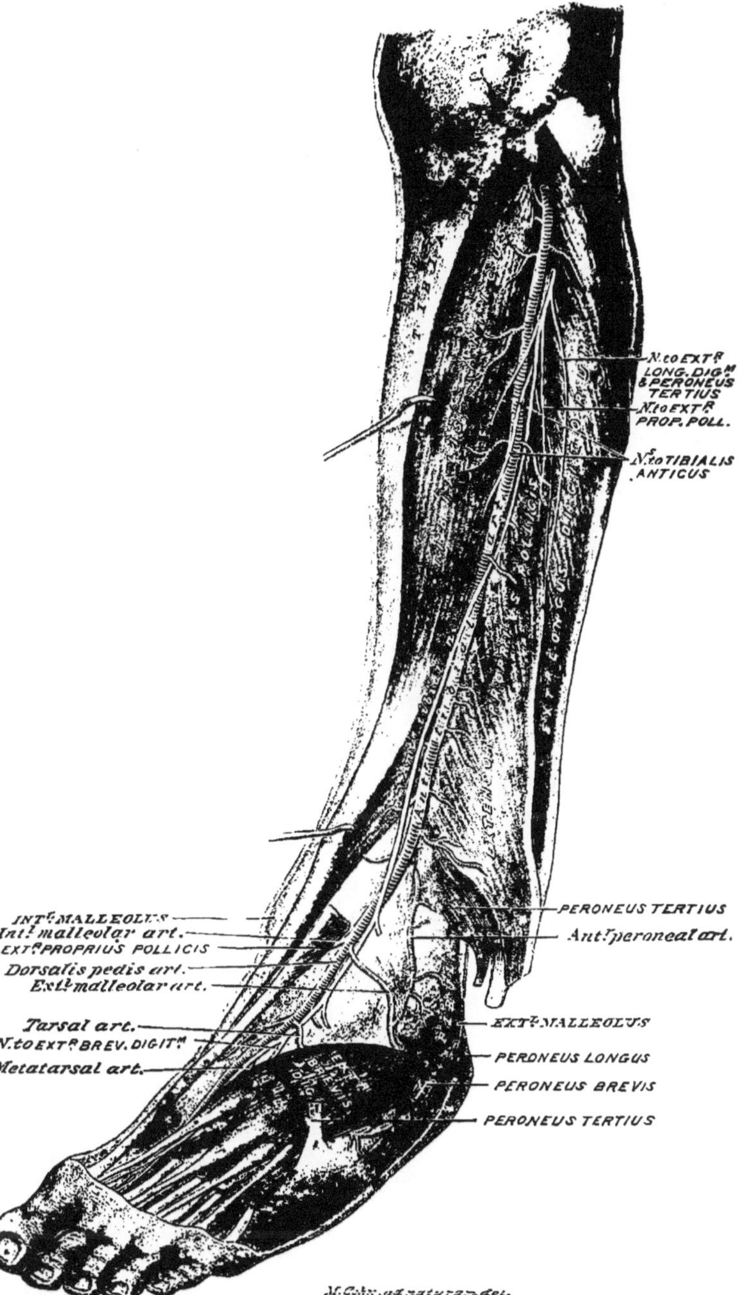
N. to EXT.R LONG. DIG.M & PERONEUS TERTIUS
N. to EXT.R PROP. POLL.
N.S to TIBIALIS ANTICUS
INT.L MALLEOLUS
Int.l malleolar art.
EXT.R PROPRIUS POLLICIS
Dorsalis pedis art.
Ext.l malleolar art.
Tarsal art.
N. to EXT.R BREV. DIGIT.M
Metatarsal art.
PERONEUS TERTIUS
Ant.r peroneal art.
EXT.L MALLEOLUS
PERONEUS LONGUS
PERONEUS BREVIS
PERONEUS TERTIUS

M. Cohn, ad naturam del.

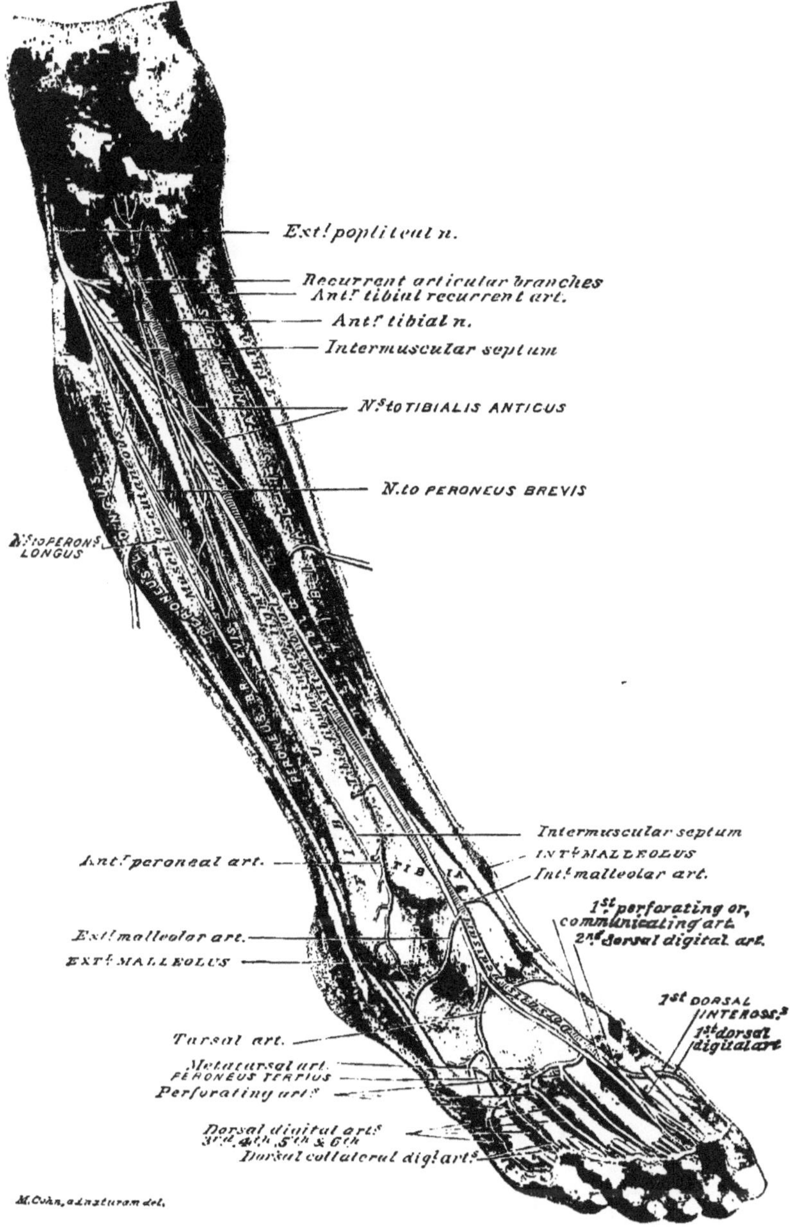
Ext! popliteal n.
Recurrent articular branches
Antr tibial recurrent art.
Antr tibial n.
Intermuscular septum
Ns to TIBIALIS ANTICUS
N. to PERONEUS BREVIS
Ns to PERONs LONGUS
Intermuscular septum
INTl MALLEOLUS
Antr peroneal art.
Intl malleolar art.
1st perforating or, communicating art.
Extl malleolar art.
2nd dorsal digital art.
EXTl MALLEOLUS
1st DORSAL INTEROSS.s
1st dorsal digital art
Tarsal art.
Metatarsal art.
PERONEUS TERTIUS
Perforating arts
Dorsal digital arts 3rd, 4th, 5th & 6th
Dorsal collateral digl arts
TIBIA
M. Cohn, ad naturam del.

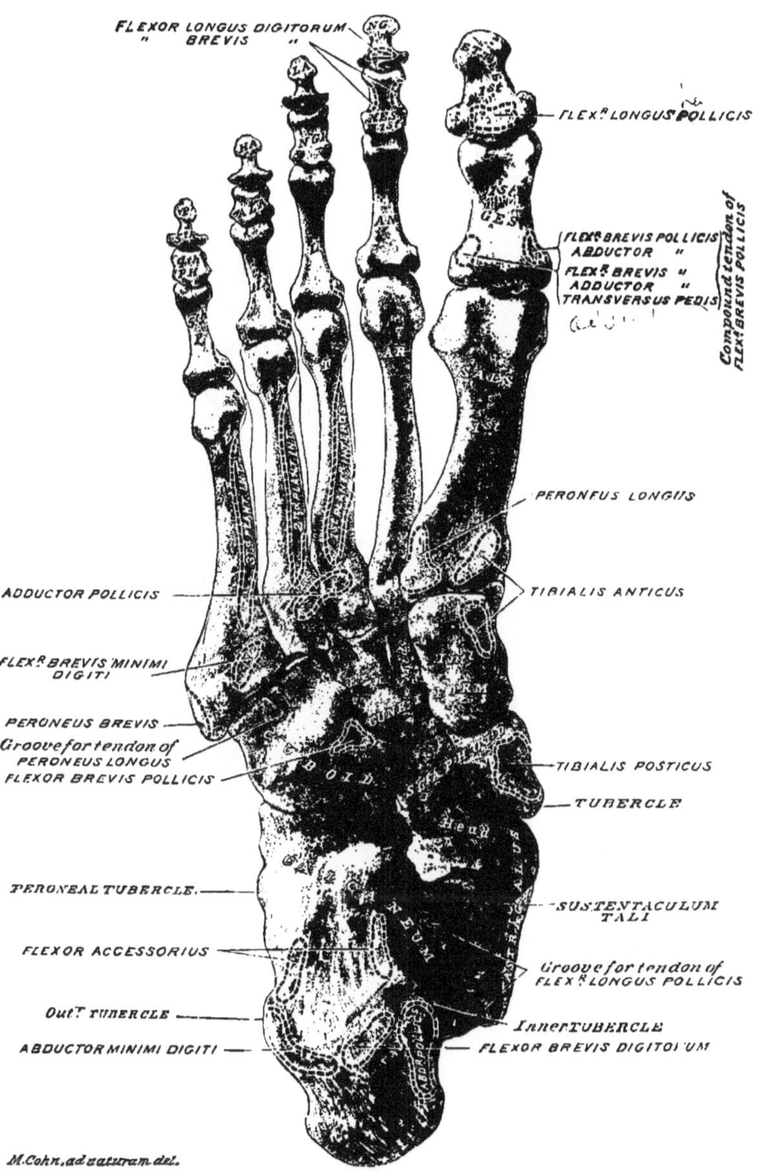
FLEXOR LONGUS DIGITORUM
" BREVIS "
FLEX.R LONGUS POLLICIS
FLEX.R BREVIS POLLICIS
ABDUCTOR "
FLEX.R BREVIS "
ADDUCTOR "
TRANSVERSUS PEDIS
Compound tendon of FLEX.R BREVIS POLLICIS
PERONEUS LONGUS
ADDUCTOR POLLICIS
TIBIALIS ANTICUS
FLEX.R BREVIS MINIMI DIGITI
PERONEUS BREVIS
Groove for tendon of PERONEUS LONGUS
FLEXOR BREVIS POLLICIS
TIBIALIS POSTICUS
TUBERCLE
PERONEAL TUBERCLE.
SUSTENTACULUM TALI
FLEXOR ACCESSORIUS
Groove for tendon of FLEX.R LONGUS POLLICIS
Out.r TUBERCLE
Inner TUBERCLE
ABDUCTOR MINIMI DIGITI
FLEXOR BREVIS DIGITORUM
M.Cohn, ad naturam del.

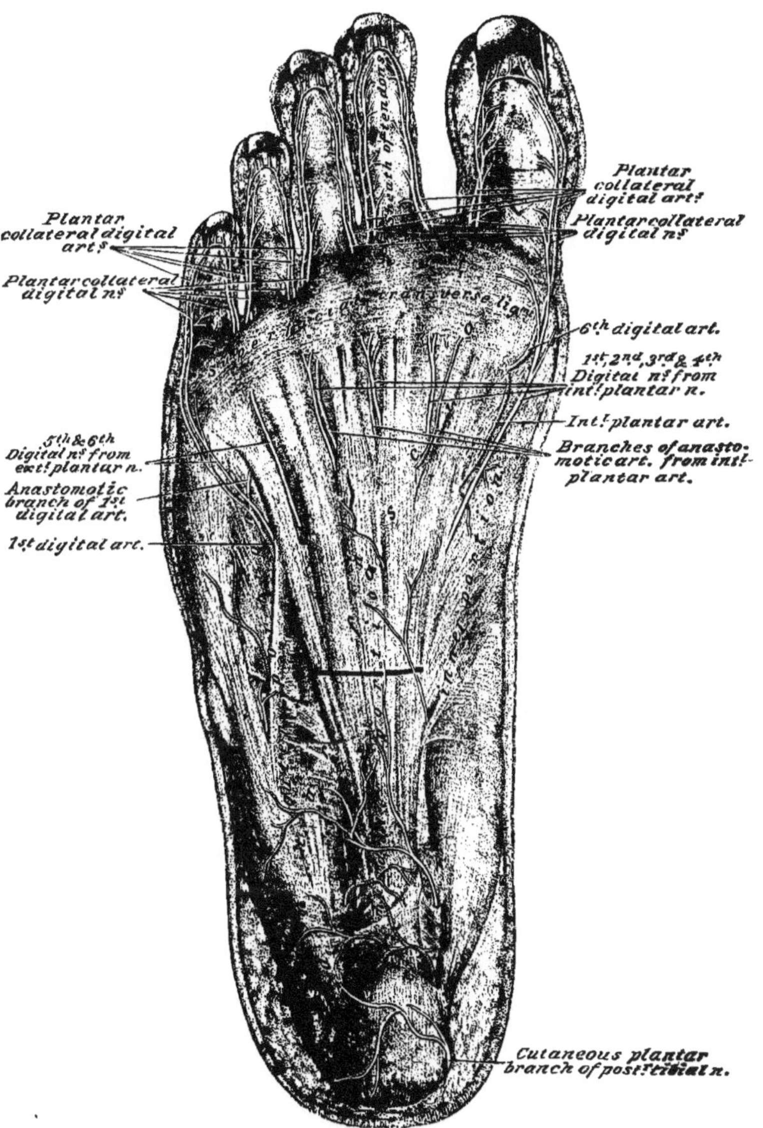
Plantar collateral digital art.s
Plantar collateral digital n.s
Plantar collateral digital art.s
Plantar collateral digital n.s
6th digital art.
1st, 2nd, 3rd & 4th Digital n.s from int.l plantar n.
Int.l plantar art.
Branches of anastomotic art. from int.l plantar art.
5th & 6th Digital n.s from ext.l plantar n.
Anastomotic branch of 1st digital art.
1st digital art.
Cutaneous plantar branch of post.r tibial n.

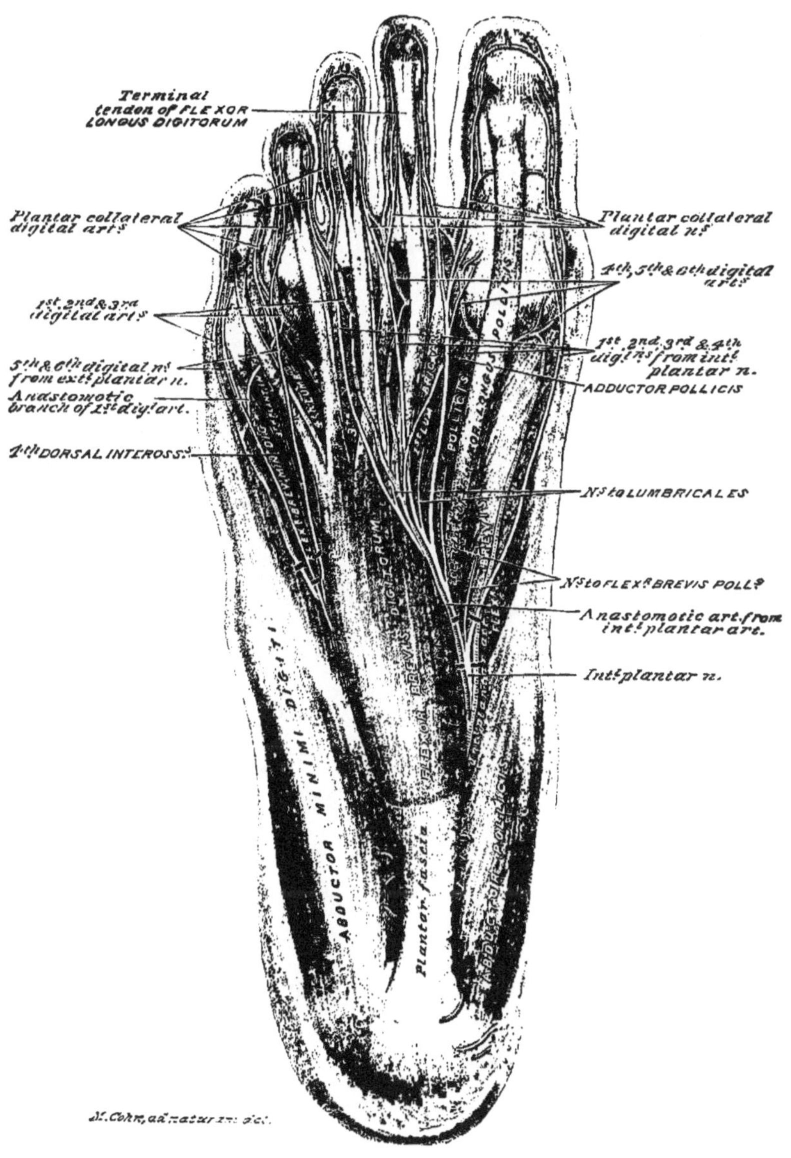

Terminal tendon of FLEXOR LONGUS DIGITORUM
Plantar collateral digital arts
Plantar collateral digital ns
4th, 5th & 6th digital arts
1st, 2nd & 3rd digital arts
1st, 2nd, 3rd & 4th digl ns from intl plantar n.
5th & 6th digital ns from extl plantar n.
ADDUCTOR POLLICIS
Anastomotic branch of 1st digl art.
4th DORSAL INTEROSS.s
Ns to LUMBRICALES
Ns to FLEXr BREVIS POLL.s
Anastomotic art. from intl plantar art.
Intl plantar n.
ABDUCTOR MINIMI DIGITI
Plantar fascia
M. Cohn, ad natur. del.

PLATE 78

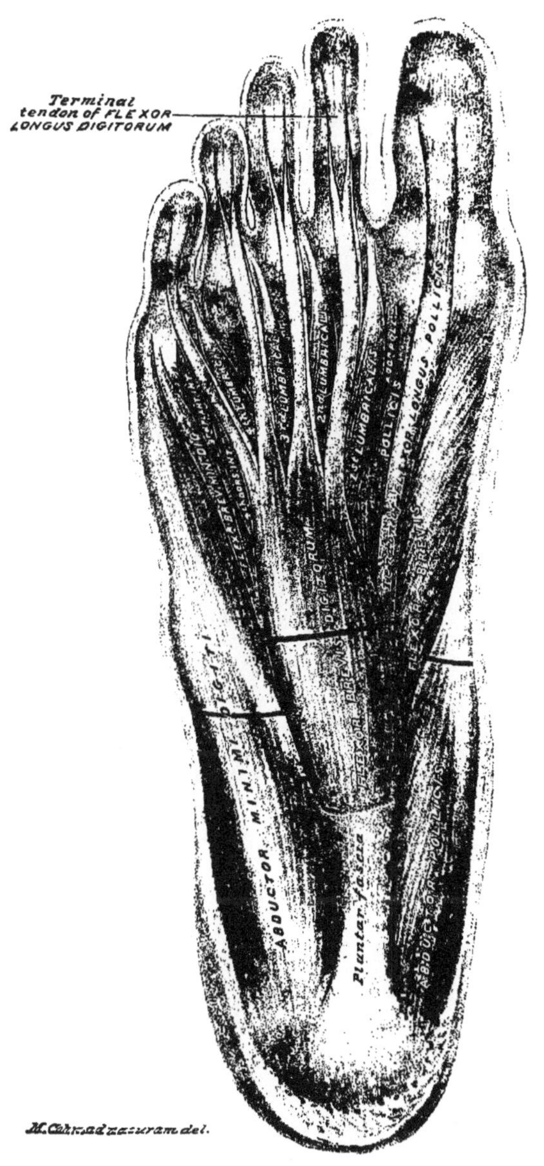
Terminal
tendon of FLEXOR
LONGUS DIGITORUM
2nd LUMBRICALIS
1st LUMBRICALIS
FLEXOR LONGUS POLLICIS
ABDUCTOR MINIMI DIGITI
Plantar fascia

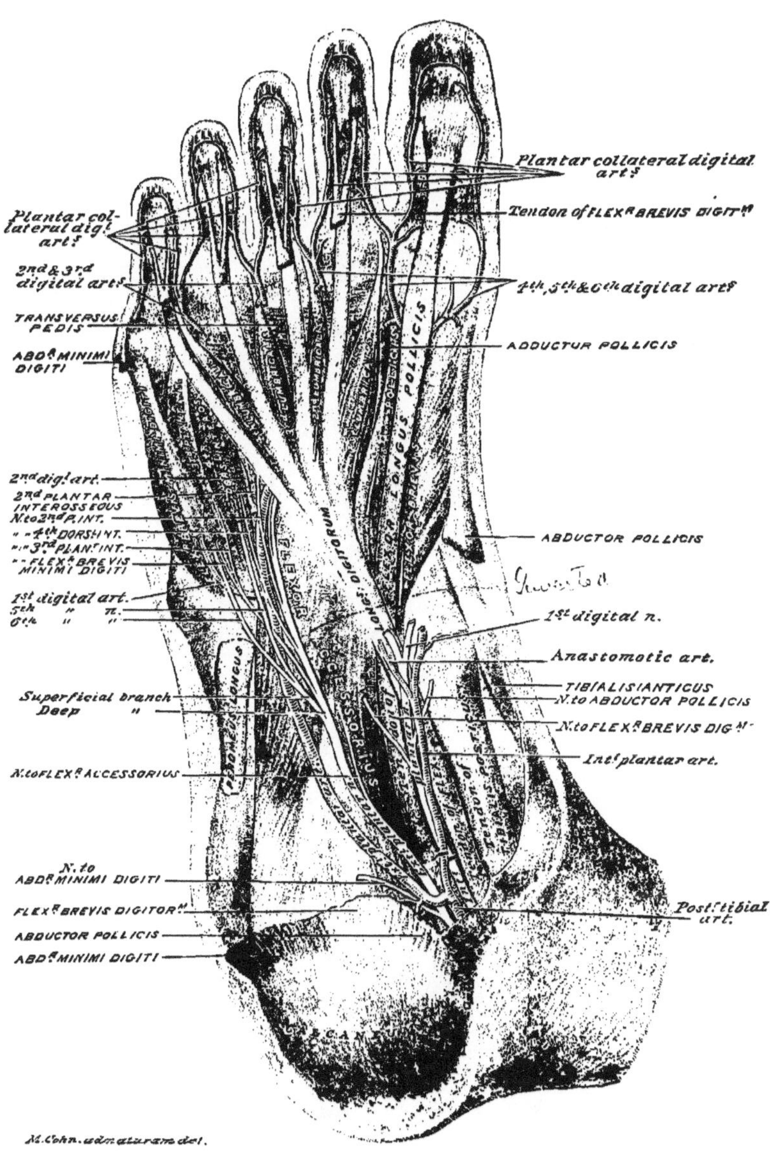
Plantar collateral digital art.s
Tendon of FLEX.R BREVIS DIGIT.M
4th, 5th & 6th digital art.s
ADDUCTOR POLLICIS
ABDUCTOR POLLICIS
1st digital n.
Anastomotic art.
TIBIALIS ANTICUS
N. to ABDUCTOR POLLICIS
N. to FLEX.R BREVIS DIG.M
Int.l plantar art.
Post.r tibial art.
Plantar collateral dig.l art.s
2nd & 3rd digital art.s
TRANSVERSUS PEDIS
ABD.R MINIMI DIGITI
2nd dig.l art.
2nd PLANTAR INTEROSSEOUS
N. to 2nd P. INT.
" " 4th DORS. INT.
" " 3rd PLAN.T INT.
" " FLEX.R BREVIS MINIMI DIGITI
1st digital art.
5th " n.
6th " "
Superficial branch
Deep "
N. to FLEX.R ACCESSORIUS
N. to ABD.R MINIMI DIGITI
FLEX.R BREVIS DIGITOR.M
ABDUCTOR POLLICIS
ABD.R MINIMI DIGITI
FLEXOR LONGUS POLLICIS
LONG.S DIGITORUM
FLEXOR ACCESSORIUS
PERONEUS LONGUS
Tendon of TIBIALIS POSTICUS
OS CALCIS
M. Cohn. ad naturam del.

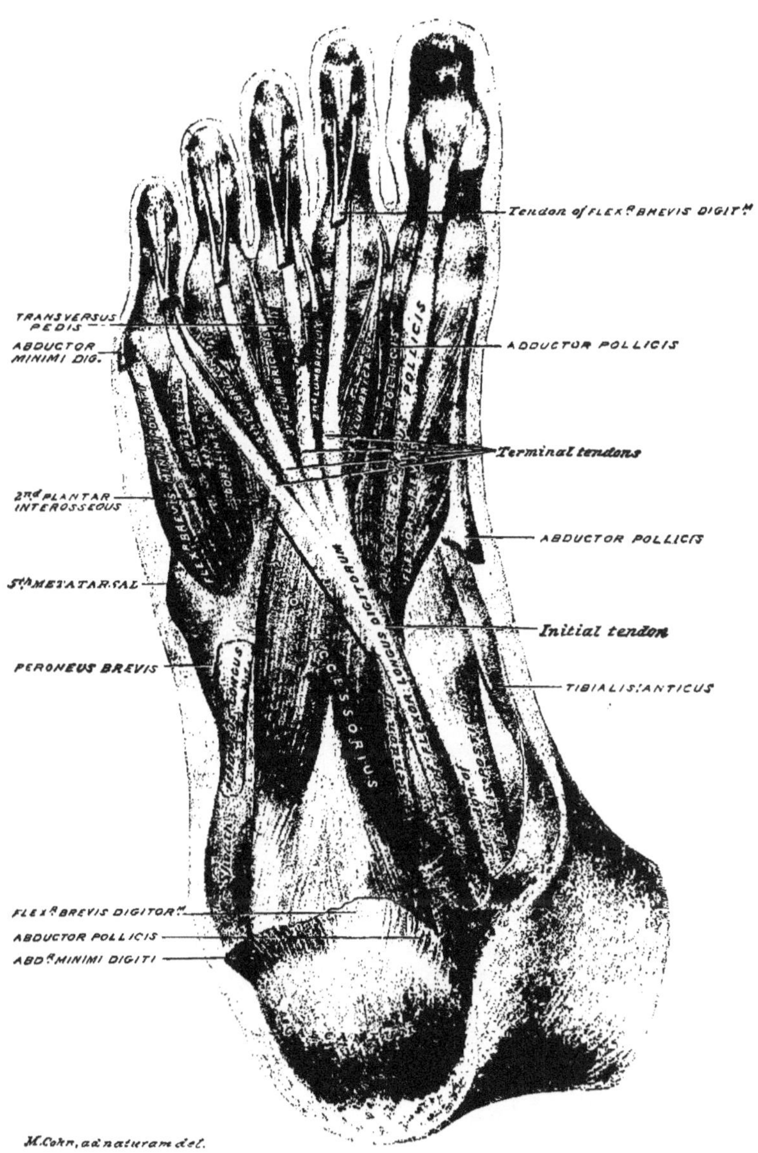
Tendon of FLEXR BREVIS DIGITM
TRANSVERSUS PEDIS
ABDUCTOR MINIMI DIG.
ADDUCTOR POLLICIS
Terminal tendons
2nd PLANTAR INTEROSSEOUS
ABDUCTOR POLLICIS
5th METATARSAL
Initial tendon
PERONEUS BREVIS
TIBIALIS ANTICUS
FLEXR BREVIS DIGITORM
ABDUCTOR POLLICIS
ABDR MINIMI DIGITI
FLEXOR LONGUS DIGITORUM
ACCESSORIUS
POLLICIS

M. Cohn, ad naturam del.

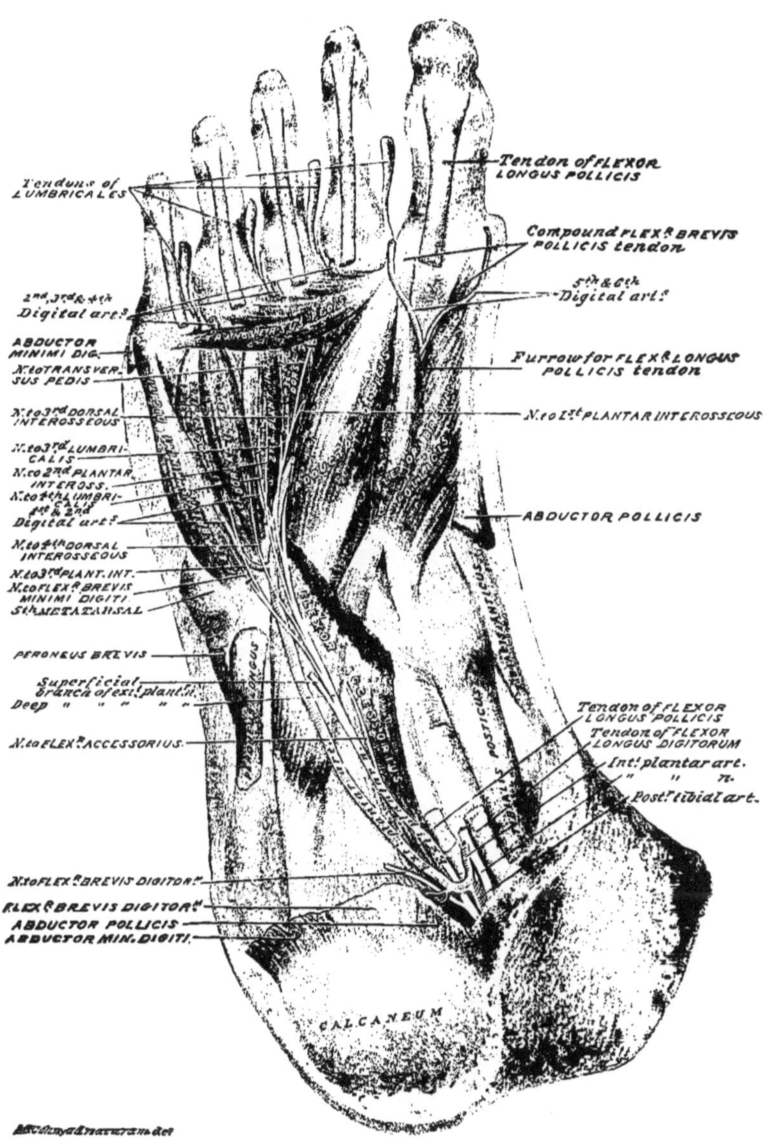
Tendons of LUMBRICALES
Tendon of FLEXOR LONGUS POLLICIS
Compound FLEXR BREVIS POLLICIS tendon
2nd, 3rd & 4th Digital arts
5th & 6th Digital arts
ABDUCTOR MINIMI DIG.
N. to TRANSVERSUS PEDIS
Furrow for FLEXR LONGUS POLLICIS tendon
N. to 3rd DORSAL INTEROSSEOUS
N. to 1st PLANTAR INTEROSSEOUS
N. to 3rd LUMBRICALIS
N. to 2nd PLANTAR INTEROSS.
N. to 4th LUMBRICALIS
1st & 2nd Digital arts
ABDUCTOR POLLICIS
N. to 4th DORSAL INTEROSSEOUS
N. to 3rd PLANT. INT.
N. to FLEXR BREVIS MINIMI DIGITI
5th METATARSAL
PERONEUS BREVIS
Superficial branch of extl plantr n.
Deep " " " " "
N. to FLEXR ACCESSORIUS
Tendon of FLEXOR LONGUS POLLICIS
Tendon of FLEXOR LONGUS DIGITORUM
Intl plantar art.
" " n.
Postr tibial art.
N. to FLEXR BREVIS DIGITORM
FLEXR BREVIS DIGITORM
ABDUCTOR POLLICIS
ABDUCTOR MIN. DIGITI
CALCANEUM

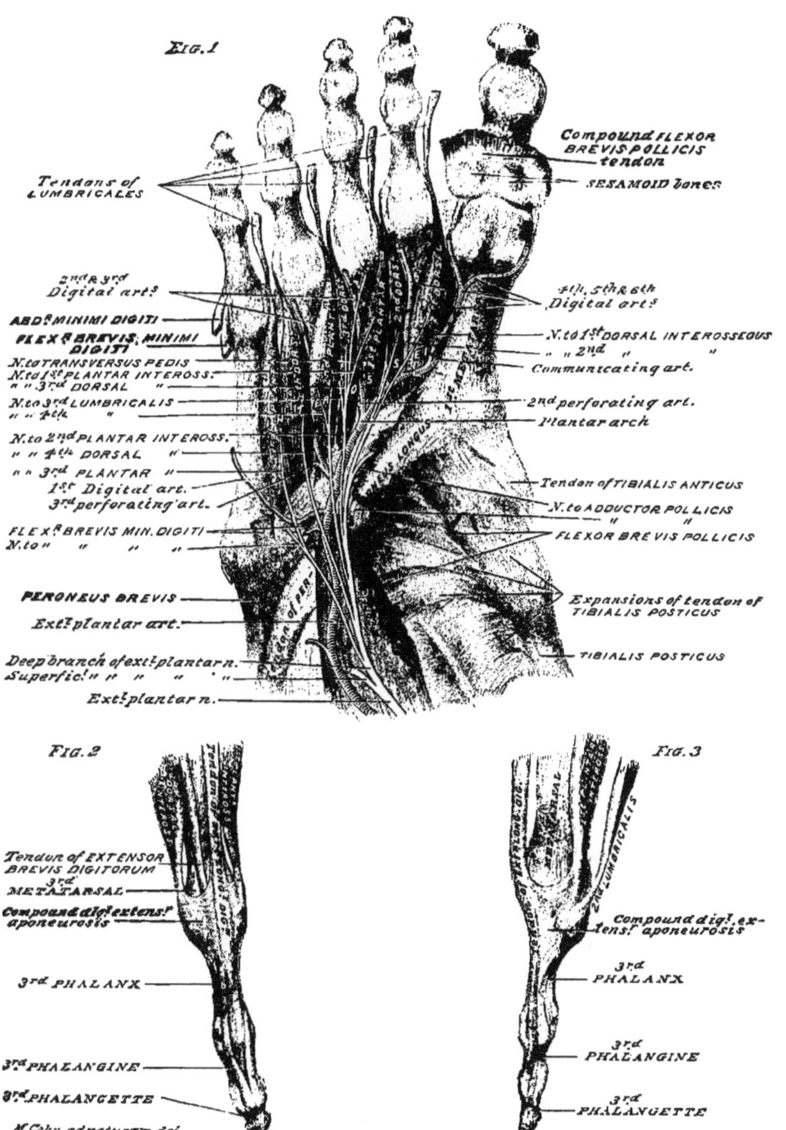
Fig. 1
Compound FLEXOR BREVIS POLLICIS tendon
SESAMOID bones
Tendons of LUMBRICALES
2nd & 3rd Digital arts.
4th, 5th & 6th Digital arts.
ABDr. MINIMI DIGITI
FLEXR. BREVIS MINIMI DIGITI
N. to 1st DORSAL INTEROSSEOUS
" " 2nd " "
N. to TRANSVERSUS PEDIS
Communicating art.
N. to 1st PLANTAR INTEROSS.
" " 3rd DORSAL "
N. to 3rd LUMBRICALIS
" " 4th "
2nd perforating art.
Plantar arch
N. to 2nd PLANTAR INTEROSS.
" " 4th DORSAL "
" " 3rd PLANTAR "
Tendon of TIBIALIS ANTICUS
1st Digital art.
3rd perforating art.
N. to ADDUCTOR POLLICIS
" "
FLEXR. BREVIS MIN. DIGITI
N. to " " "
FLEXOR BREVIS POLLICIS
PERONEUS BREVIS
Expansions of tendon of TIBIALIS POSTICUS
Extl. plantar art.
TIBIALIS POSTICUS
Deep branch of extl. plantar n.
Superficl. " " "
Extl. plantar n.
Fig. 2
Tendon of EXTENSOR BREVIS DIGITORUM
3rd METATARSAL
Compound digl. extensr. aponeurosis
3rd PHALANX
3rd PHALANGINE
3rd PHALANGETTE
M. Cohn. ad naturam del.
Fig. 3
2nd LUMBRICALIS
Compound digl. extensr. aponeurosis
3rd PHALANX
3rd PHALANGINE
3rd PHALANGETTE

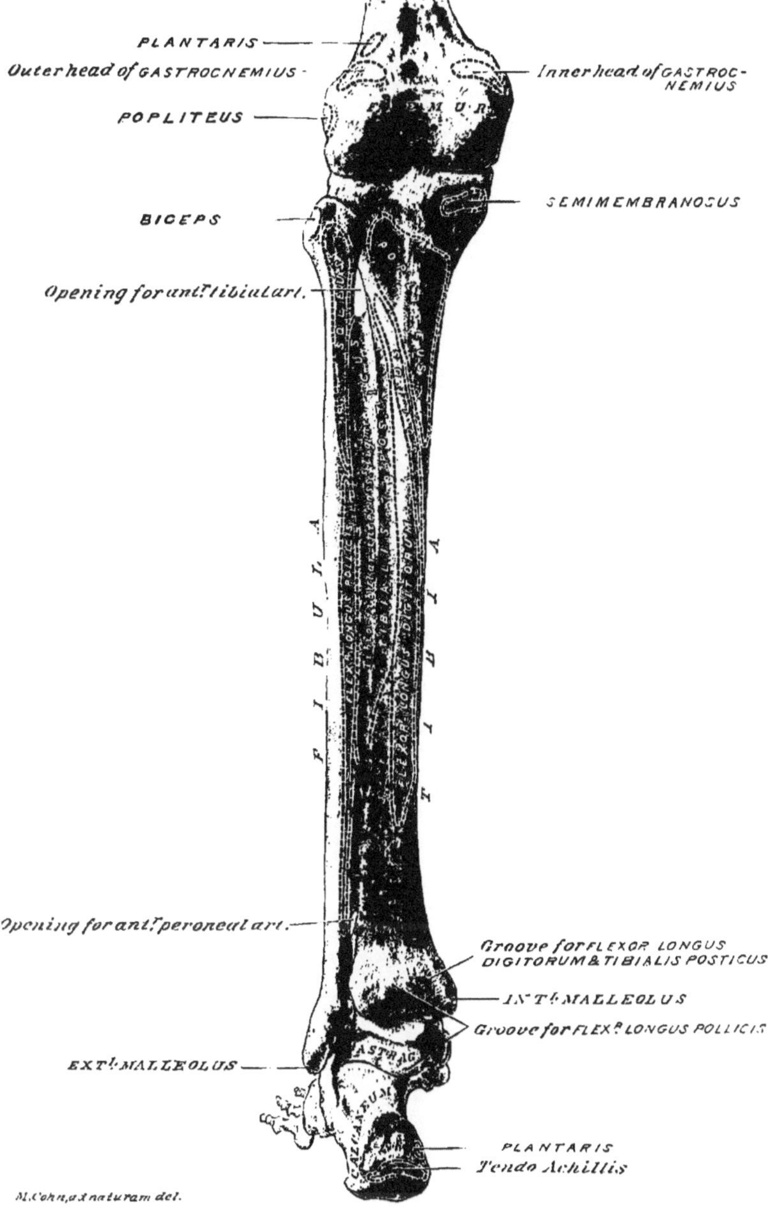
PLANTARIS
Outer head of GASTROCNEMIUS
Inner head of GASTROCNEMIUS
FEMUR
POPLITEUS
SEMIMEMBRANOSUS
BICEPS
Opening for ant.r tibial art.
FIBULA
TIBIA
FLEXOR LONGUS POLLICIS
FLEXOR LONGUS DIGITORUM
Opening for ant.r peroneal art.
Groove for FLEXOR LONGUS DIGITORUM & TIBIALIS POSTICUS
INT.L MALLEOLUS
Groove for FLEX.R LONGUS POLLICIS
EXT.L MALLEOLUS
ASTRAG.
CALCANEUM
PLANTARIS
Tendo Achillis

M. Cohn, ad naturam del.

PLATE 84

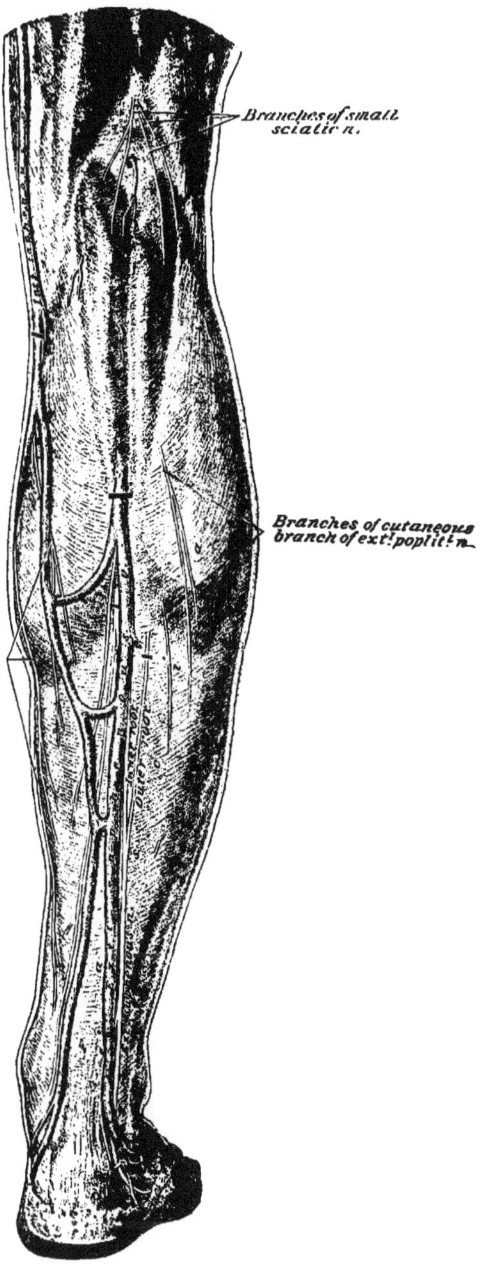
Branches of small sciatic n.
Branches of cutaneous branch of ext! poplit! n.
Branches of int! saphen! n.

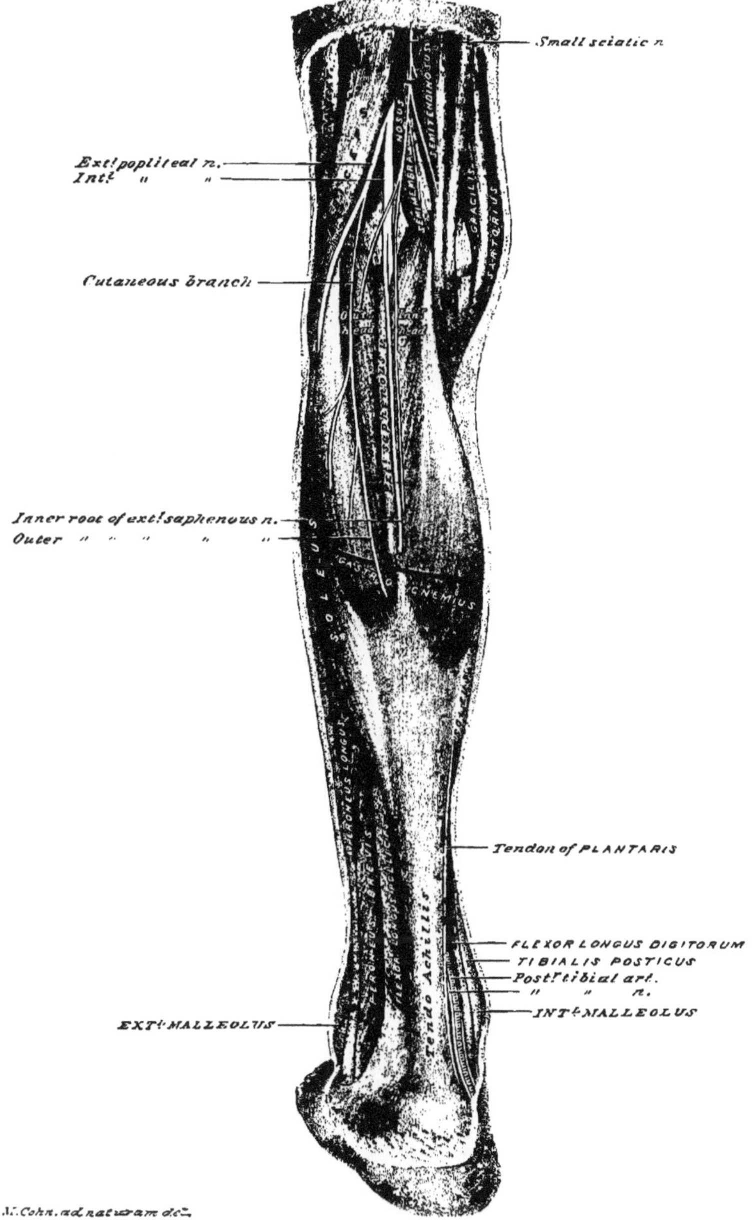
Small sciatic n.
Ext! popliteal n.
Int! " "
Cutaneous branch
Inner root of ext! saphenous n.
Outer " " " " "
SOLEUS
GASTROCNEMIUS
GRACILIS
SARTORIUS
Tendon of PLANTARIS
FLEXOR LONGUS DIGITORUM
TIBIALIS POSTICUS
Post! tibial art.
" " n.
INT! MALLEOLUS
EXT! MALLEOLUS
Tendo Achillis
M. Cohn. ad naturam del.

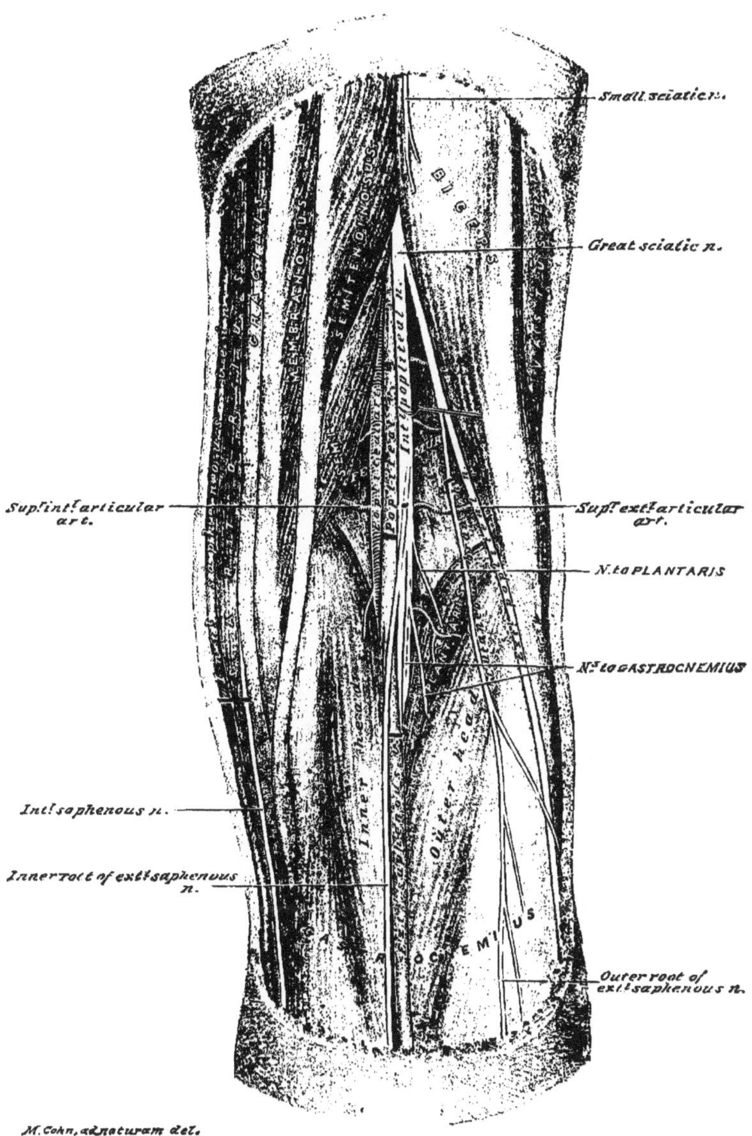
Small sciatic n.
Great sciatic n.
Supr intl articular art.
Supr extl articular art.
N. to PLANTARIS
Ns to GASTROCNEMIUS
Intl saphenous n.
Inner root of extl saphenous n.
Outer root of extl saphenous n.
BICEPS
SEMITENDINOSUS
SEMIMEMBRANOSUS
GRACILIS
Popliteal
Intl popliteal n.
Inner head
Outer head
GASTROCNEMIUS

M. Cohn, ad naturam del.

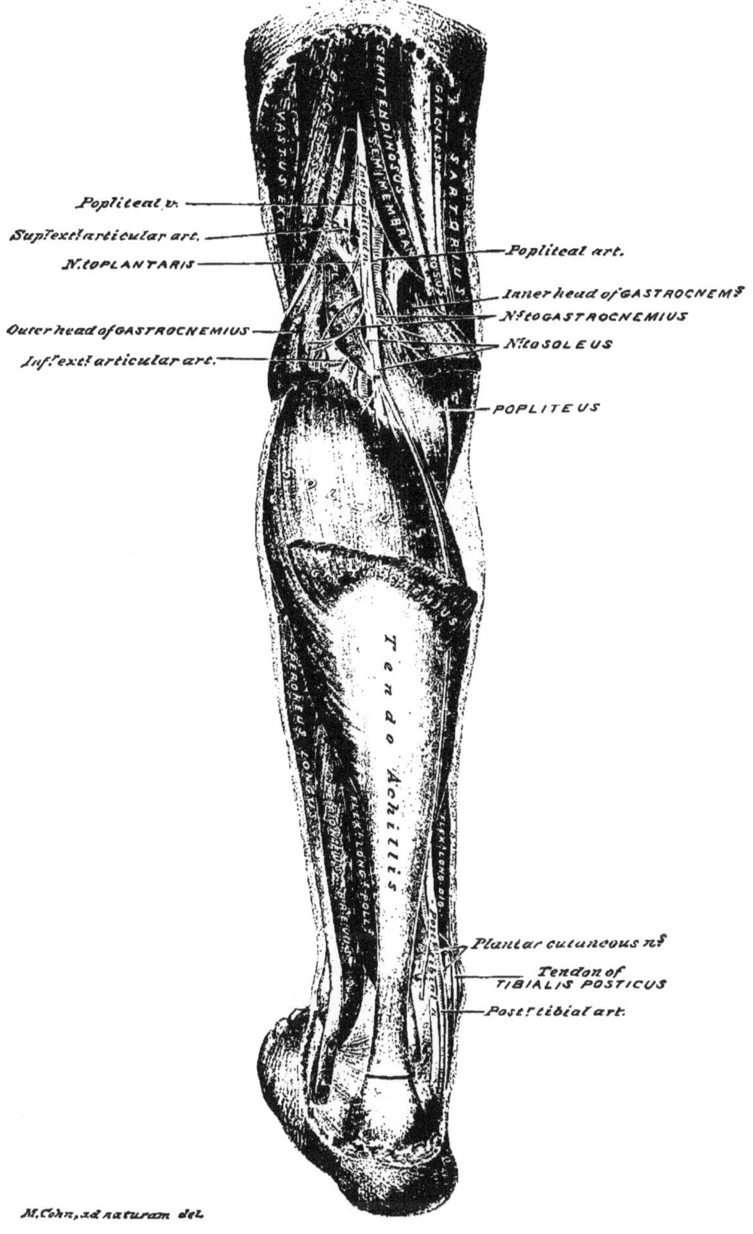
Popliteal v.
Supr. ext. articular art.
N. to PLANTARIS
Outer head of GASTROCNEMIUS
Infr. ext. articular art.
Popliteal art.
Inner head of GASTROCNEMS.
Ns. to GASTROCNEMIUS
Ns. to SOLEUS
POPLITEUS
Plantar cutaneous ns.
Tendon of TIBIALIS POSTICUS
Postr. tibial art.
VASTUS EXT.
SEMITENDINOSUS
SARTORIUS
Tendo Achillis
M. Cohn, ad naturam del.

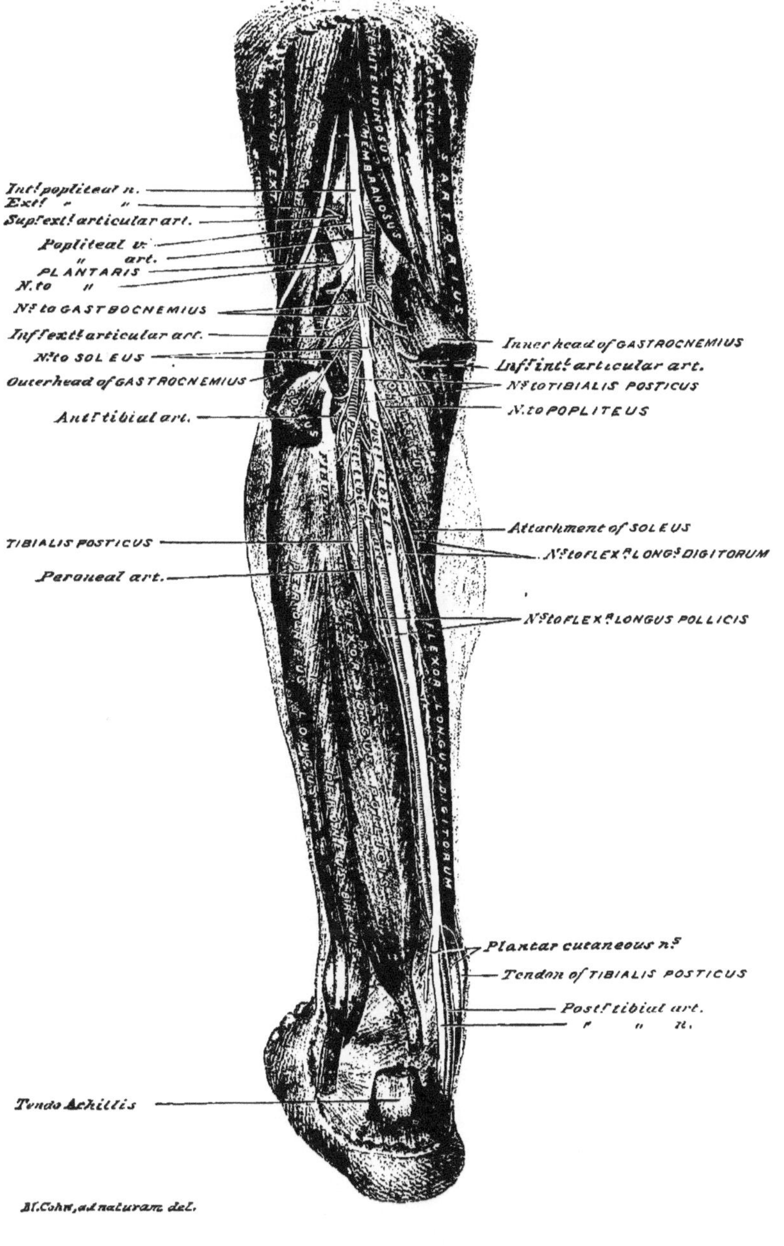
Intl popliteal n.
Extl " "
Supr extl articular art.
Popliteal v.
" art.
PLANTARIS
N. to "
Ns to GASTROCNEMIUS
Infr extl articular art.
N. to SOLEUS
Outer head of GASTROCNEMIUS
Antr tibial art.
TIBIALIS POSTICUS
Peroneal art.
Tendo Achillis
Inner head of GASTROCNEMIUS
Infr intl articular art.
Ns to TIBIALIS POSTICUS
N. to POPLITEUS
Attachment of SOLEUS
Ns to FLEXR LONGS DIGITORUM
Ns to FLEXR LONGUS POLLICIS
Plantar cutaneous ns
Tendon of TIBIALIS POSTICUS
Postr tibial art.
" " n.
VASTUS EXT.
SEMITENDINOSUS
SEMIMEMBRANOSUS
GRACILIS
SARTORIUS
FLEXOR LONGUS DIGITORUM
PERONEUS LONGUS

Bl. Cohn, ad naturam del.

PLATE 89

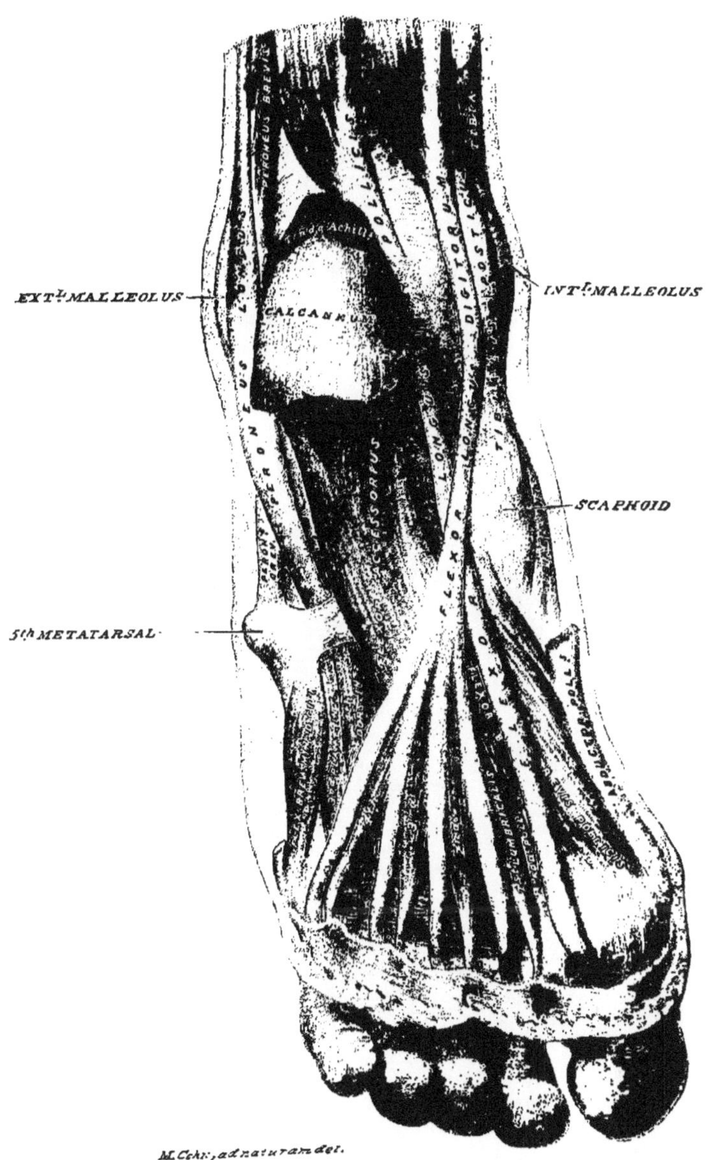

PERONEUS BREVIS
Tendo Achillis
CALCANEUM
EXT^L MALLEOLUS
INT^L MALLEOLUS
DIGITORUM
TIBIA
PERONEUS
ACCESSORIUS
FLEXOR
SCAPHOID
5th METATARSAL
FLEXOR

M. Cohn, ad naturam del.

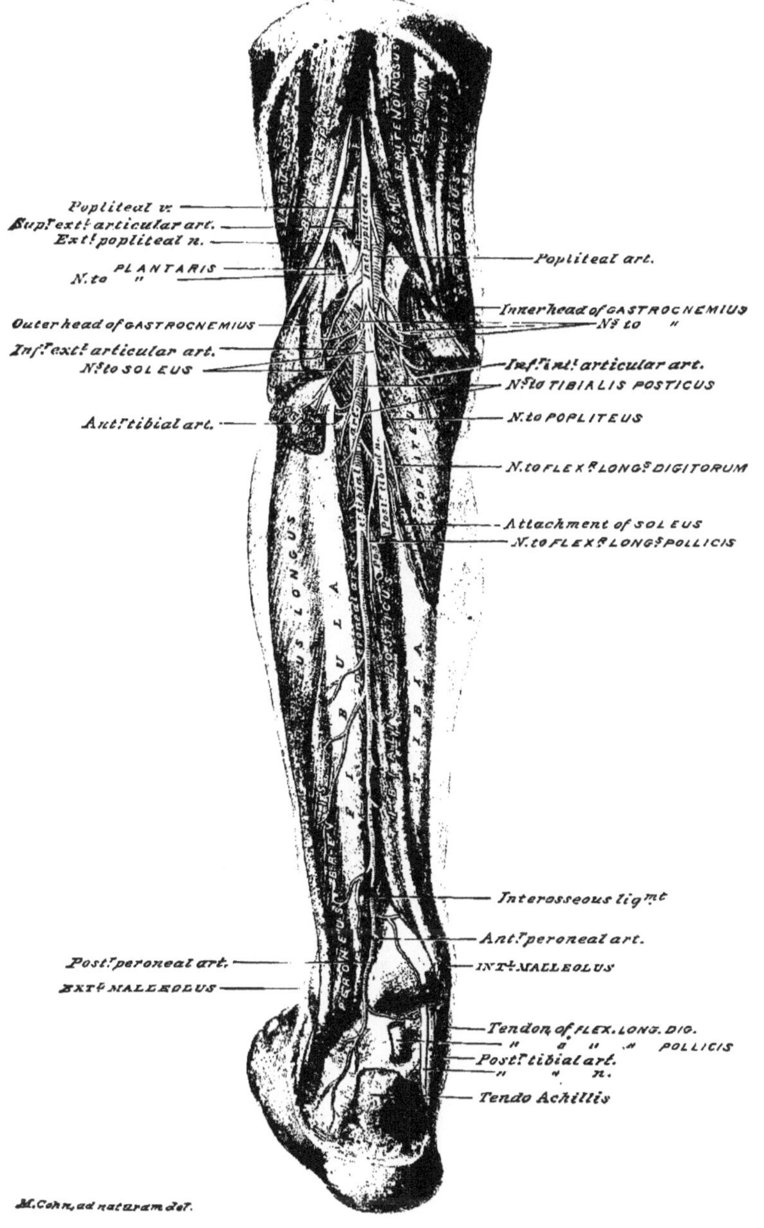
Popliteal v.
Supr. extl. articular art.
Extl. popliteal n.
PLANTARIS
N. to "
Outer head of GASTROCNEMIUS
Infr. extl. articular art.
Ns. to SOLEUS
Antr. tibial art.
Popliteal art.
Inner head of GASTROCNEMIUS
Ns. to "
Infr. intl. articular art.
Ns. to TIBIALIS POSTICUS
N. to POPLITEUS
N. to FLEXR. LONGS. DIGITORUM
Attachment of SOLEUS
N. to FLEXR. LONGS. POLLICIS
Interosseous ligmt
Antr. peroneal art.
Postr. peroneal art.
INTL. MALLEOLUS
EXTL. MALLEOLUS
Tendon of FLEX. LONG. DIG.
" " " POLLICIS
Postr. tibial art.
" " n.
Tendo Achillis
VASTUS EXT.
BICEPS
SEMITENDINOSUS
SEMIMEMBRAN.
GRACILIS
SARTORIUS
POPLITEUS
PERONEUS LONGUS
FIBULA
TIBIA
TIBIALIS POSTICUS
PERONEUS BREVIS
M. Cohn, ad naturam del.

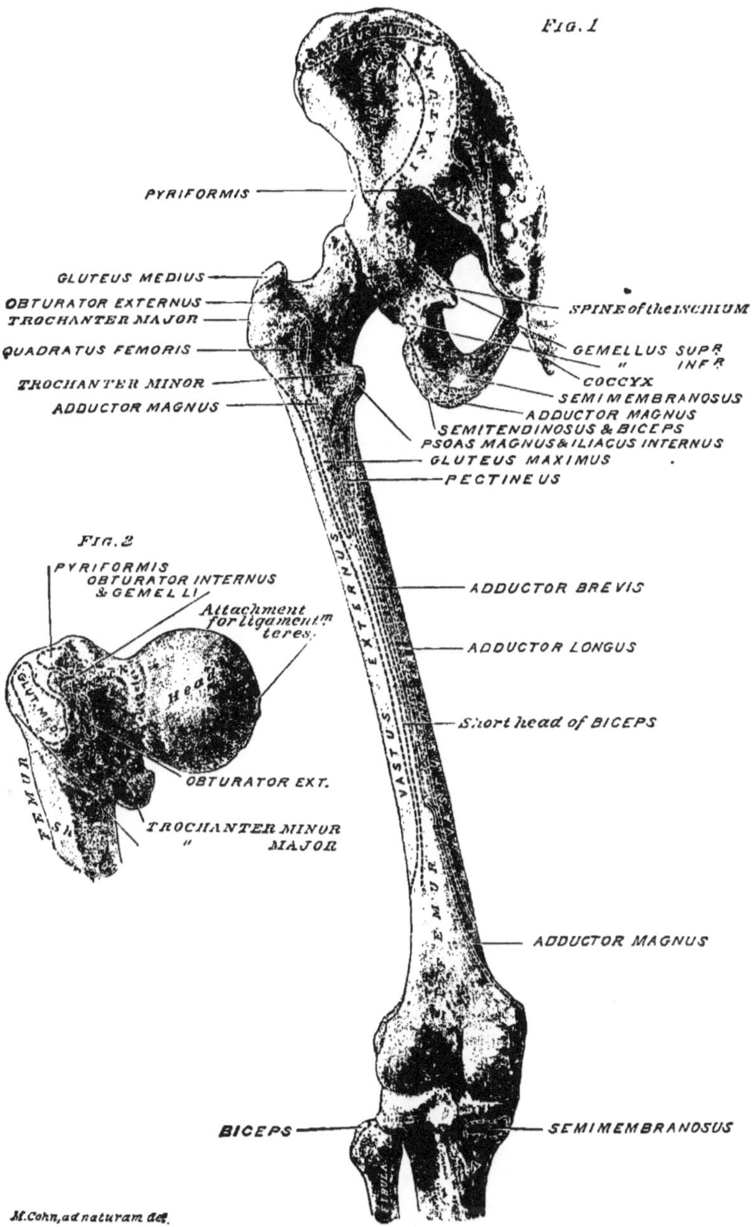
FIG. 1
PYRIFORMIS
GLUTEUS MEDIUS
OBTURATOR EXTERNUS
TROCHANTER MAJOR
QUADRATUS FEMORIS
TROCHANTER MINOR
ADDUCTOR MAGNUS
SPINE of the ISCHIUM
GEMELLUS SUPR.
" INFR.
COCCYX
SEMIMEMBRANOSUS
ADDUCTOR MAGNUS
SEMITENDINOSUS & BICEPS
PSOAS MAGNUS & ILIACUS INTERNUS
GLUTEUS MAXIMUS
PECTINEUS
ADDUCTOR BREVIS
ADDUCTOR LONGUS
Short head of BICEPS
VASTUS EXTERNUS
FEMUR
ADDUCTOR MAGNUS
BICEPS
SEMIMEMBRANOSUS
FIG. 2
PYRIFORMIS
OBTURATOR INTERNUS
& GEMELLI
Attachment
for ligament.m
teres.
Head
GLUT. MED.
FEMUR
OBTURATOR EXT.
TROCHANTER MINOR
" MAJOR
M. Cohn, ad naturam del.

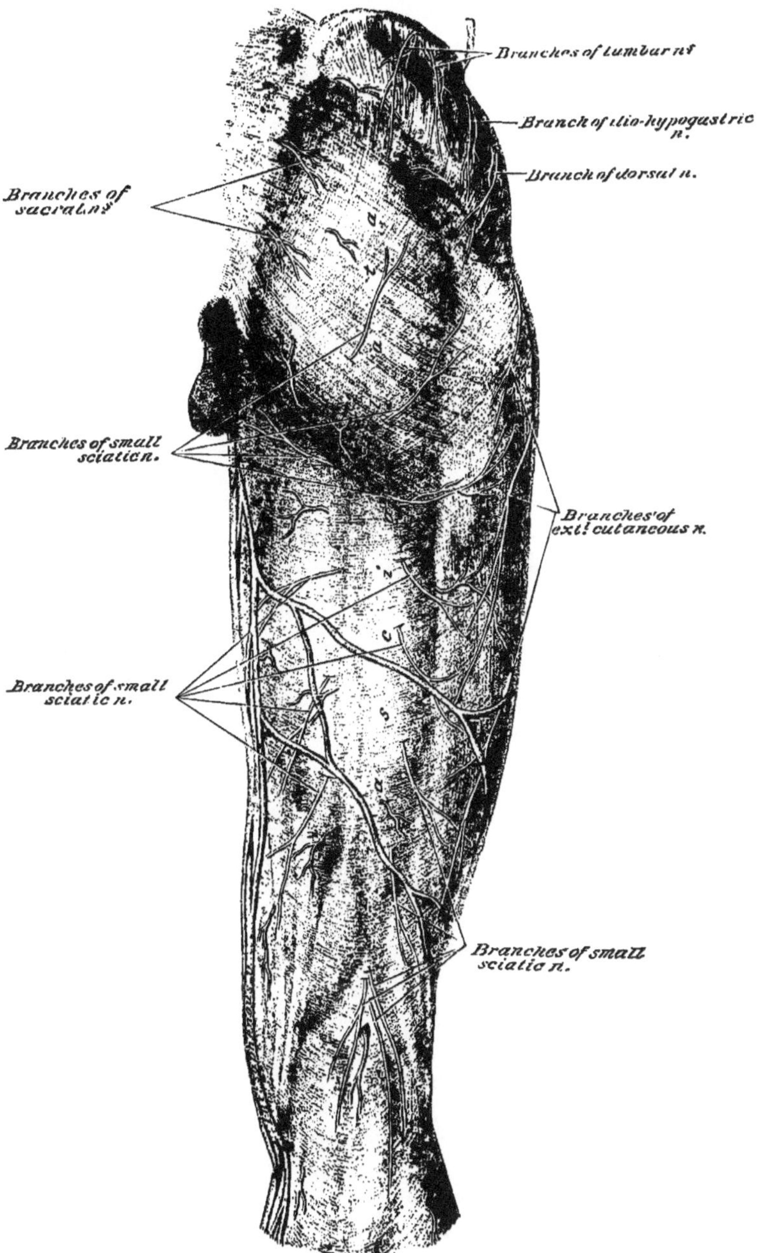
Branches of lumbar nf
Branch of ilio-hypogastric n.
Branch of dorsal n.
Branches of sacral nf
Branches of small sciatic n.
Branches of ext! cutaneous n.
Branches of small sciatic n.
Branches of small sciatic n.

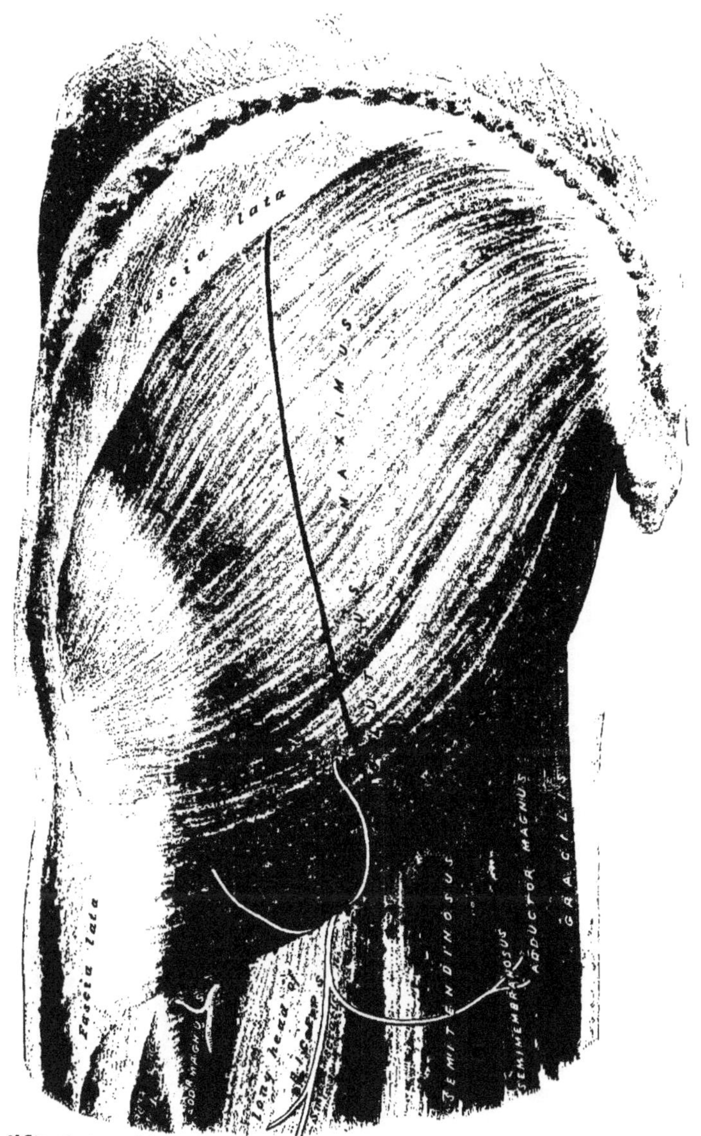
Fascia lata
MAXIMUS
Fascia lata
long head of
SEMITENDINOSUS
SEMIMEMBRANOSUS
ADDUCTOR MAGNUS
GRACILIS

M. Cohn, ad naturam del.

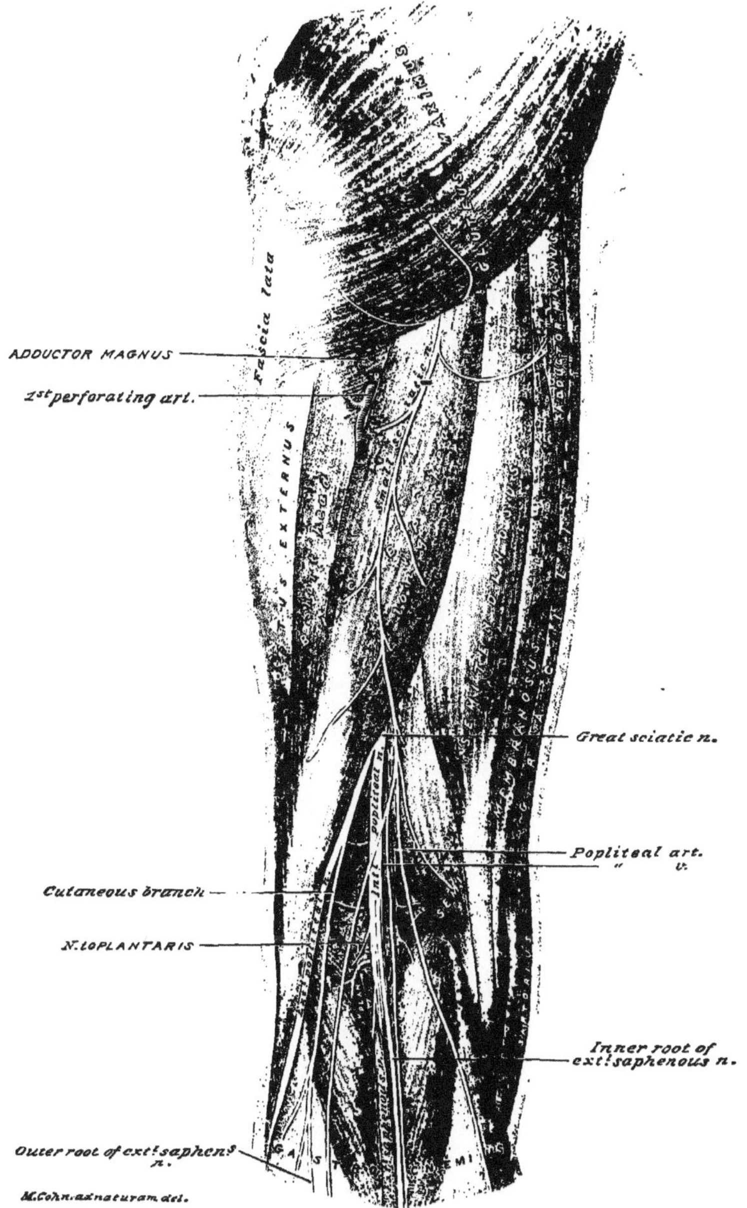
ADDUCTOR MAGNUS
2nd perforating art.
Fascia lata
Great sciatic n.
Popliteal art.
" v.
Cutaneous branch
N. to PLANTARIS
Inner root of ext! saphenous n.
Outer root of ext! saphen! n.
M. Cohn. ad naturam del.

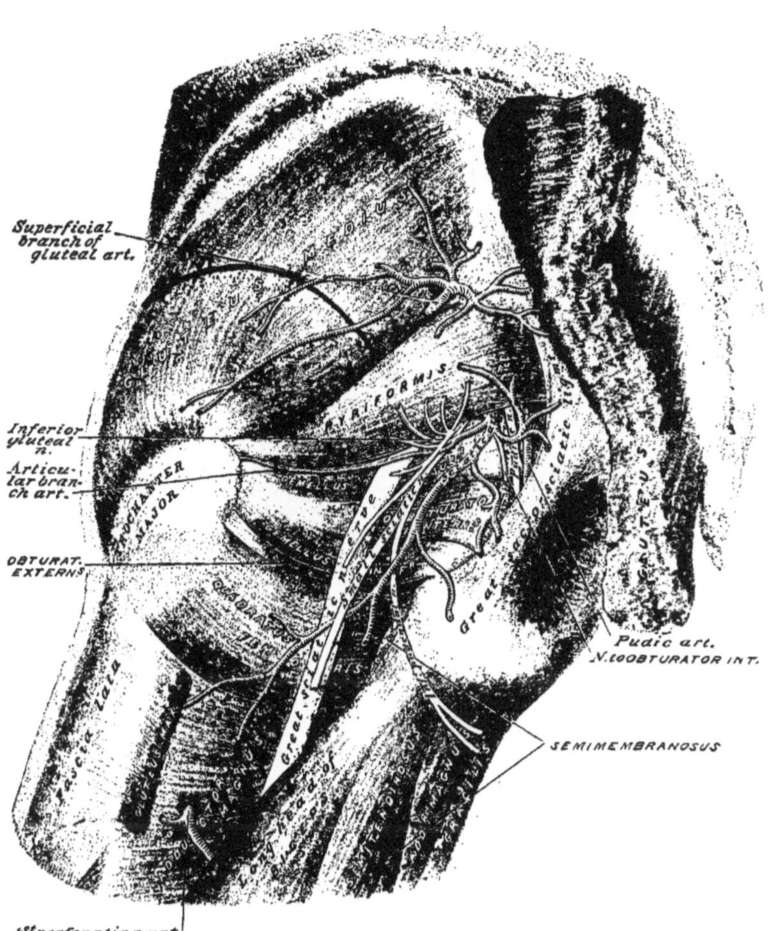
Superficial branch of gluteal art.
Inferior gluteal n.
Articular branch art.
OBTURAT. EXTERNS
TROCHANTER MAJOR
PYRIFORMIS
Great sciatic nerve
Great sacro sciatic lig.
Pudic art.
N. to OBTURATOR INT.
SEMIMEMBRANOSUS
Fascia lata
Long head of biceps
1st perforating art.

M. Cohn, ad naturam del.

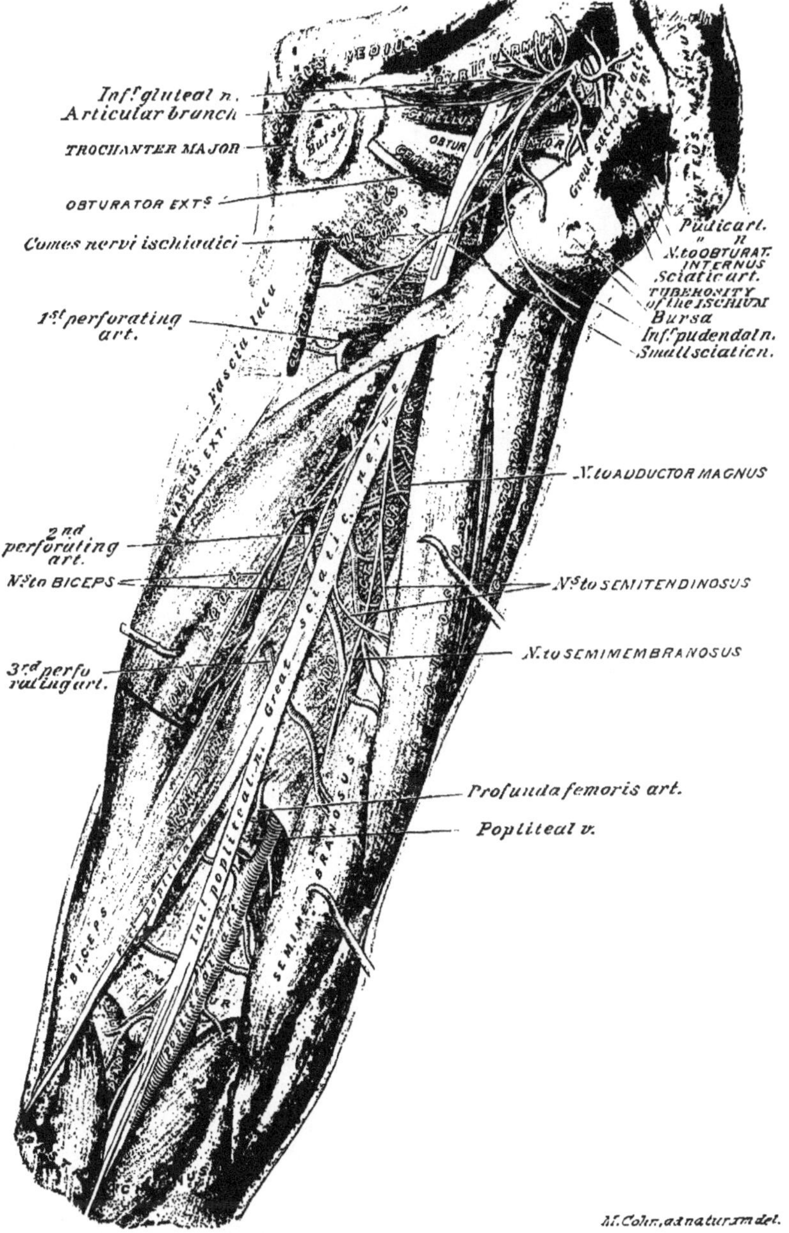
Infr. gluteal n.
Articular branch
TROCHANTER MAJOR
OBTURATOR EXTS.
Comes nervi ischiadici
1st perforating art.
2nd perforating art.
Ns. to BICEPS
3rd perforating art.
Pudic art.
" n.
N. to OBTURAT. INTERNUS
Sciatic art.
TUBEROSITY of the ISCHIUM
Bursa
Infr. pudendal n.
Small sciatic n.
N. to ADDUCTOR MAGNUS
Ns. to SEMITENDINOSUS
N. to SEMIMEMBRANOSUS
Profunda femoris art.
Popliteal v.
Great sacro-sciatic ligmt.
Great sciatic nerve
Fascia lata
VASTUS EXT.
BICEPS
SEMIMEMBRANOSUS
Bursa
GEMELLUS
OBTUR
PYRIFORMIS
MEDIUS
Infr. popliteal n.

M. Cohn, ad naturam del.

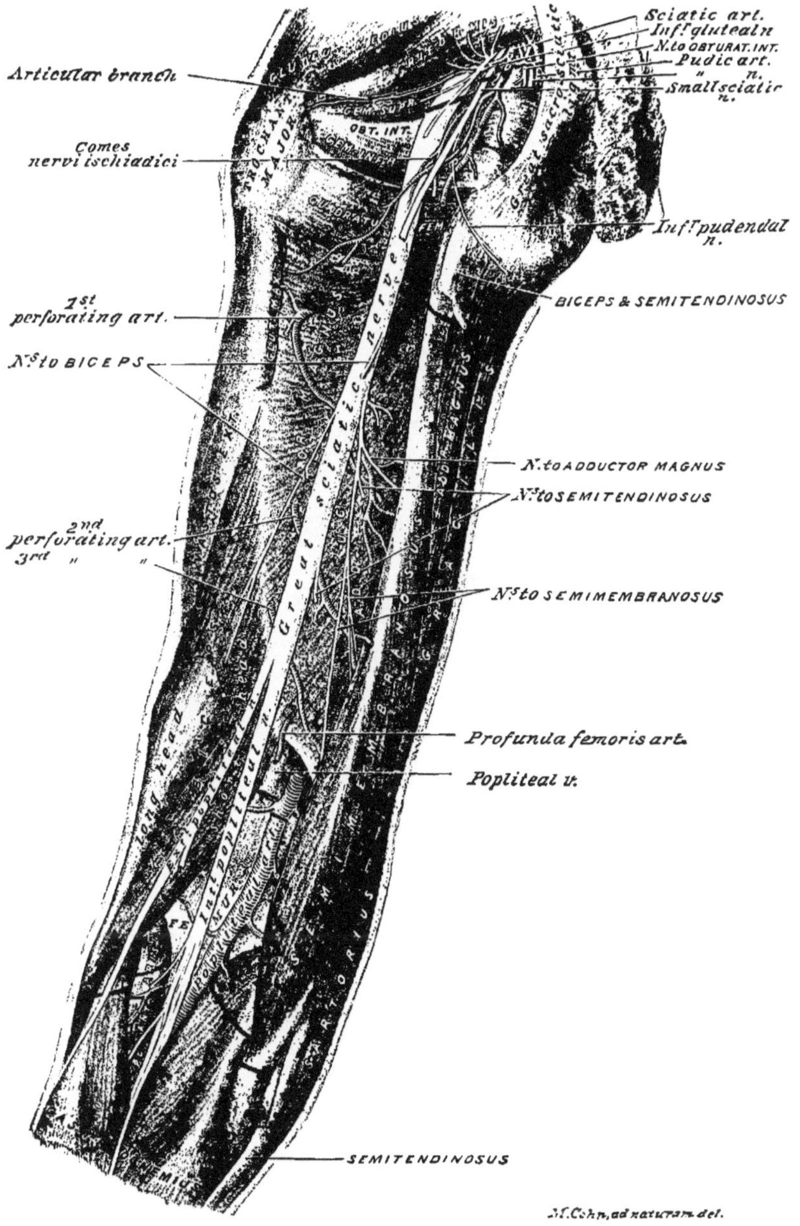
Sciatic art.
Infr gluteal n
N. to OBTURAT. INT.
Pudic art.
" n.
Small sciatic n.
Articular branch
Comes nervi ischiadici
OBT. INT.
Infr pudendal n.
BICEPS & SEMITENDINOSUS
1st perforating art.
Ns to BICEPS
Great sciatic nerve
N. to ADDUCTOR MAGNUS
Ns to SEMITENDINOSUS
2nd perforating art.
3rd " "
Ns to SEMIMEMBRANOSUS
Profunda femoris art.
Popliteal v.
Long head
Intl popliteal n.
SEMITENDINOSUS
SARTORIUS

M. Cohn, ad naturam del.

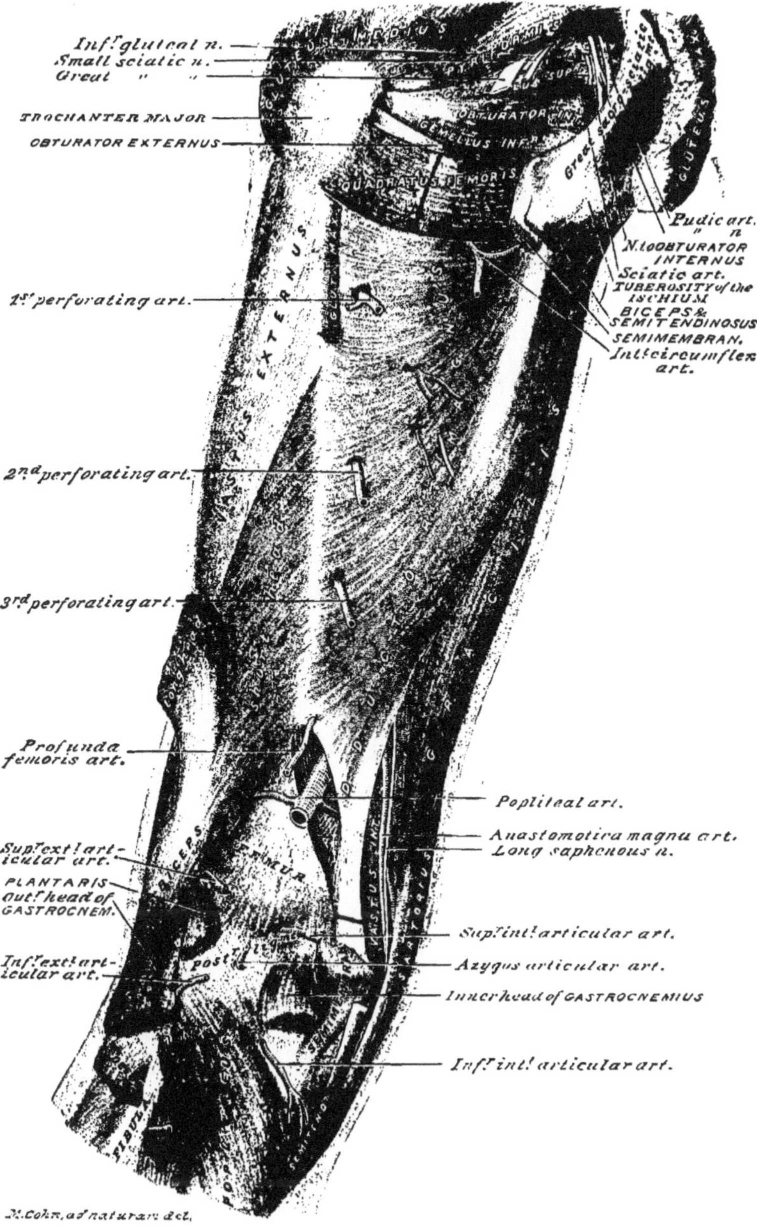
Infr. gluteal n.
Small sciatic n.
Great " "
TROCHANTER MAJOR
OBTURATOR EXTERNUS
1st perforating art.
2nd perforating art.
3rd perforating art.
Profunda femoris art.
Supr. extl. articular art.
PLANTARIS
Outr. head of GASTROCNEM.
Infr. extl. articular art.
GLUTEUS MEDIUS
PYRIFORMIS
GEMELLUS SUP.
OBTURATOR INT.
GEMELLUS INFR.
QUADRATUS FEMORIS
VASTUS EXTERNUS
GRACILIS
BICEPS
FEMUR
FIBULA
Great sacro-sciatic lig.
GLUTEUS MAX.
Pudic art.
" n
N. to OBTURATOR INTERNUS
Sciatic art.
TUBEROSITY of the ISCHIUM
BICEPS & SEMITENDINOSUS
SEMIMEMBRAN.
Intl. circumflex art.
Popliteal art.
Anastomotica magna art.
Long saphenous n.
Supr. intl. articular art.
Azygos articular art.
Inner head of GASTROCNEMIUS
Infr. intl. articular art.
M. Cohn, ad naturam del.

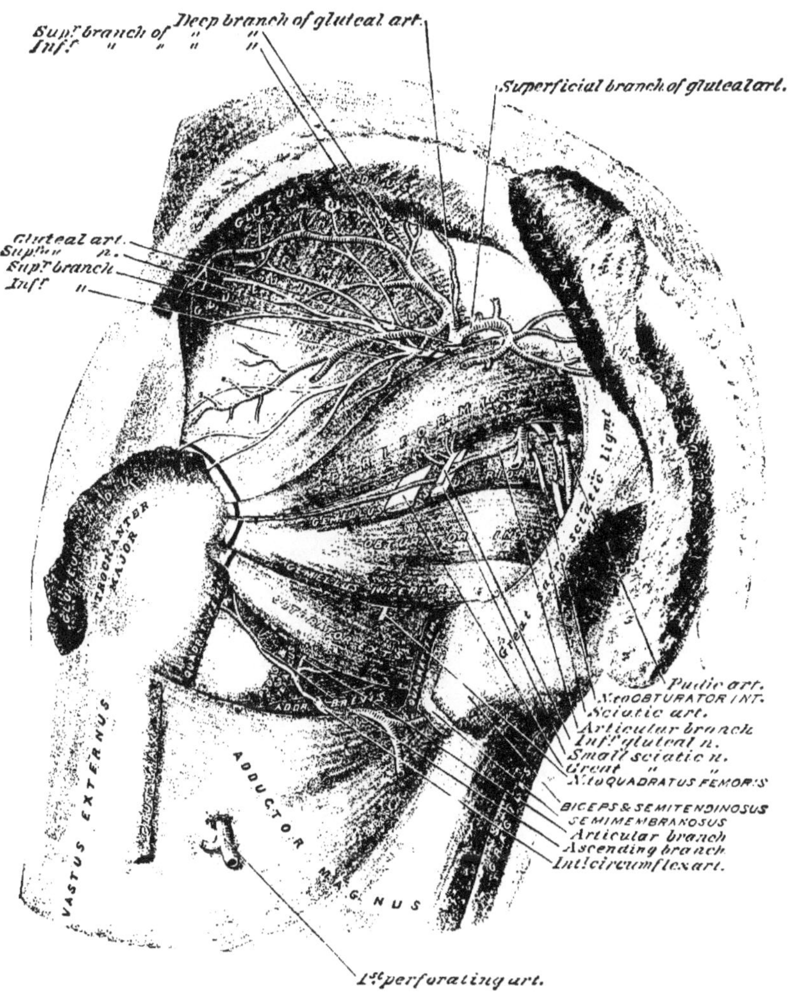
Deep branch of gluteal art.
Supr. branch of " "
Infr. " " " "
Superficial branch of gluteal art.
Gluteal art.
Supr. " n.
Supr. branch
Infr. "
GLUTEUS MEDIUS
GLUTEUS MAXIMUS
PYRIFORMIS
GEMELLUS
OBTURATOR INT.
GEMELLUS INFERIOR
OBTURATOR EXT.
QUADRATUS
ADDUCTOR BREVIS
TROCHANTER MAJOR
GLUTEUS MEDIUS
Great Sacro sciatic ligmt.
Pudic art.
N. to OBTURATOR INT.
Sciatic art.
Articular branch
Infr. gluteal n.
Small sciatic n.
Great " "
N. to QUADRATUS FEMORIS
BICEPS & SEMITENDINOSUS
SEMIMEMBRANOSUS
Articular branch
Ascending branch
Intl. circumflex art.
VASTUS EXTERNUS
ADDUCTOR MAGNUS
1st perforating art.

M. Cohn, ad naturam del.

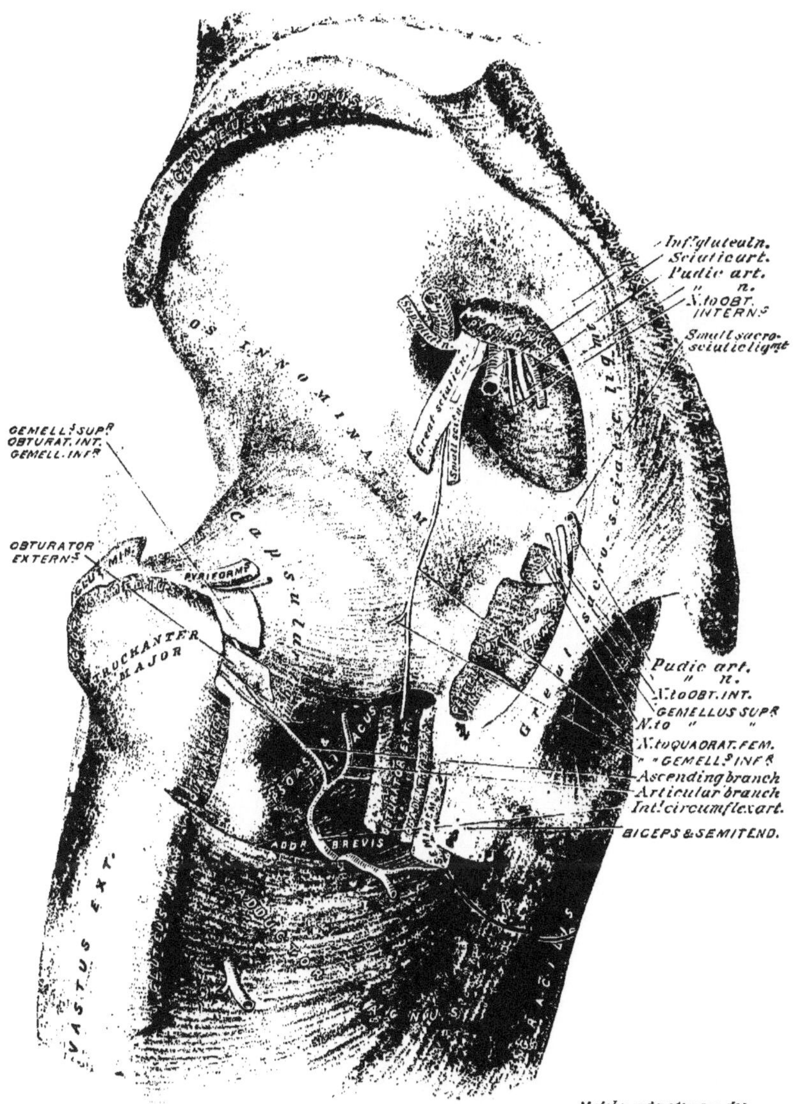
Infr gluteal n.
Sciatic art.
Pudic art.
" n.
N. to OBT. INTERNS
Small sacro-sciatic ligmt
GEMELLS SUPR
OBTURAT. INT.
GEMELL. INFR
OBTURATOR EXTERNS
PYRIFORMIS
TROCHANTER MAJOR
OS INNOMINATUM
Great sciatic n.
Small sciatic n.
Pudic art.
" n.
N. to OBT. INT.
GEMELLUS SUPR
N. to " "
N. to QUADRAT. FEM.
" GEMELLS INFR
Ascending branch
Articular branch
Intl circumflex art.
BICEPS & SEMITEND.
ADDR BREVIS
PSOAS & ILIACUS
VASTUS EXT.
GRACILIS

M. John. ad naturam del.

PLATE 101

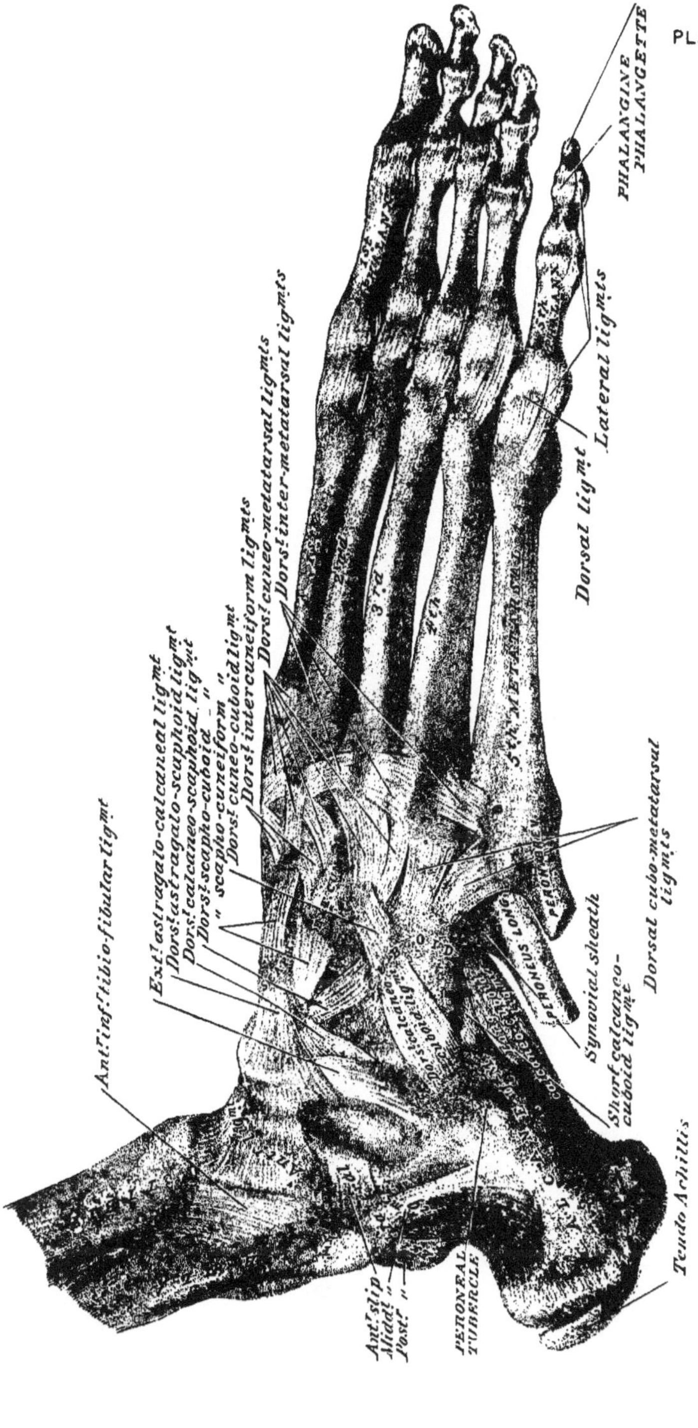

Ant^{r} inf^{r} tibio-fibular lig^{mt}
Ext^{l} astragalo-calcaneal lig^{mt}
Dors^{l} astragalo-scaphoid lig^{mt}
Dors^{l} calcaneo-scaphoid lig^{mt}
Dors^{l} scapho-cuboid " "
" scapho-cuneiform "
Dors^{l} cuneo-cuboid lig^{mt}
Dors^{l} intercuneiform lig^{mts}
Dors^{l} cuneo-metatarsal lig^{mts}
Dors^{l} inter-metatarsal lig^{mts}
Ant^{r} slip
Midd^{l} "
Post^{r} "
PERONEAL TUBERCLE
Dors^{l} calcaneo-cuboid lig^{mt}
3rd
4th
5th METATARSAL
PERONEUS LONG
5th PHALANX
Synovial sheath
Short calcaneo-cuboid lig^{mt}
Dorsal cubo-metatarsal lig^{mts}
Tendo Achillis
Dorsal lig^{mt}
Lateral lig^{mts}
PHALANGINE
PHALANGETTE

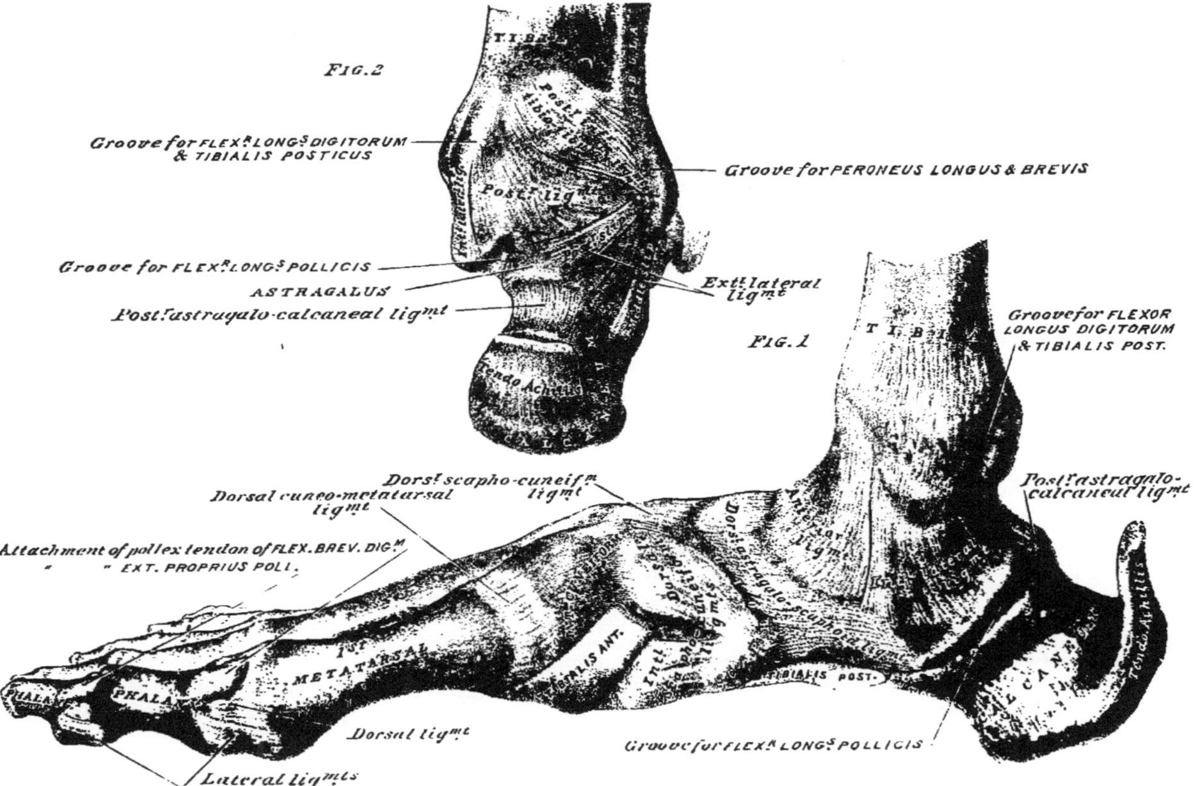
FIG. 2
Groove for FLEX.R LONG.S DIGITORUM & TIBIALIS POSTICUS
Groove for PERONEUS LONGUS & BREVIS
Groove for FLEX.R LONG.S POLLICIS
ASTRAGALUS
Post! astragalo-calcaneal ligmt
Ext! lateral ligmt
FIG. 1
Groove for FLEXOR LONGUS DIGITORUM & TIBIALIS POST.
Dors! scapho-cuneifm ligmt
Dorsal cuneo-metatarsal ligmt
Post! astragalo-calcaneal ligmt
Attachment of pollex tendon of FLEX. BREV. DIG.M
" " EXT. PROPRIUS POLL.
Dorsal ligmt
Groove for FLEX.R LONG.S POLLICIS
Lateral ligmts
M. Cohn, ad naturam del.

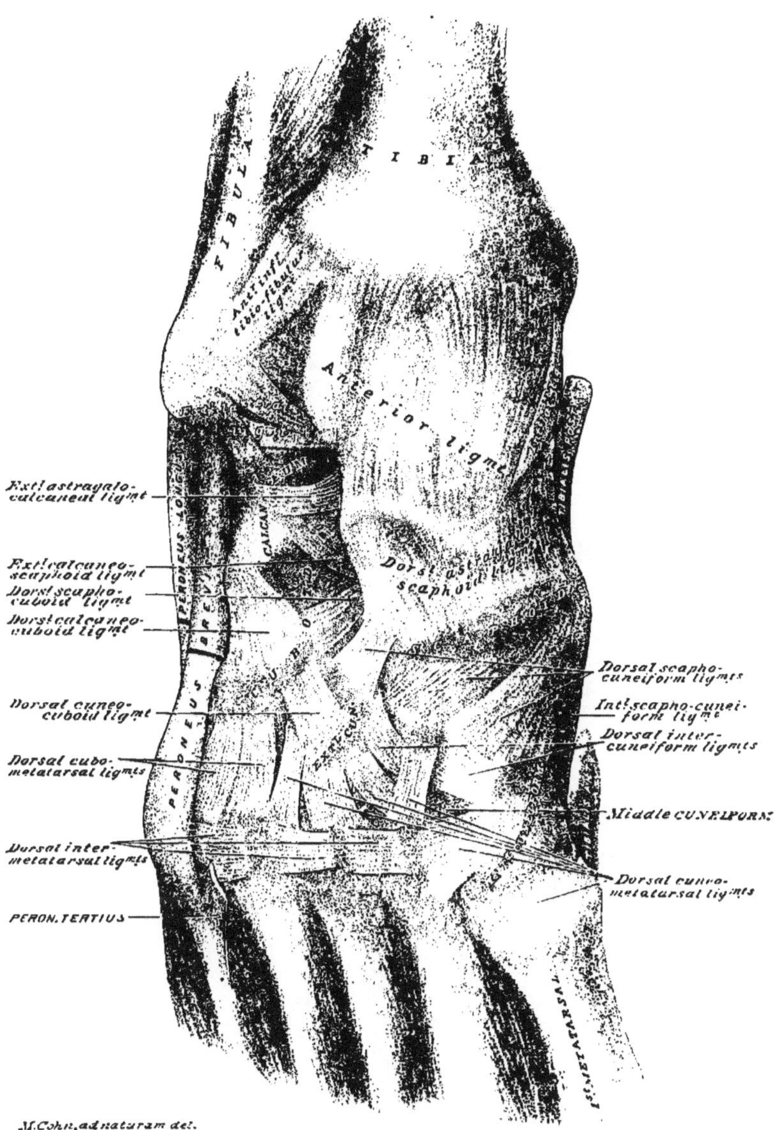
TIBIA
FIBULA
Ant! inf! tibio-fibular ligt
Anterior ligmt
Ext! astragalo-calcaneal ligmt
Ext! calcaneo-scaphoid ligmt
Dors! scapho-cuboid ligmt
Dors! calcaneo-cuboid ligmt
Dorsal cuneo-cuboid ligmt
Dorsal cubo-metatarsal ligmts
Dorsal inter-metatarsal ligmts
PERON. TERTIUS
Dors! astrag. scaphoid ligt
Dorsal scapho-cuneiform ligmts
Int! scapho-cunei-form ligmt
Dorsal inter-cuneiform ligmts
Middle CUNEIFORM
Dorsal cuneo-metatarsal ligmts
PERONEUS LONGUS
PERONEUS BREVIS
CUBOID
CALCAN
EXT. CUN.
TIBIALIS
1st METATARSAL

M. Cohn, ad naturam del.

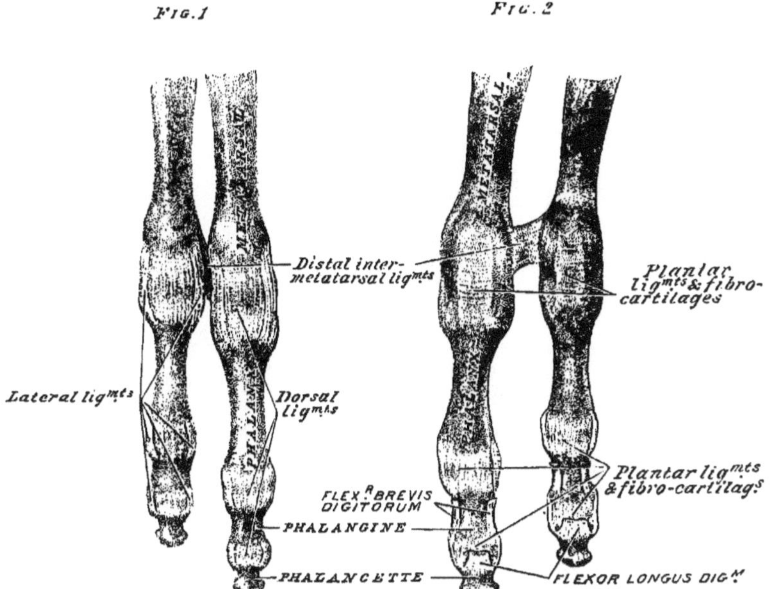
FIG. 1
FIG. 2
METATARSAL
PHALANX
METATARSAL
PHALANX
Distal inter-metatarsal ligmts
Plantar ligmts & fibro-cartilages
Lateral ligmts
Dorsal ligmts
Plantar ligmts & fibro-cartilags
FLEXR BREVIS DIGITORUM
PHALANGINE
PHALANGETTE
FLEXOR LONGUS DIGM

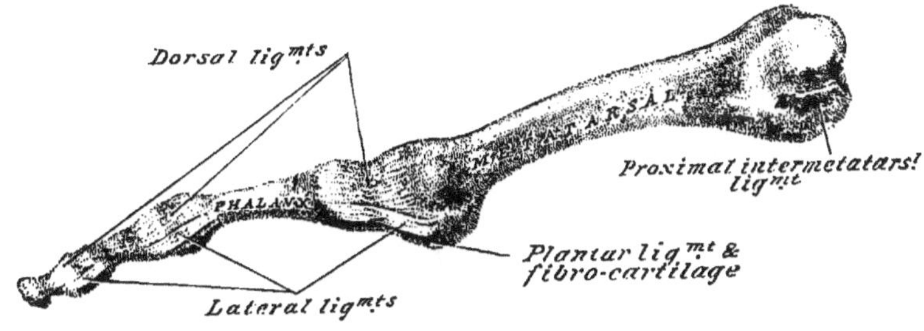
FIG. 3
Dorsal ligmts
METATARSAL
PHALANX
Proximal intermetatarsl ligmt
Plantar ligmt & fibro-cartilage
Lateral ligmts

M. Cohn, ad naturam del.

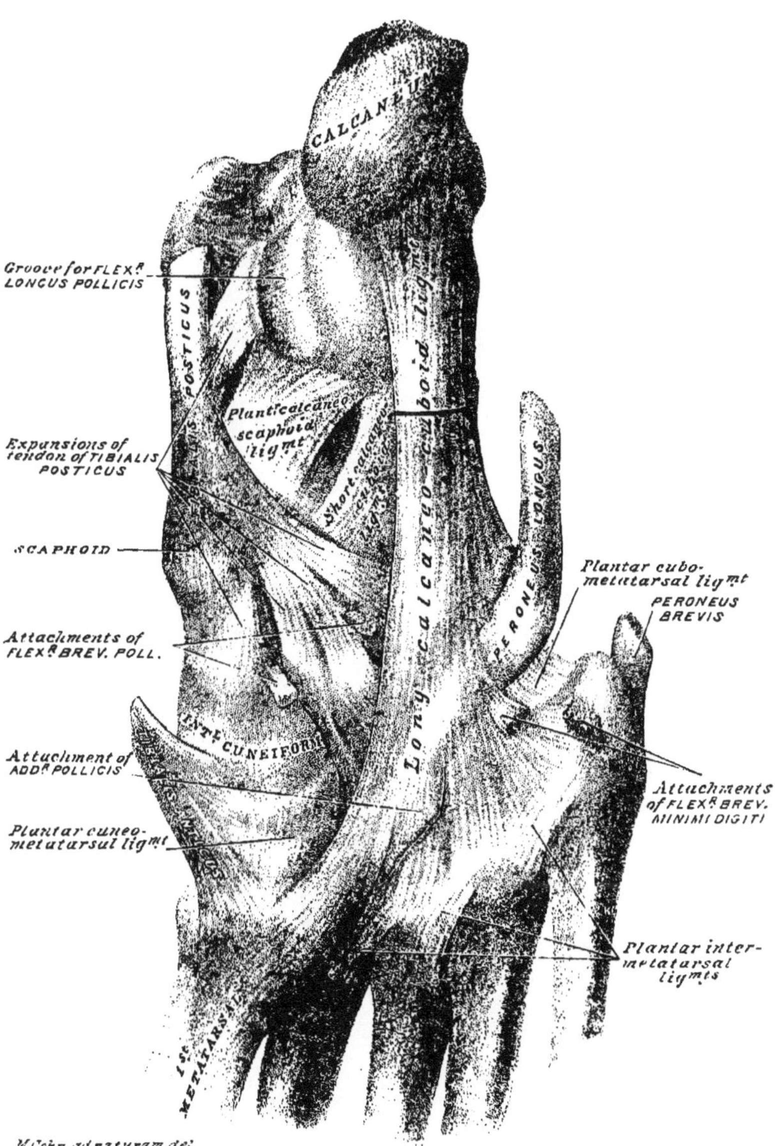
CALCANEUM
Groove for FLEX.R LONGUS POLLICIS
POSTICUS
Plant.r calcaneo-scaphoid ligmt
Short calcaneo-cuboid ligmt
Long calcaneo-cuboid Ligmt
Expansions of tendon of TIBIALIS POSTICUS
PERONEUS LONGUS
SCAPHOID
Plantar cubo-metatarsal ligmt
PERONEUS BREVIS
Attachments of FLEX.R BREV. POLL.
INT.L CUNEIFORM
Attachment of ADD.R POLLICIS
Attachments of FLEX.R BREV. MINIMI DIGITI
Plantar cuneo-metatarsal ligmt
Plantar inter-metatarsal ligmts
1st METATARSAL
M. Cohn, ad naturam del.

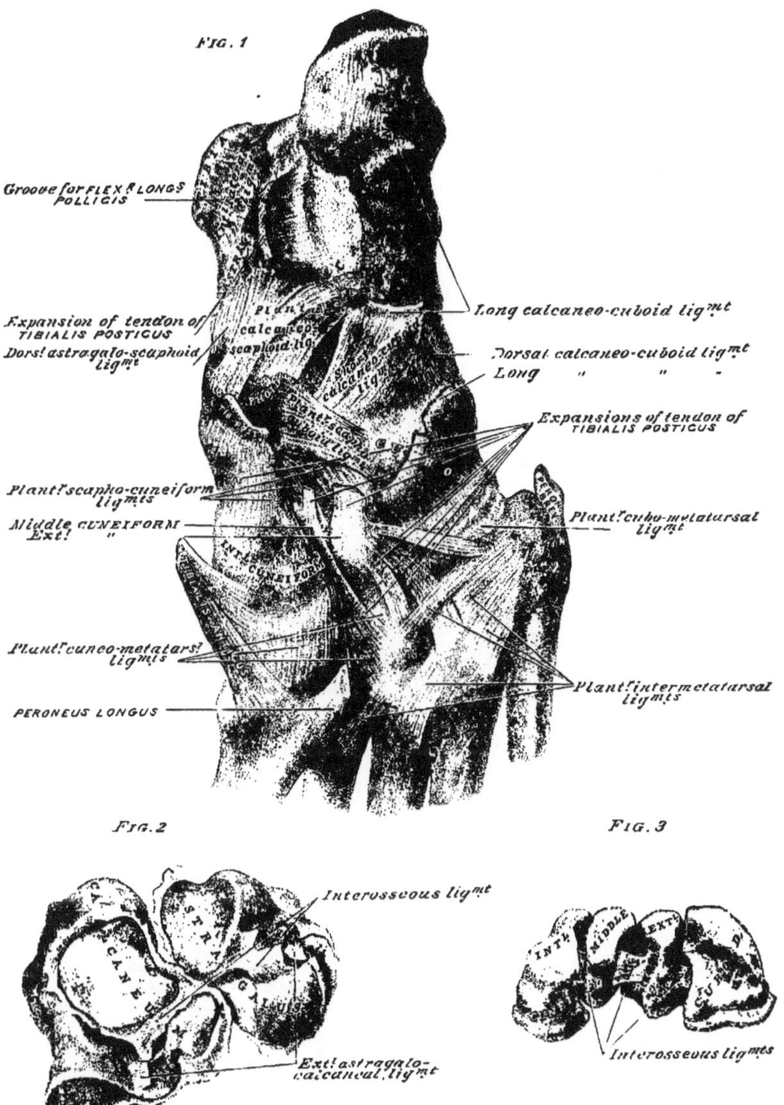
FIG. 1
Groove for FLEX.r LONGS POLLICIS
Long calcaneo-cuboid lig.mt
Expansion of tendon of TIBIALIS POSTICUS
Dors.l astragalo-scaphoid lig.mt
Dorsal calcaneo-cuboid lig.mt
Long " " "
Expansions of tendon of TIBIALIS POSTICUS
Plant.r scapho-cuneiform lig.mts
Middle CUNEIFORM
Ext.l "
Plant.r cubo-metatarsal lig.mt
Plant.r cuneo-metatars.l lig.mts
Plant.r intermetatarsal lig.mts
PERONEUS LONGUS
FIG. 2
Interosseous lig.mt
Ext.l astragalo-calcaneal lig.mt
FIG. 3
INT.L
MIDDLE
EXT.L
Interosseous lig.mes

M. Colin, ad naturam del.

Fig. 1

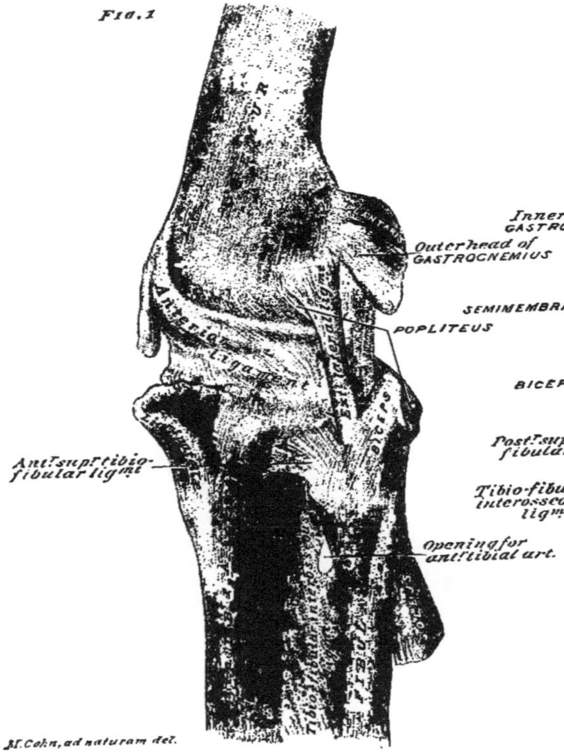

Outer head of GASTROCNEMIUS

POPLITEUS

Ant.r supr tibio-fibular ligmt

Opening for ant.r tibial art.

Anterior Ligament

Ext. lateral ligmt

BICEPS

FIBULA

M. Cohn, ad naturam del.

Fig. 2

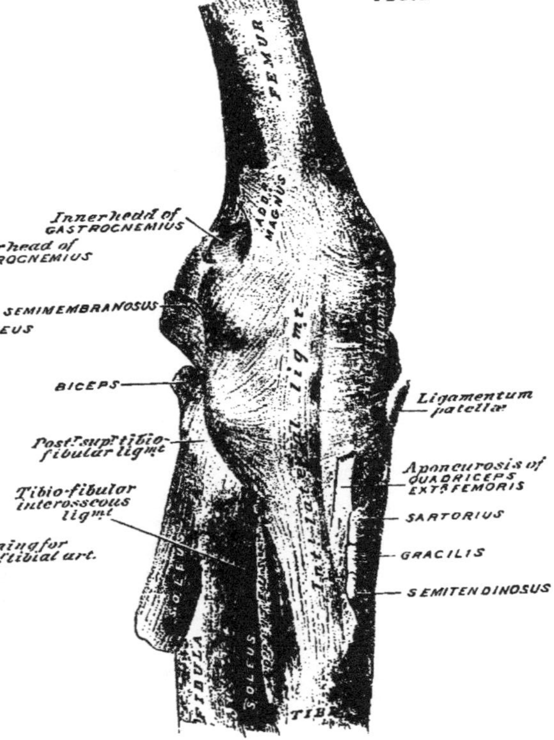

FEMUR

ADDR MAGNUS

Inner head of GASTROCNEMIUS

SEMIMEMBRANOSUS

BICEPS

Post.r supr tibio-fibular ligmt

Tibio-fibular interosseous ligmt

Int. lateral ligmt

Ligamentum patellæ

Aponeurosis of QUADRICEPS EXT.R FEMORIS

SARTORIUS

GRACILIS

SEMITENDINOSUS

SOLEUS

FIBULA

TIB.

Fig. 1

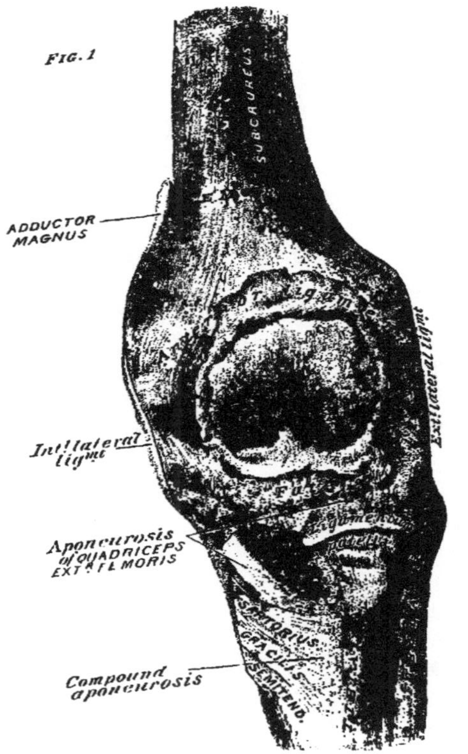
SUBCRUREUS
ADDUCTOR MAGNUS
Intl lateral ligmt
Extl lateral ligmt
Aponeurosis of QUADRICEPS EXTR FEMORIS
SARTORIUS
GRACILIS
SEMITEND.
Compound aponeurosis

Fig. 2

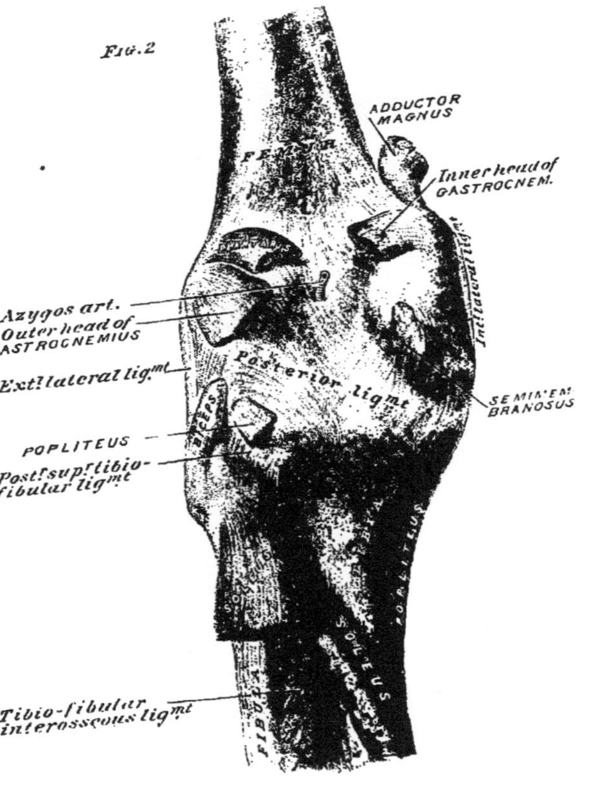
ADDUCTOR MAGNUS
Inner head of GASTROCNEM.
Intl lateral ligmt
Azygos art.
Outer head of GASTROCNEMIUS
Extl lateral ligmt
Posterior ligmt
SEMIMEMBRANOSUS
POPLITEUS
Postr supr tibio-fibular ligmt
Tibio-fibular interosseous ligmt

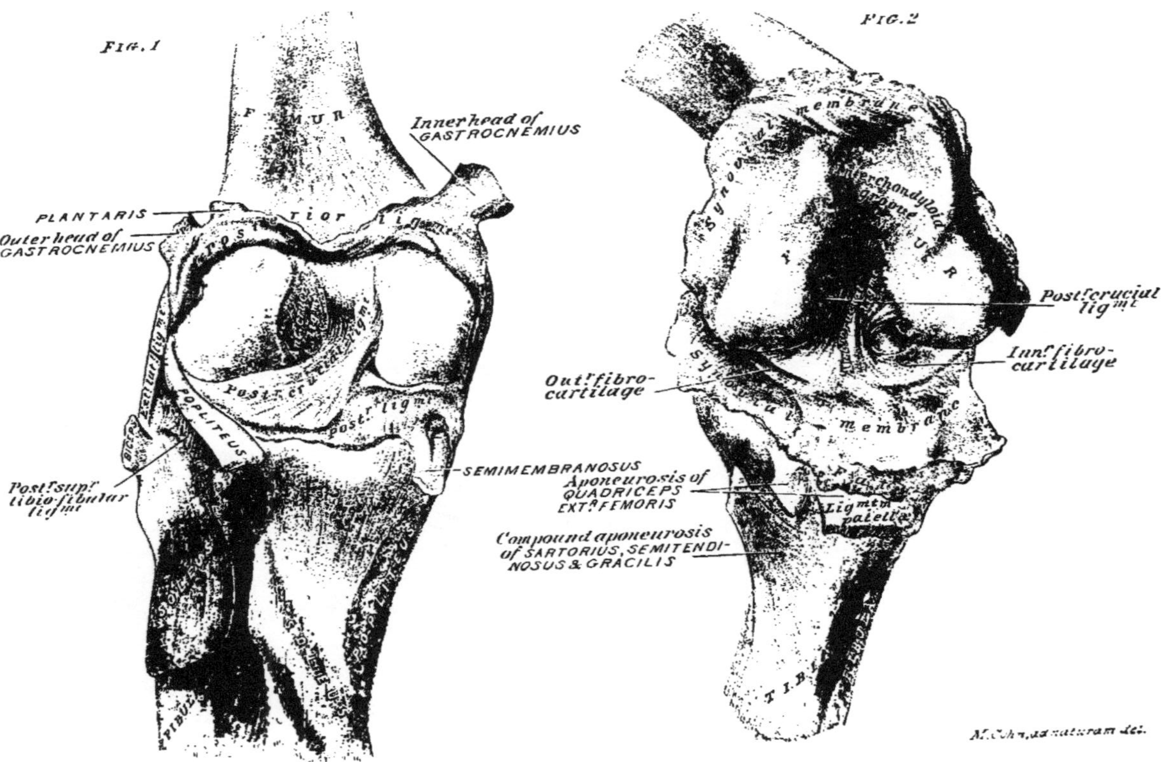
FIG. 1
FEMUR
Inner head of GASTROCNEMIUS
PLANTARIS
Outer head of GASTROCNEMIUS
Posterior ligm.t
Post.r crucial ligm.t
Post.r ligm.t
POPLITEUS
BICEPS
SEMIMEMBRANOSUS
Post.r sup.r tibio-fibular ligm.t
FIG. 2
Synovial membrane
Interchondyloid groove
FEMUR
Post.r crucial ligm.t
Inn.r fibro-cartilage
Out.r fibro-cartilage
Aponeurosis of QUADRICEPS EXT.R FEMORIS
Ligm.t patellæ
Compound aponeurosis of SARTORIUS, SEMITENDINOSUS & GRACILIS
TIBIA
M. Cohn, ad naturam del.

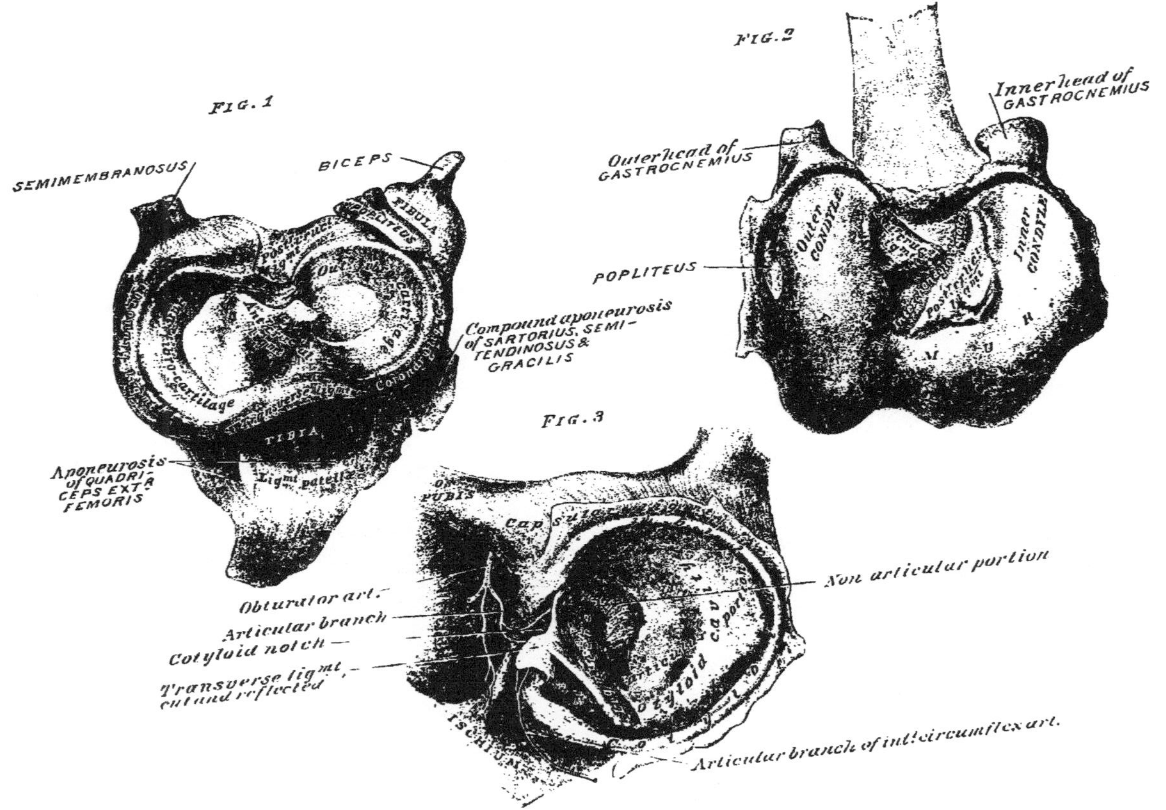
FIG. 1
SEMIMEMBRANOSUS
BICEPS
FIBULA
POPLITEUS
TIBIA
Ligmt patellæ
Compound aponeurosis of SARTORIUS, SEMI-TENDINOSUS & GRACILIS
Aponeurosis of QUADRICEPS EXTR FEMORIS
FIG. 2
Outer head of GASTROCNEMIUS
Inner head of GASTROCNEMIUS
Outer CONDYLE
Inner CONDYLE
POPLITEUS
FIG. 3
PUBIS
ISCHIUM
Obturator art.
Articular branch
Cotyloid notch
Transverse ligmt, cut and reflected
Non articular portion
Articular branch of intl circumflex art.

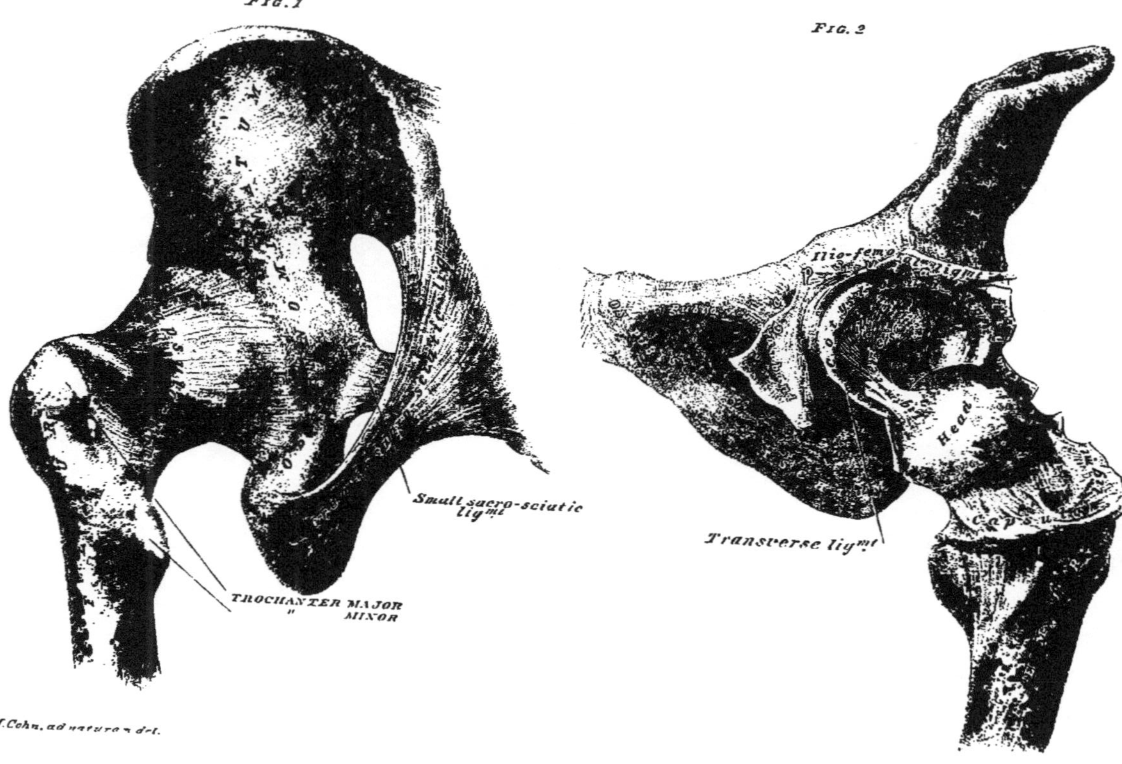
FIG. 1
Small sacro-sciatic ligmt
TROCHANTER MAJOR
" MINOR
M. Cohn, ad naturam del.
FIG. 2
Ilio-femoral
Head
Transverse ligmt

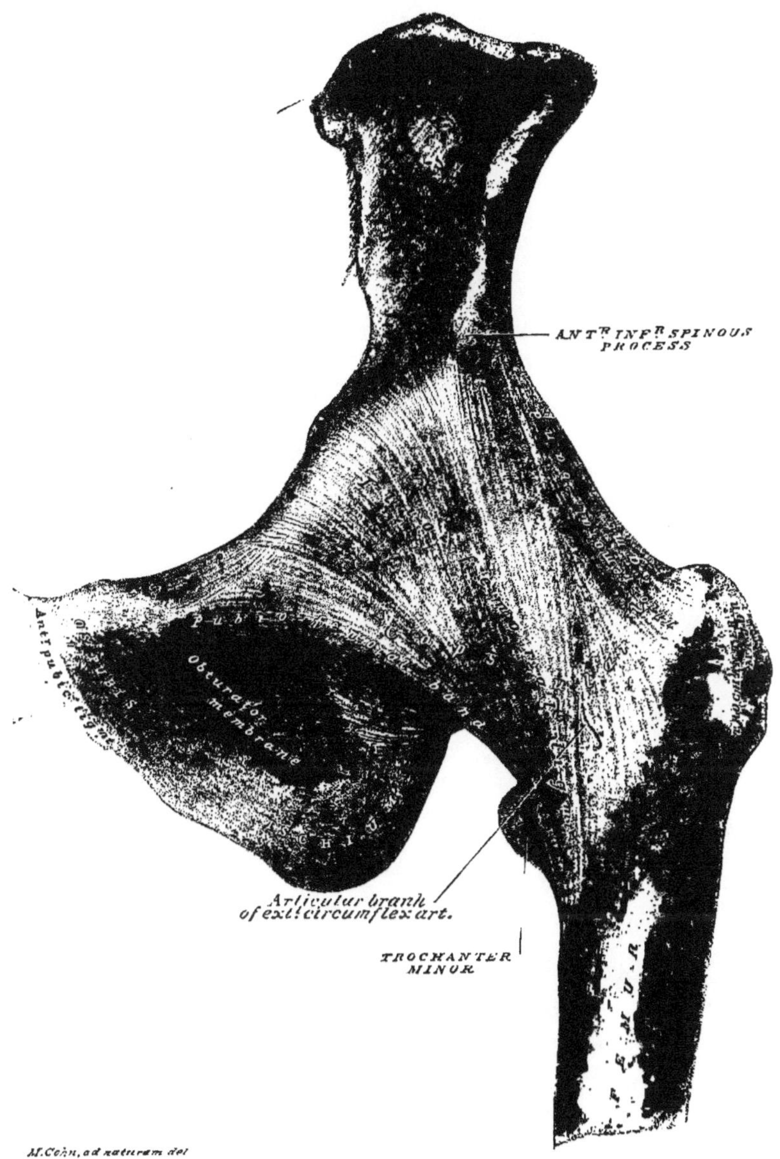
ANTR INFR SPINOUS PROCESS
Obturator membrane
Antr pubic ligt.
Articular branh of extl circumflex art.
TROCHANTER MINOR
FEMUR

M. Cohn, ad naturam del

PLATE 113

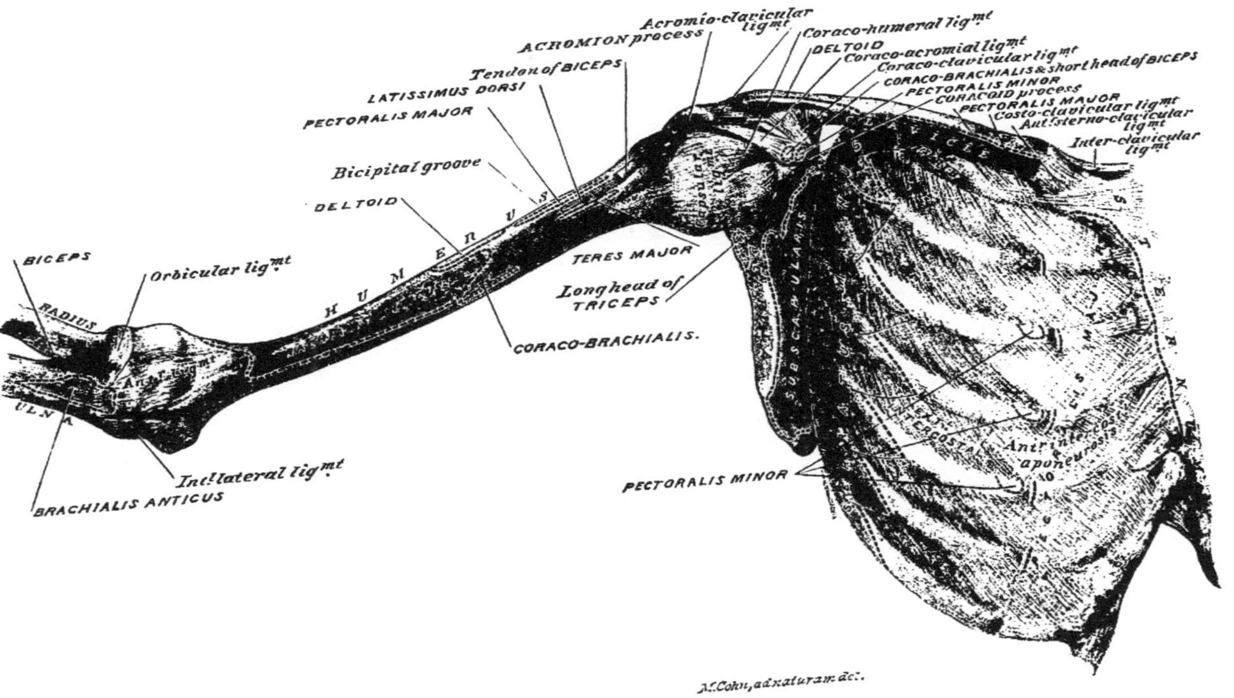

Acromio-clavicular ligmt
Coraco-humeral ligmt
ACROMION process
DELTOID
Coraco-acromial ligmt
Coraco-clavicular ligmt
Tendon of BICEPS
CORACO-BRACHIALIS & short head of BICEPS
PECTORALIS MINOR
LATISSIMUS DORSI
CORACOID process
PECTORALIS MAJOR
PECTORALIS MAJOR
Costo-clavicular ligmt
Antl sterno-clavicular ligmt
Inter-clavicular ligmt
Bicipital groove
DELTOID
HUMERUS
CLAVICLE
STERNUM
BICEPS
Orbicular ligmt
TERES MAJOR
Long head of TRICEPS
RADIUS
SUBSCAPULARIS
CORACO-BRACHIALIS.
ULNA
EXT. INTERCOSTAL
Antr intercostal aponeurosis
PECTORALIS MINOR
Intl lateral ligmt
BRACHIALIS ANTICUS
M.Cohn, ad naturam del.

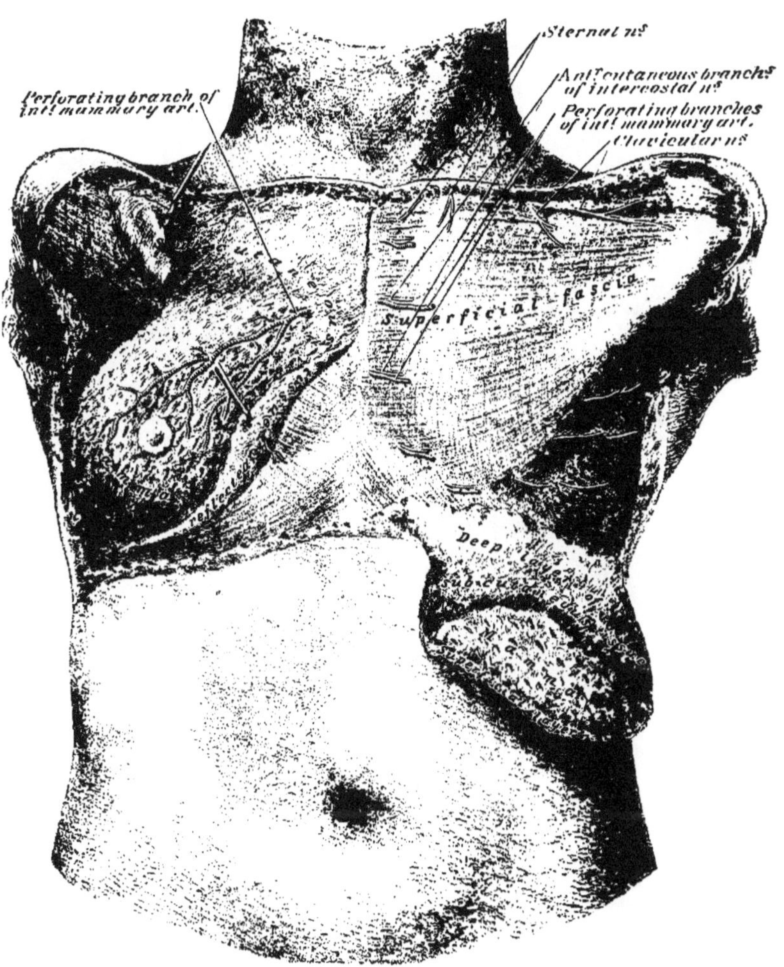
Sternal n.s
Ant. cutaneous branch.s of intercostal n.s
Perforating branches of int. mammary art.
Clavicular n.s
Perforating branch of int. mammary art.
Superficial fascia
Deep layer of

M.Cohn, ad naturam del.

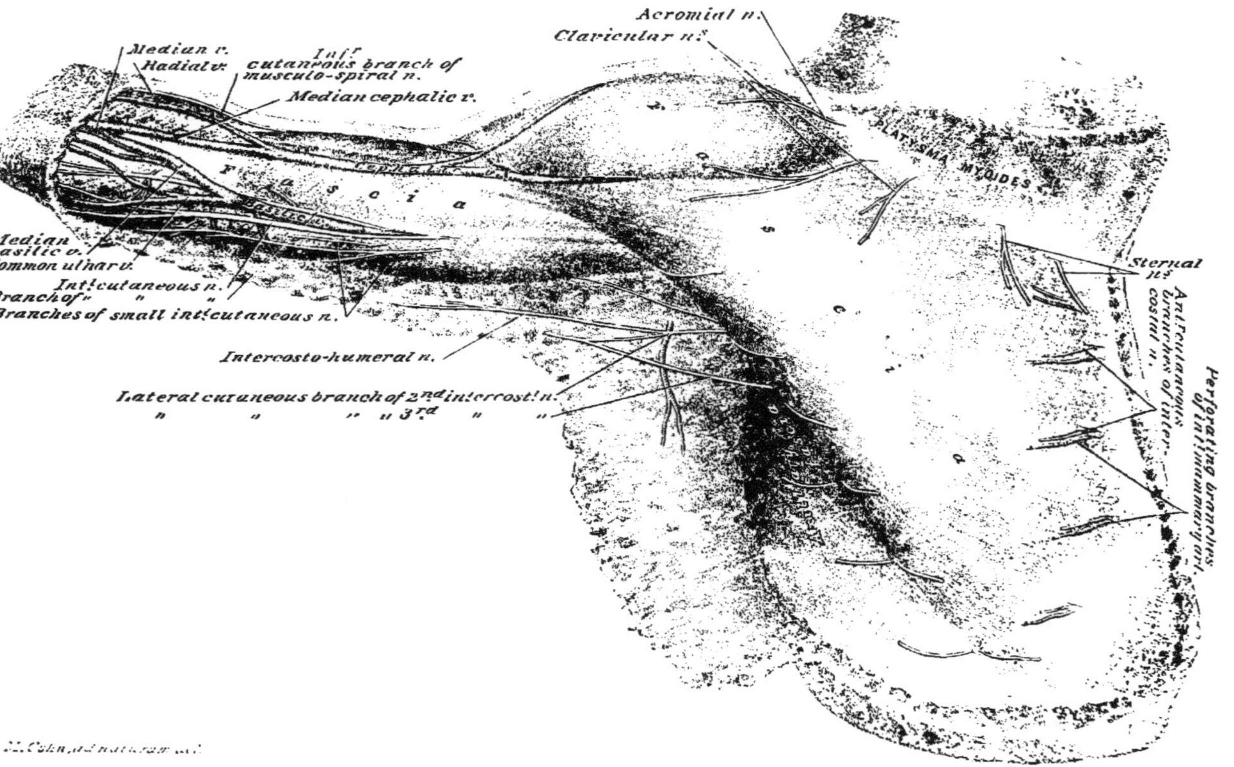
Acromial n.
Clavicular n.s
Median v.
Radial v.
Inf! cutaneous branch of musculo-spiral n.
Median cephalic v.
PLATYSMA MYOIDES
Median basilic v.
Common ulnar v.
Int! cutaneous n.
Branch of " "
Branches of small int! cutaneous n.
Intercosto-humeral n.
Lateral cutaneous branch of 2nd intercost! n.
" " " " 3rd " "
Sternal n.s
Ant! cutaneous branches of inter-costal n.
Perforating branches of int! mammary art.

PLATE 116

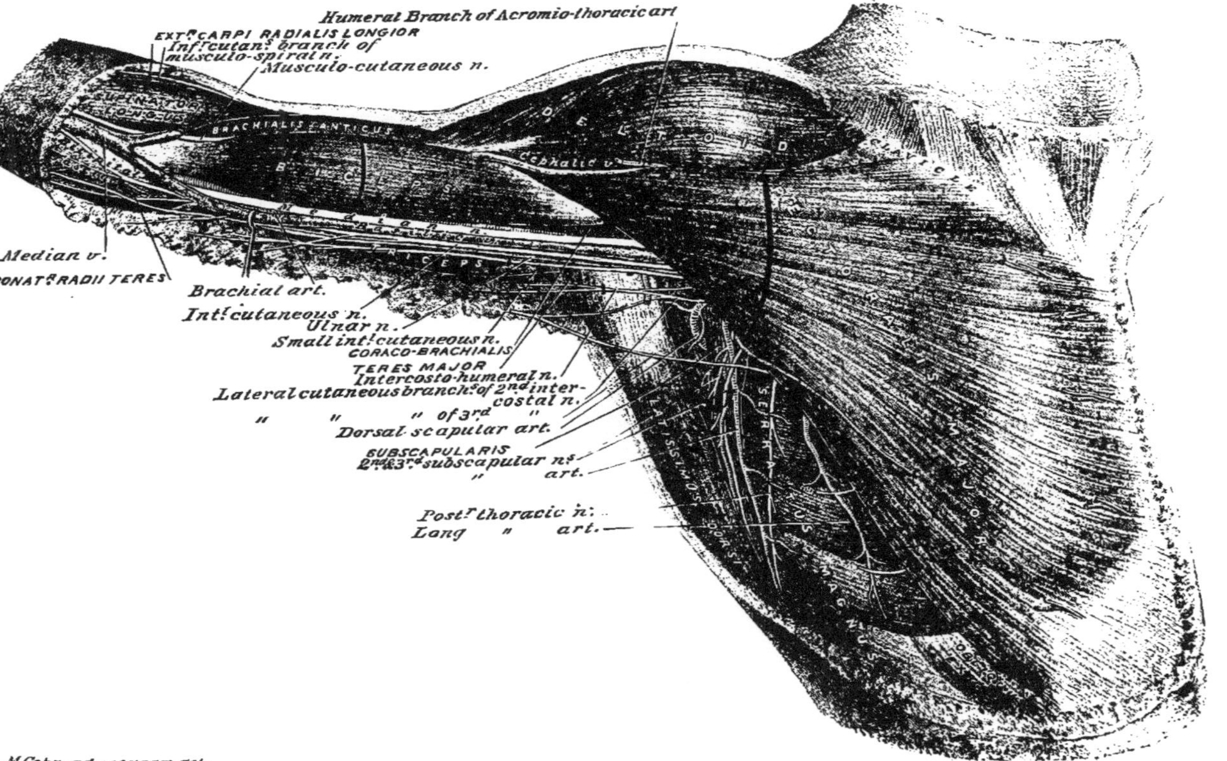

Humeral Branch of Acromio-thoracic art
EXT.R CARPI RADIALIS LONGIOR
Int.l cutan.s branch of musculo-spiral n.
Musculo-cutaneous n.
SUPINATOR LONGUS
BRACHIALIS ANTICUS
DELTOID
Cephalic v.
BICEPS
TRICEPS
Median v.
PRONAT.R RADII TERES
Brachial art.
Int.l cutaneous n.
Ulnar n.
Small int.l cutaneous n.
CORACO-BRACHIALIS
TERES MAJOR
Intercosto-humeral n.
Lateral cutaneous branch.s of 2nd intercostal n.
" " " of 3rd "
Dorsal scapular art.
SUBSCAPULARIS
2nd & 3rd subscapular n.s
" art.
Post.r thoracic n.
Long " art.
SERRATUS MAGNUS
LATISSIMUS DORSI
M. Cohn, ad naturam del.

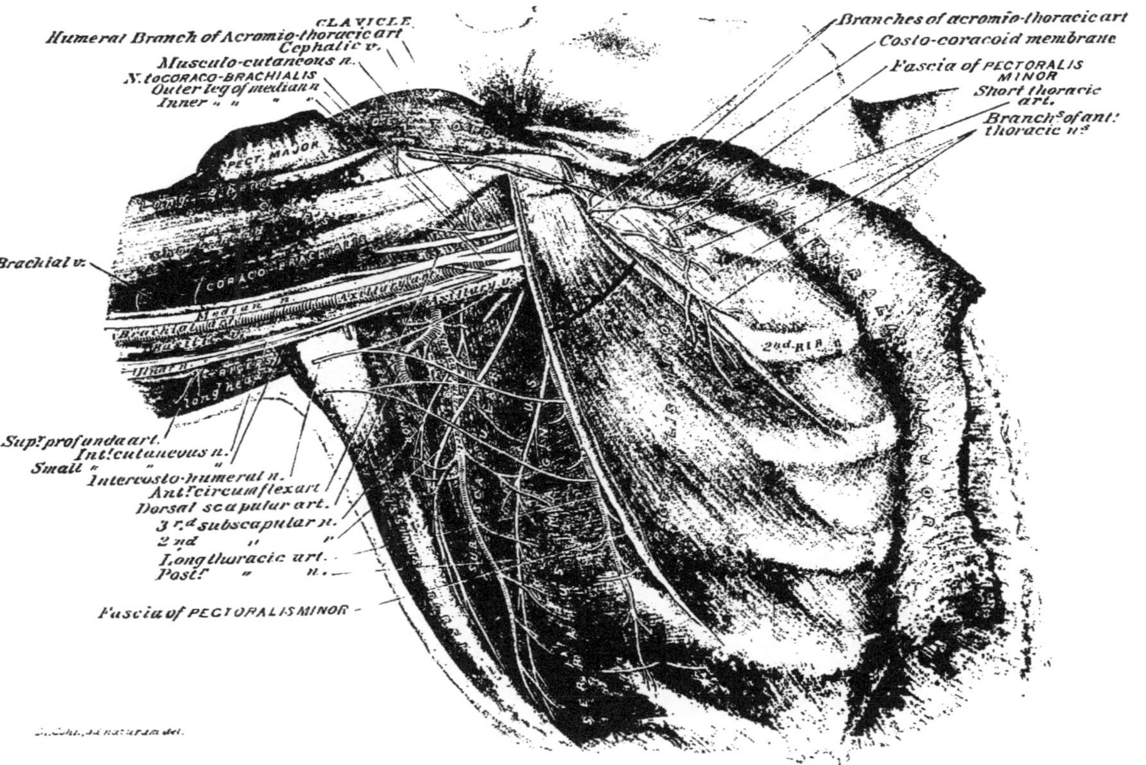
CLAVICLE
Humeral Branch of Acromio-thoracic art
Cephalic v.
Musculo-cutaneous n.
N. to CORACO-BRACHIALIS
Outer leg of median n
Inner " " "
Brachial v.
PECT. MAJOR
CORACO-BRACHIALIS
Median n.
Axillary v.
2nd RIB
Sup! profunda art.
Int! cutaneous n.
Small " "
Intercosto-humeral n.
Ant! circumflex art
Dorsal scapular art.
3rd subscapular n.
2nd " "
Long thoracic art.
Post! " n.
Fascia of PECTORALIS MINOR
Branches of acromio-thoracic art
Costo-coracoid membrane
Fascia of PECTORALIS MINOR
Short thoracic art.
Branch.s of ant! thoracic n.s

PLATE 118

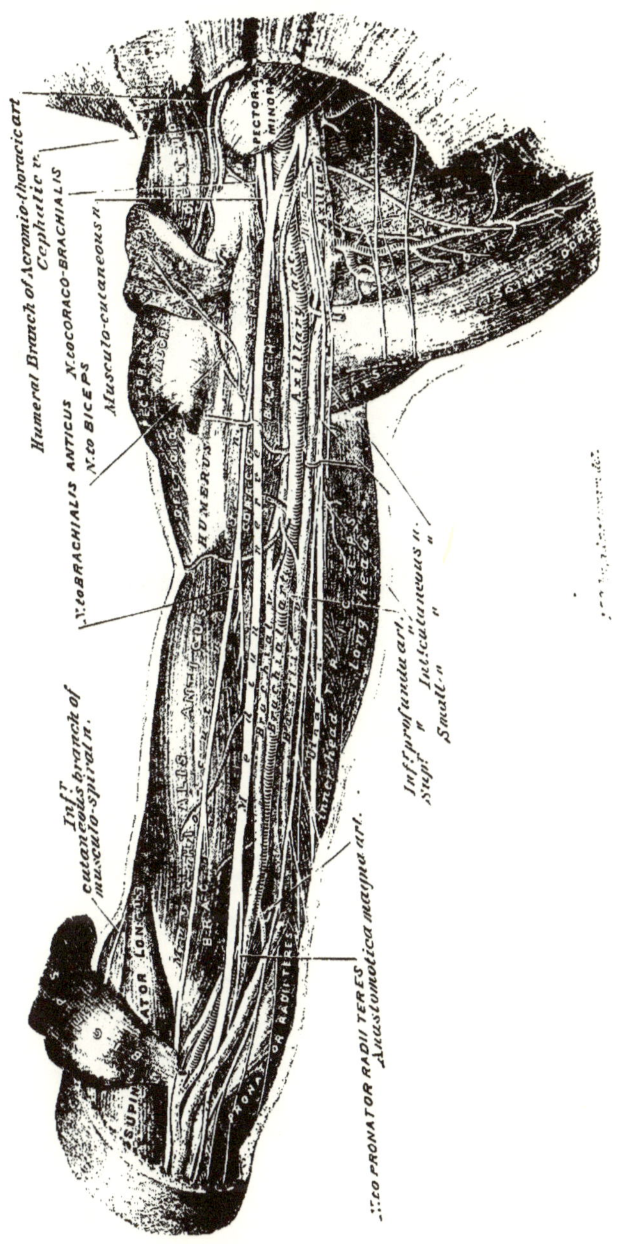
Humeral Branch of Acromio-thoracic art
Cephalic v.
N. to CORACO-BRACHIALIS
Musculo-cutaneous n.
BRACHIALIS ANTICUS
N. to BICEPS
N. to BRACHIALIS
Inf.r cutaneous branch of musculo-spiral n.
N. to PRONATOR RADII TERES
Anastomotica magna art.
Inf.r profunda art.
Sup.r " "
Internal cutaneous n.
Small " "
PECTORAL MINOR
Axillary art.
HUMERUS
Long head
PRONATOR RADII TERES
SUPINATOR LONGUS

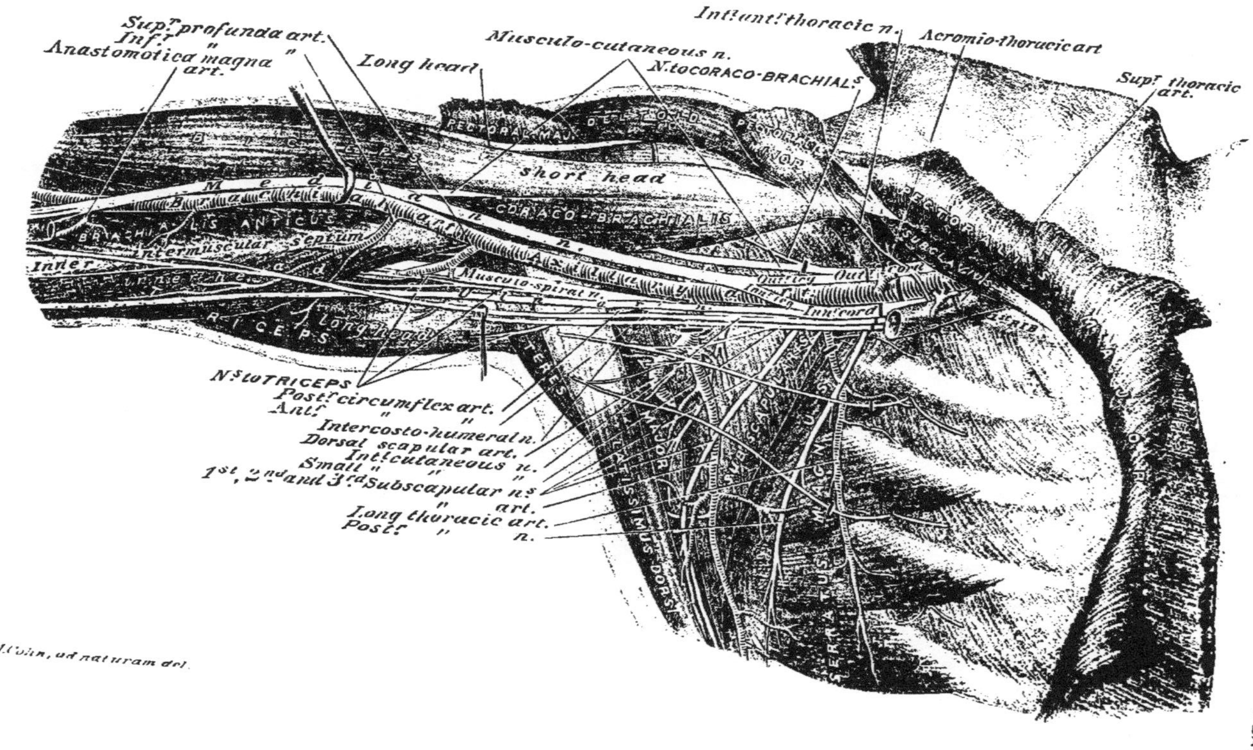
Supr profunda art.
Infr "
Anastomotica magna art.
Long head
Musculo-cutaneous n.
N. to CORACO-BRACHIALIS
Int. ant. thoracic n.
Acromio-thoracic art
Supr thoracic art.
DELTOID
PECTORAL. MAJ.
short head
CORACO-BRACHIALIS
BRACHIALIS ANTICUS
Intermuscular septum
SUBCLAVIUS
Musculo-spiral n.
Inn. cord
TRICEPS
Long head
TERES MAJOR
LATISSIMUS DORSI
SERRATUS MAGNUS
Ns to TRICEPS
Postr circumflex art.
Antr " "
Intercosto-humeral n.
Dorsal scapular art.
Intr cutaneous n.
Small " "
1st, 2nd and 3rd Subscapular ns
" art.
Long thoracic art.
Postr " n.
M. Cohn, ad naturam del.

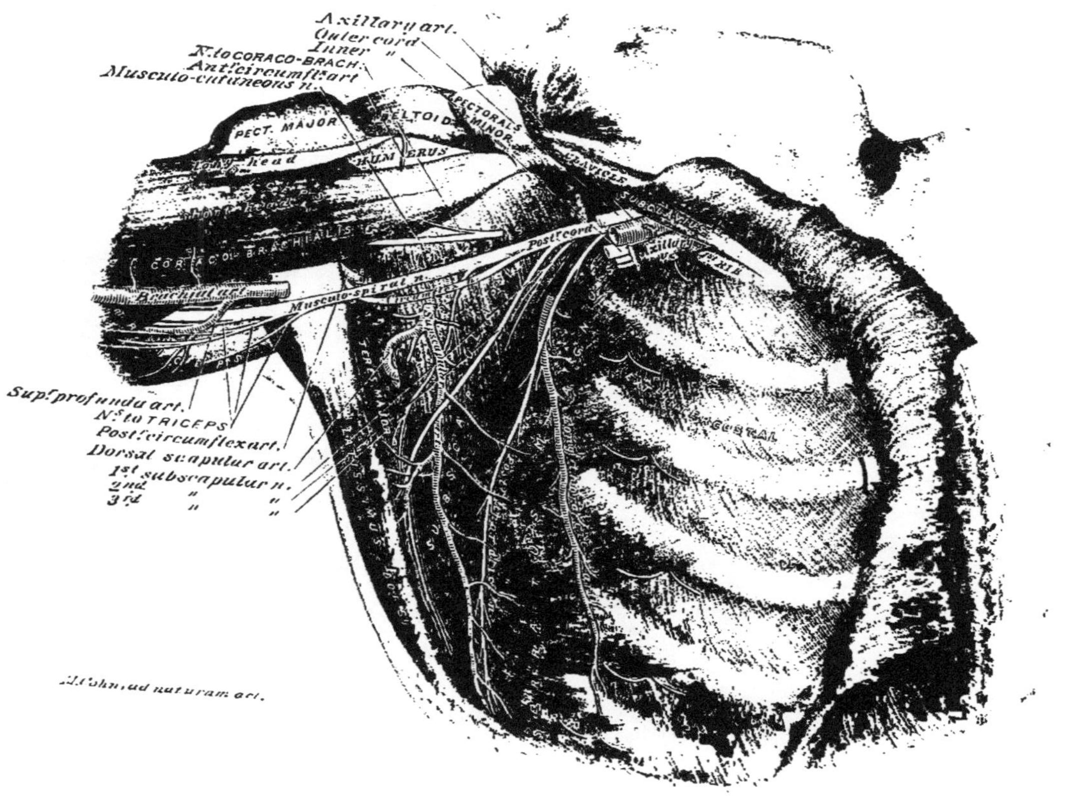
Axillary art.
Outer cord
Inner "
N. to CORACO-BRACH.
Ant. circumfl. art.
Musculo-cutaneous n.
PECT. MAJOR
DELTOID
PECTORALIS MINOR
HUMERUS
CLAVICLE
Long head
CORACO-BRACHIALIS
Post. cord
Brachial art.
Musculo-spiral n.
Sup. profunda art.
N. to TRICEPS
Post. circumflex art.
Dorsal scapular art.
1st subscapular n.
2nd " "
3rd " "
TERES MAJOR
LATISSIMUS DORSI
M. Cohn ad naturam del.

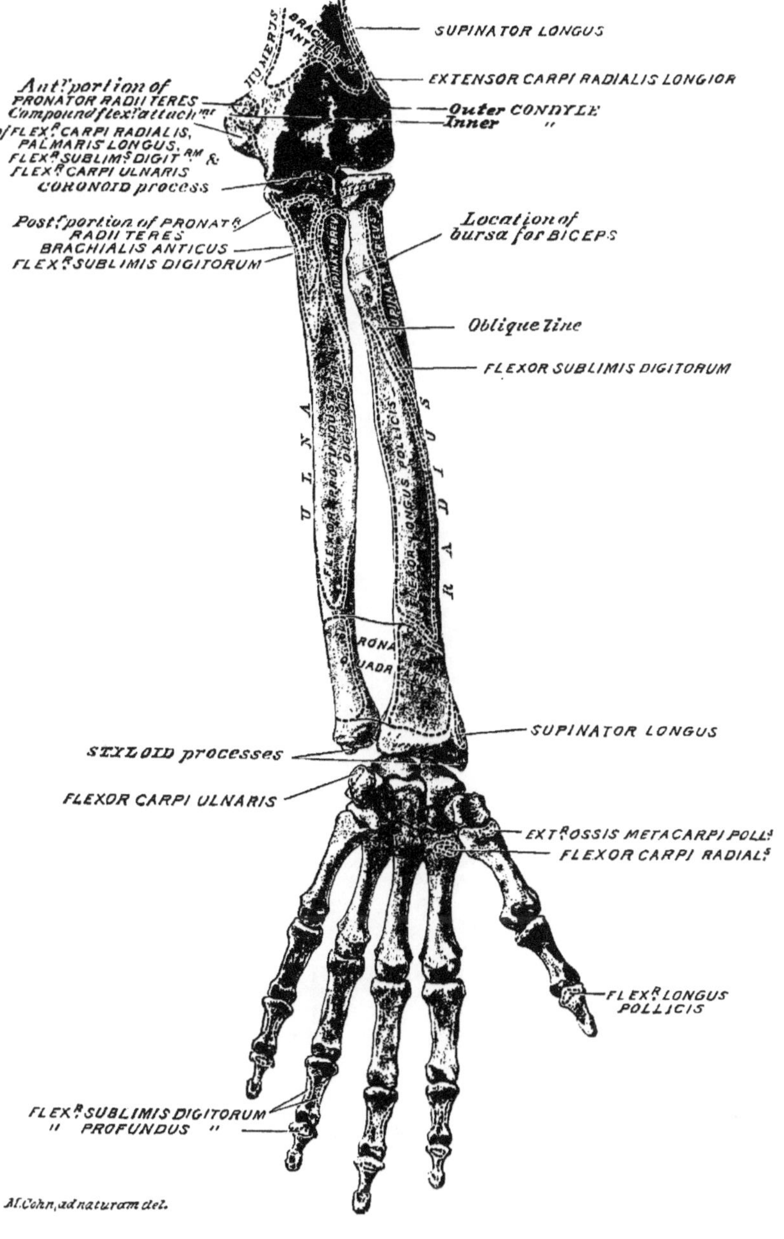
SUPINATOR LONGUS
EXTENSOR CARPI RADIALIS LONGIOR
Outer CONDYLE
Inner "
Antr. portion of PRONATOR RADII TERES
Compound flexr. attachmt of FLEXR. CARPI RADIALIS, PALMARIS LONGUS, FLEXR. SUBLIMS. DIGITRM & FLEXR. CARPI ULNARIS
CORONOID process
Postr. portion of PRONATR. RADII TERES
BRACHIALIS ANTICUS
FLEXR. SUBLIMIS DIGITORUM
Location of bursa for BICEPS
Oblique line
FLEXOR SUBLIMIS DIGITORUM
ULNA
RADIUS
SUPINATOR LONGUS
STYLOID processes
FLEXOR CARPI ULNARIS
EXTR. OSSIS METACARPI POLLs.
FLEXOR CARPI RADIALs.
FLEXR. LONGUS POLLICIS
FLEXR. SUBLIMIS DIGITORUM
" PROFUNDUS "

M. Cohn, ad naturam del.

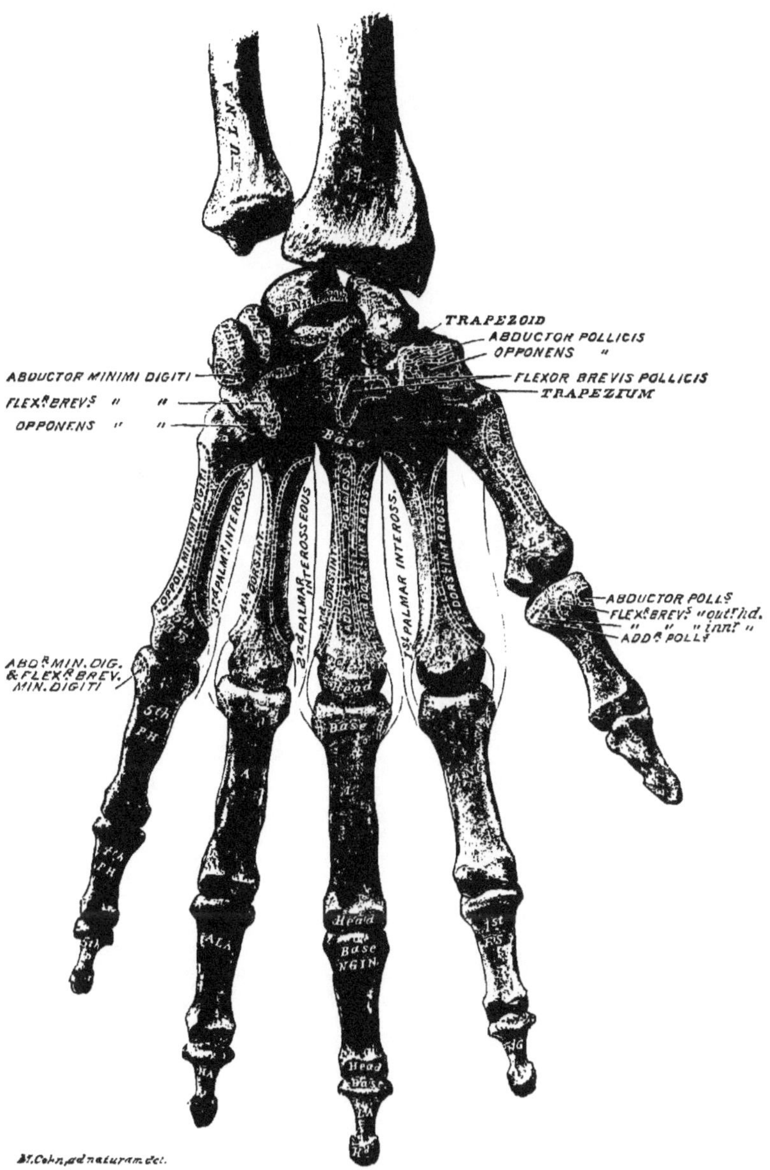
TRAPEZOID
ABDUCTOR POLLICIS
OPPONENS "
ABDUCTOR MINIMI DIGITI
FLEXOR BREVIS POLLICIS
TRAPEZIUM
FLEXR BREVS " "
OPPONENS " "
ABDUCTOR POLLS
ABDR MIN. DIG. & FLEXR BREV. MIN. DIGITI
ADDR POLLS

M. Cohn, ad naturam del.

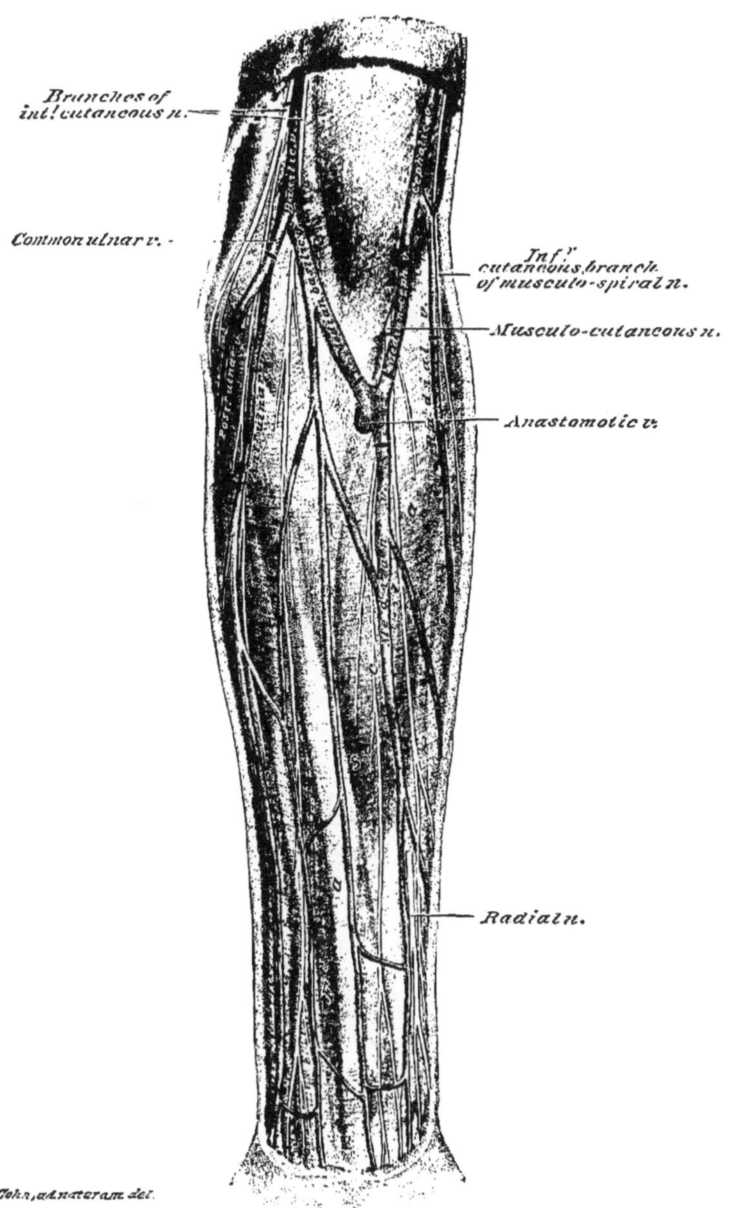
Branches of
int! cutaneous n.
Common ulnar v.
Infr
cutaneous branch
of musculo-spiral n.
Musculo-cutaneous n.
Anastomotic v.
Radial n.
M. Cohn, ad naturam del.

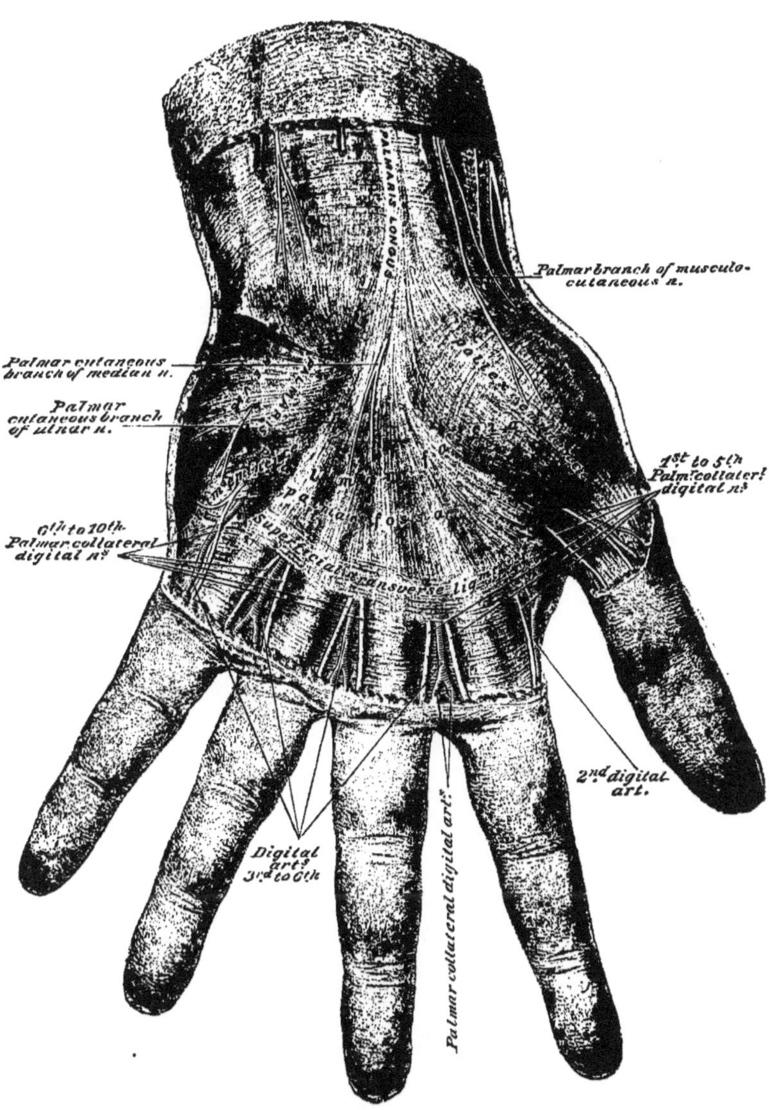
Palmar branch of musculo-cutaneous n.
Palmar cutaneous branch of median n.
Palmar cutaneous branch of ulnar n.
1st to 5th Palm. collater! digital n.s
6th to 10th Palmar collateral digital n.s
2nd digital art.
Digital art.s 3rd to 6th
Palmar collateral digital art.s
Palmaris longus
Superficial transverse lig.

M. Cohn, ad naturam del.

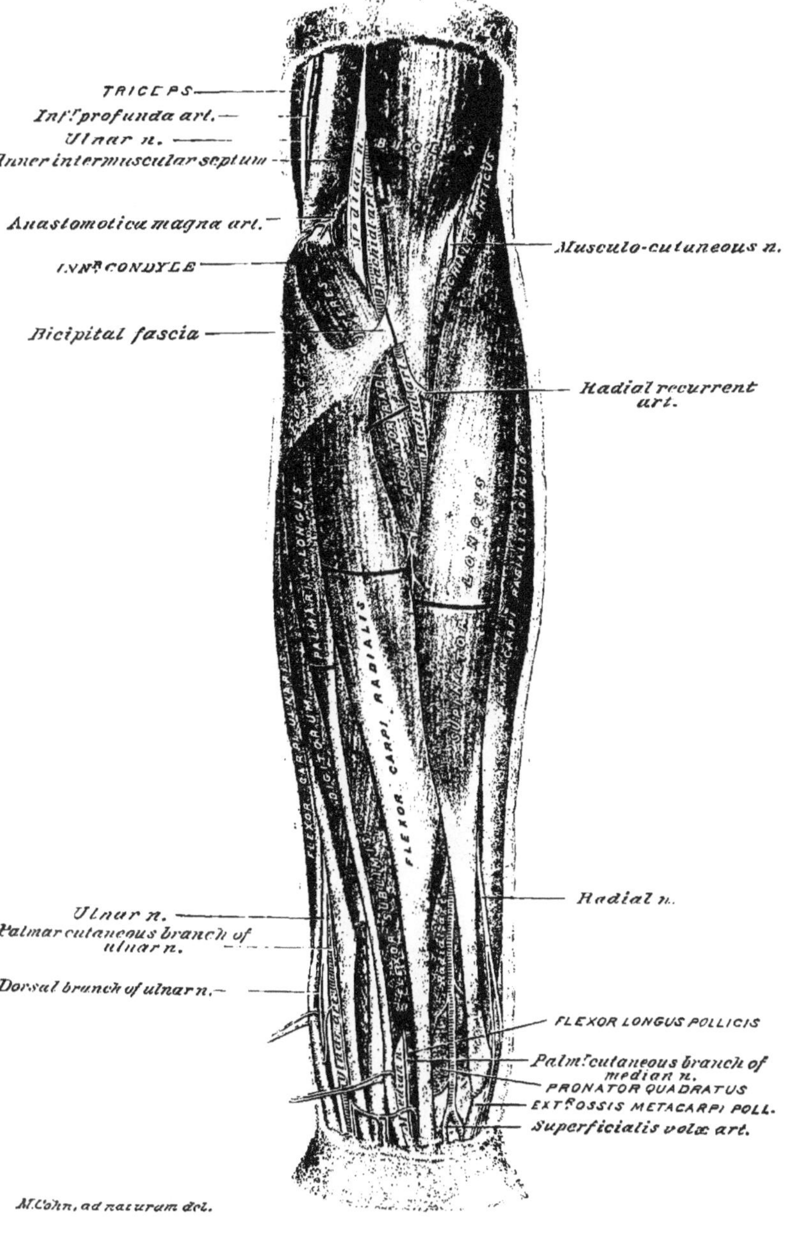
TRICEPS
Infr profunda art.
Ulnar n.
Inner intermuscular septum
Anastomotica magna art.
INNR CONDYLE
Bicipital fascia
Musculo-cutaneous n.
Radial recurrent art.
BICEPS
FLEXOR CARPI RADIALIS
Ulnar n.
Palmar cutaneous branch of ulnar n.
Dorsal branch of ulnar n.
Radial n.
FLEXOR LONGUS POLLICIS
Palmr cutaneous branch of median n.
PRONATOR QUADRATUS
EXTR OSSIS METACARPI POLL.
Superficialis volæ art.

M. Cohn, ad naturam del.

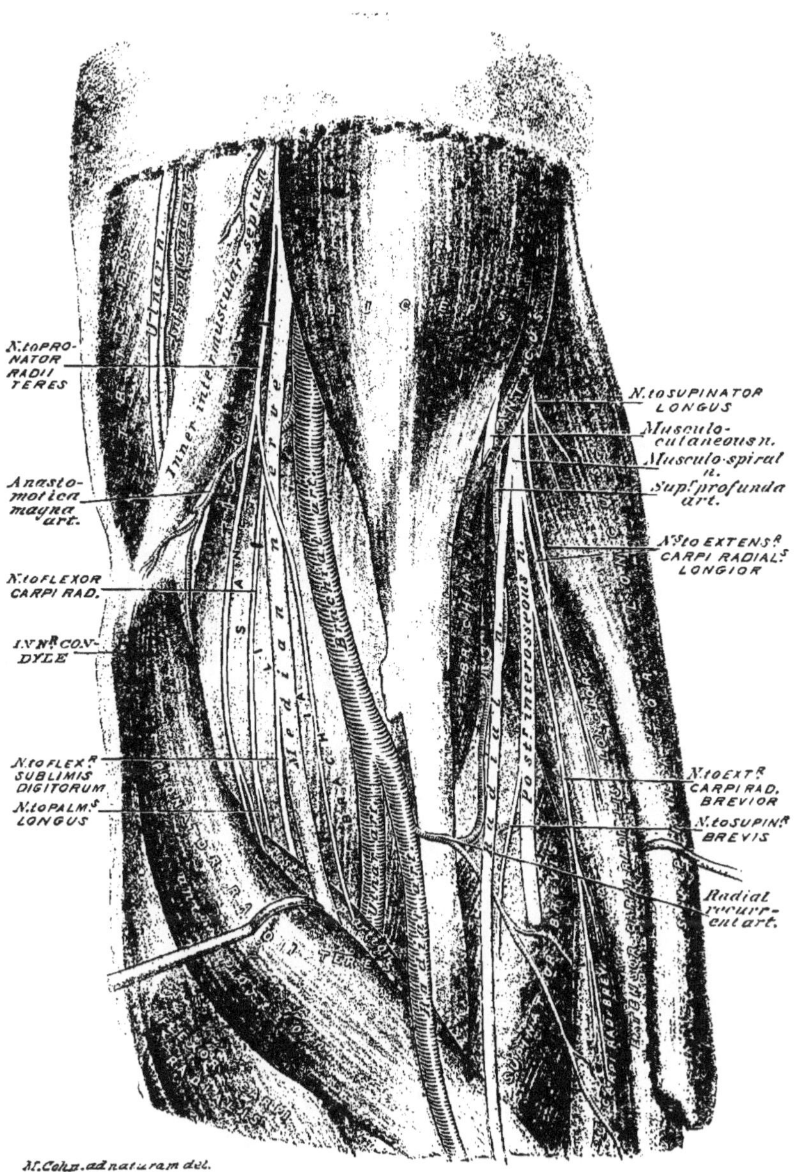
N. to PRONATOR RADII TERES
Anastomotica magna art.
N. to FLEXOR CARPI RAD.
INNR CONDYLE
N. to FLEXR SUBLIMIS DIGITORUM
N. to PALMS LONGUS
N. to SUPINATOR LONGUS
Musculo-cutaneous n.
Musculo-spiral n.
Supr profunda art.
Ns to EXTENSR CARPI RADIALS LONGIOR
N. to EXTR CARPI RAD. BREVIOR
N. to SUPINR BREVIS
Radial recurrent art.
Ulnar n.
Inner intermuscular septum
Median nerve
Post. interosseous n.
M. Cohn. ad naturam del.

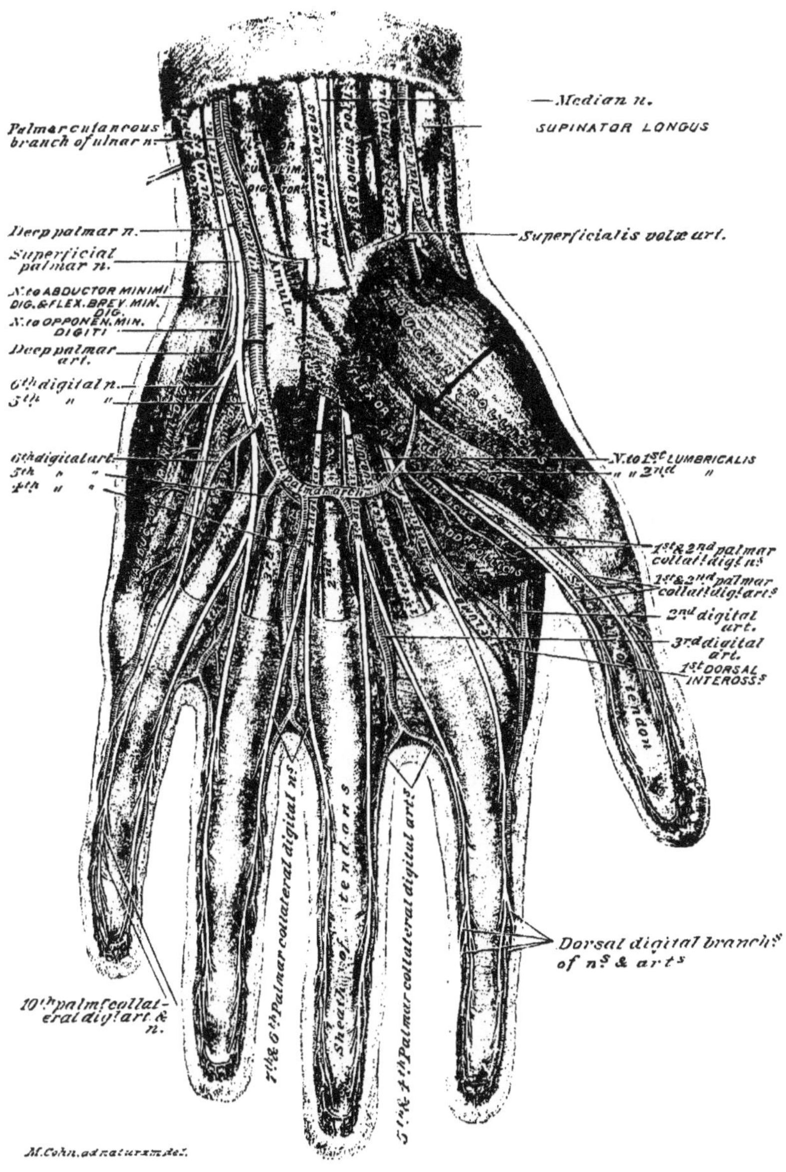
Median n.
SUPINATOR LONGUS
Palmar cutaneous branch of ulnar n.
Deep palmar n.
Superficial palmar n.
N. to ABDUCTOR MINIMI DIG. & FLEX. BREV. MIN. DIG.
N. to OPPONEN. MIN. DIGITI
Deep palmar art.
6th digital n.
5th " "
Superficialis volæ art.
6th digital art.
5th " "
4th " "
N. to 1st LUMBRICALIS
" " 2nd "
1st & 2nd palmar collat! digl ns
1st & 2nd palmar collat! digl arts
2nd digital art.
3rd digital art.
1st DORSAL INTEROSS.s
Dorsal digital branchs of ns & arts
10th palml collateral digl art. & n.
7th & 6th Palmar collateral digital ns
Sheath of tendons
5th & 4th Palmar collateral digital arts
M. Cohn. ad naturam del.

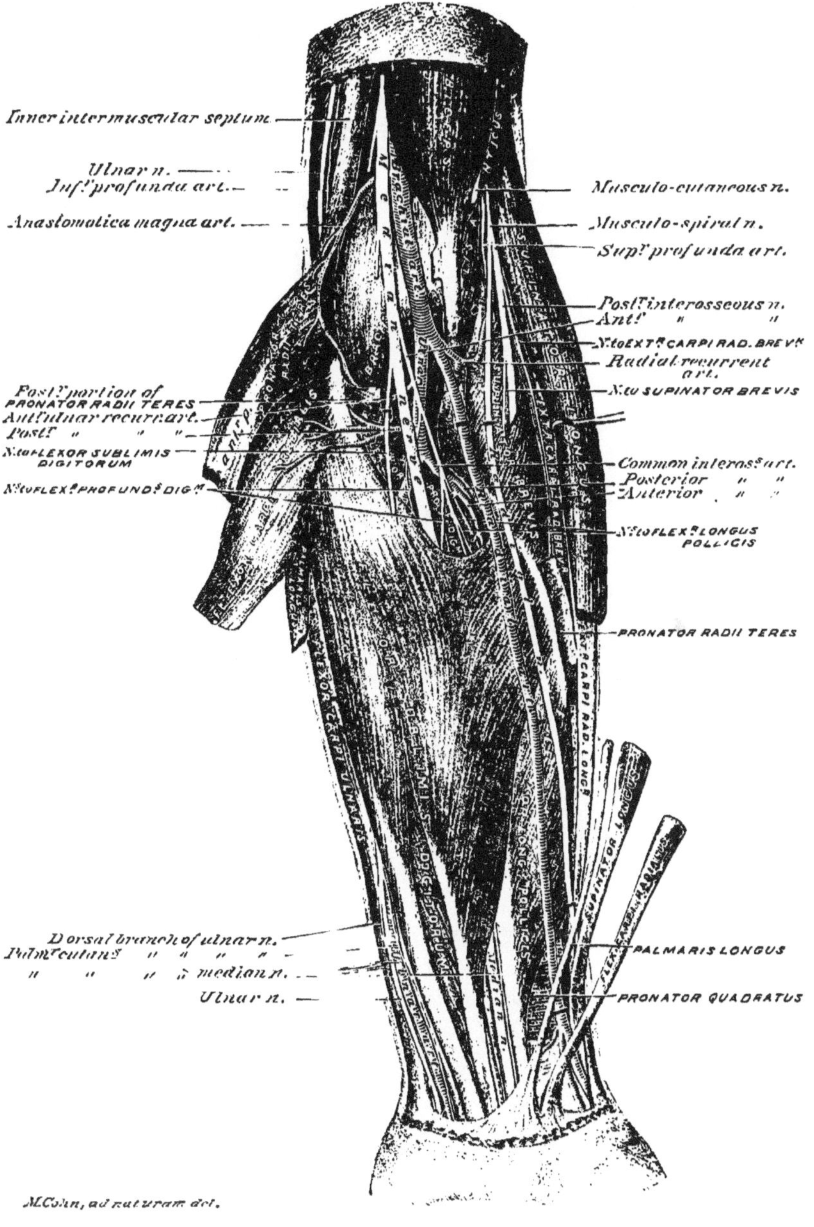
Inner intermuscular septum.
Ulnar n.
Inf.r profunda art.
Anastomotica magna art.
Musculo-cutaneous n.
Musculo-spiral n.
Sup.r profunda art.
Post.r interosseous n.
Ant.r " "
N. to EXT.R CARPI RAD. BREV.R
Radial recurrent art.
N. to SUPINATOR BREVIS
Post.r portion of PRONATOR RADII TERES
Ant.r ulnar recurr. art.
Post.r " " "
N. to FLEXOR SUBLIMIS DIGITORUM
N. to FLEX.R PROFUND.S DIG.M
Common inteross.s art.
Posterior " "
Anterior " "
N. to FLEX.R LONGUS POLLICIS
PRONATOR RADII TERES
Dorsal branch of ulnar n.
Palm.r cutan.s " " " "
" " " " median n.
Ulnar n.
PALMARIS LONGUS
PRONATOR QUADRATUS
Median nerve
SUPINATOR LONGUS
FLEX. CARPI RADIALIS
M.Cohn, ad naturam del.

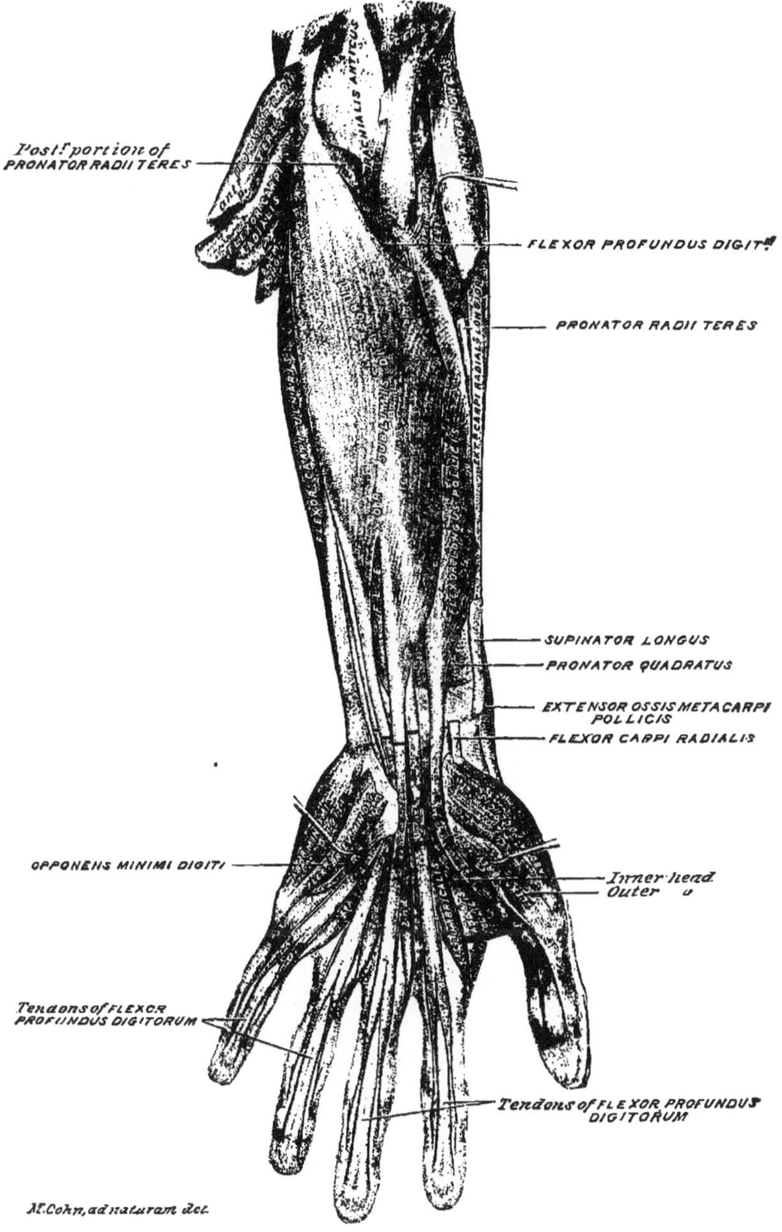
Post? portion of PRONATOR RADII TERES
FLEXOR PROFUNDUS DIGIT.
PRONATOR RADII TERES
SUPINATOR LONGUS
PRONATOR QUADRATUS
EXTENSOR OSSIS METACARPI POLLICIS
FLEXOR CARPI RADIALIS
OPPONENS MINIMI DIGITI
Inner head
Outer "
Tendons of FLEXOR PROFUNDUS DIGITORUM
Tendons of FLEXOR PROFUNDUS DIGITORUM
M. Cohn, ad naturam del.

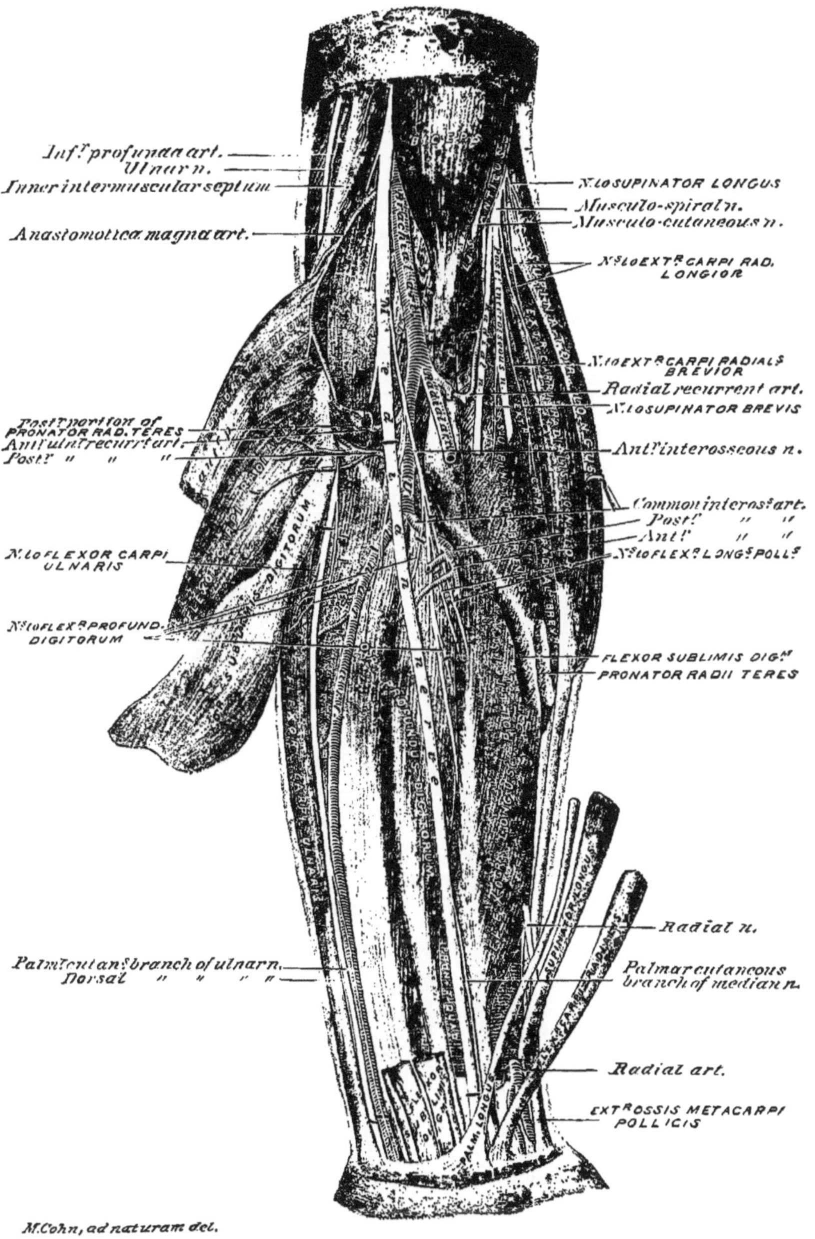
Inf.r profunda art.
Ulnar n.
Inner intermuscular septum
Anastomotica magna art.
Post.r portion of PRONATOR RAD. TERES
Ant.r uln.r recurr.t art.
Post.r " " "
N. to FLEXOR CARPI ULNARIS
N.s to FLEX.R PROFUND. DIGITORUM
Palm.r cutan.s branch of ulnar n.
Dorsal " " " "
N. to SUPINATOR LONGUS
Musculo-spiral n.
Musculo-cutaneous n.
N.s to EXT.R CARPI RAD. LONGIOR
N. to EXT.R CARPI RADIALIS BREVIOR
Radial recurrent art.
N. to SUPINATOR BREVIS
Ant.r interosseous n.
Common inteross.t art.
Post.r " "
Ant.r " "
N.s to FLEX.R LONG.S POLL.S
FLEXOR SUBLIMIS DIG.M
PRONATOR RADII TERES
Radial n.
Palmar cutaneous branch of median n.
Radial art.
EXT.R OSSIS METACARPI POLLICIS

M.Cohn, ad naturam del.

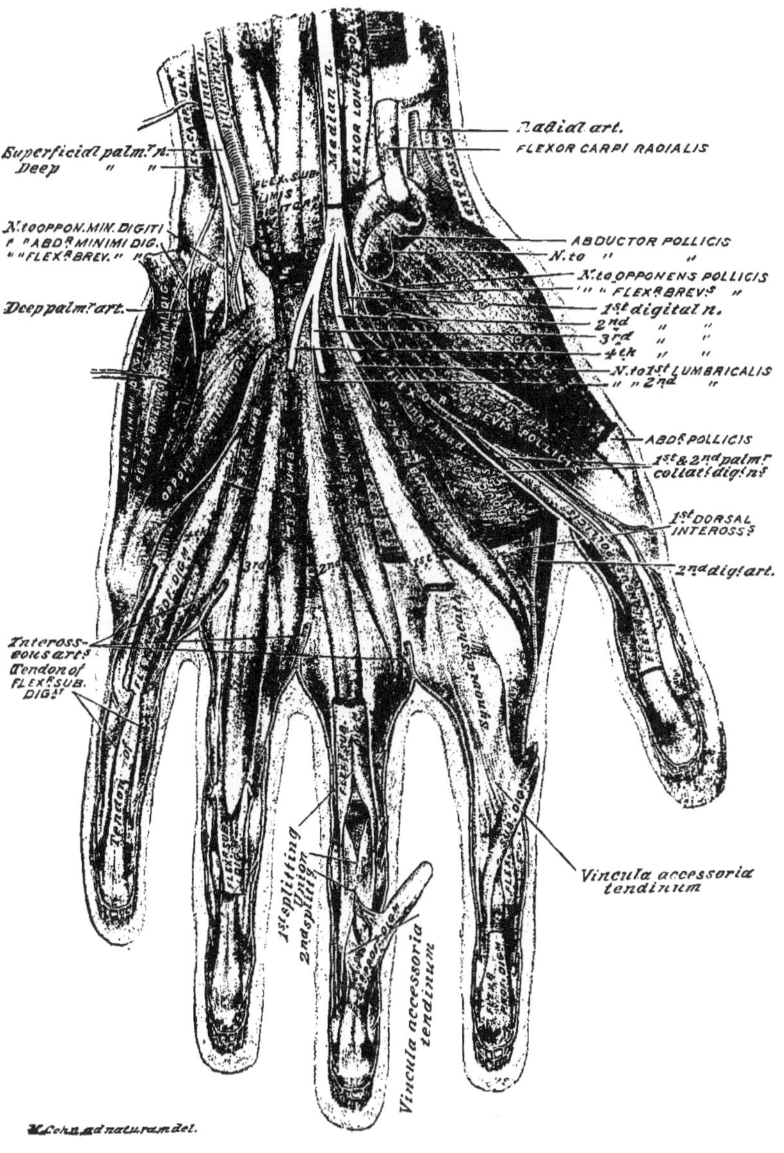
Radial art.
FLEXOR CARPI RADIALIS
Superficial palm.r n.
Deep " "
N. to OPPON. MIN. DIGITI
" " ABD.R MINIMI DIG.
" " FLEX.R BREV. "
Deep palm.r art.
ABDUCTOR POLLICIS
N. to " "
N. to OPPONENS POLLICIS
" " FLEX.R BREV.S "
1st digital n.
2nd " "
3rd " "
4th " "
N. to 1st LUMBRICALIS
" " 2nd "
ABD.R POLLICIS
1st & 2nd palm.r collat.l dig.l n.s
1st DORSAL INTEROSS.S
2nd dig.l art.
Inteross-eous arts
Tendon of FLEX.R SUB. DIG.I
Vincula accessoria tendinum
1st splitting
Union
2nd splitting
Vincula accessoria tendinum
Median n.
Ulnar n.
FLEX. SUB. LIMIS DIGITORUM
Synovial sheath

M. Cohn ad naturam del.

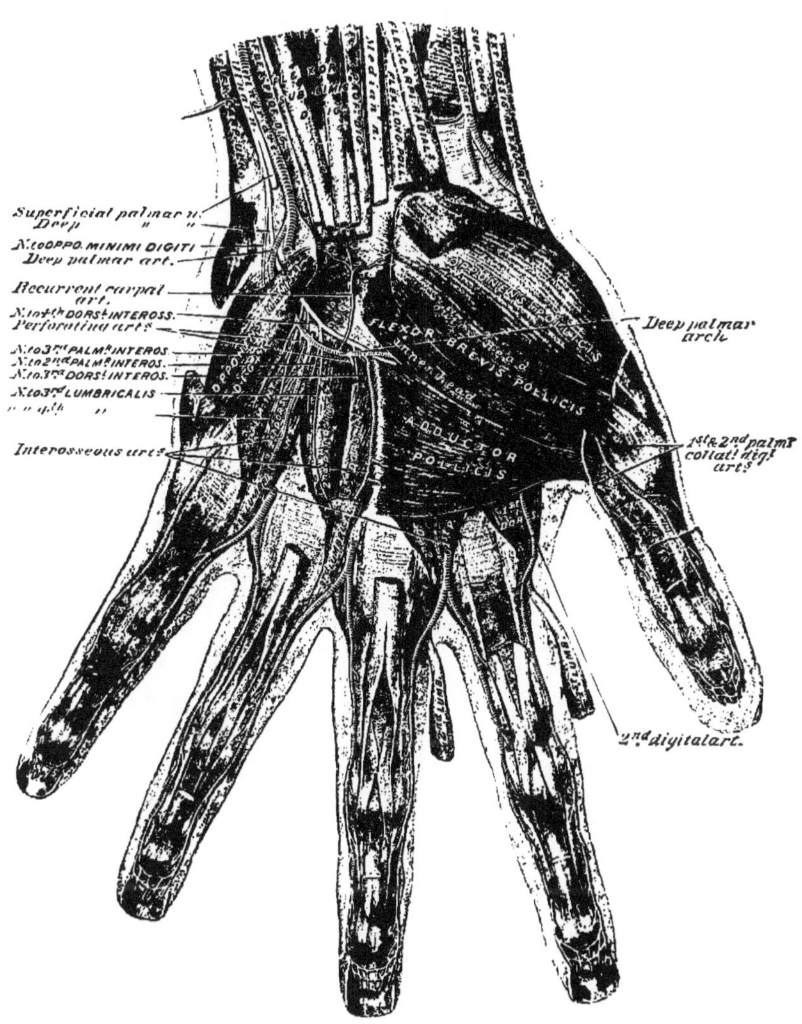
Superficial palmar n.
Deep " "
N. to OPPO. MINIMI DIGITI
Deep palmar art.
Recurrent carpal art.
N. to 4th DORS! INTEROSS.
Perforating art's
N. to 3rd PALM! INTEROS.
N. to 2nd PALM! INTEROS.
N. to 3rd DORS! INTEROS.
N. to 3rd LUMBRICALIS
" " 4th "
Interosseous art's
Median N.
FLEXOR BREVIS POLLICIS
ADDUCTOR POLLICIS
Deep palmar arch
1st & 2nd palm! collat! dig! art's
2nd digital art.

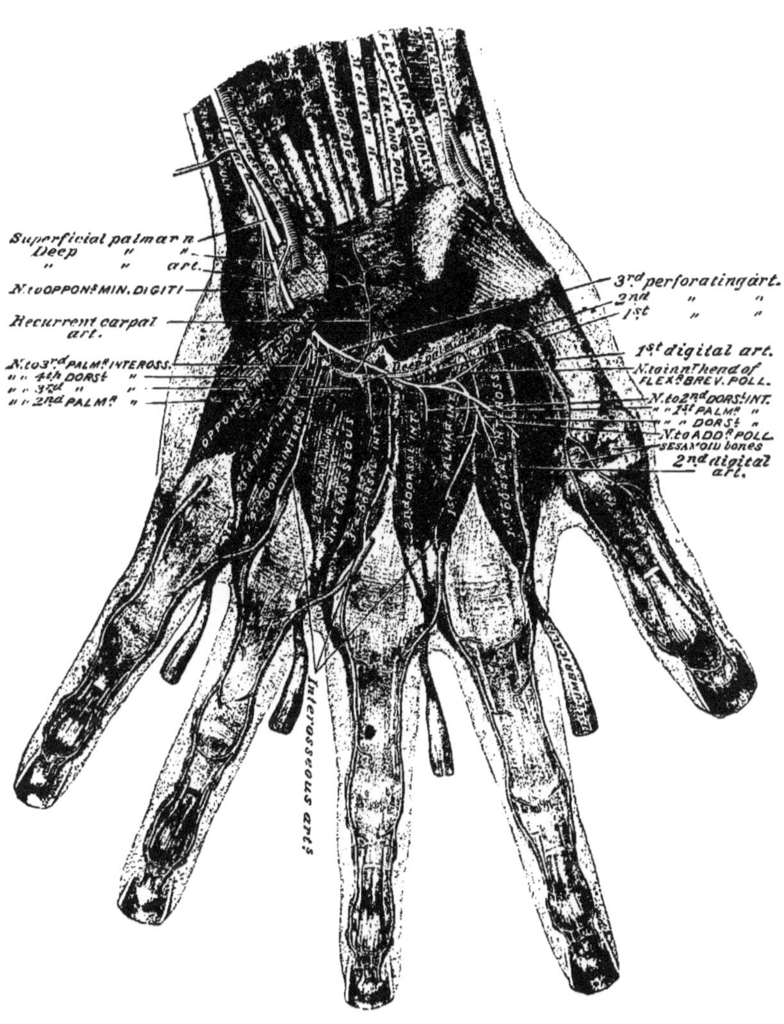
Superficial palmar n.
Deep " "
" " art.
N. to OPPONS MIN. DIGITI
Recurrent carpal art.
N. to 3rd PALMR INTEROSS.
" " 4th DORSL "
" " 3rd " "
" " 2nd PALMR "
3rd perforating art.
2nd " "
1st " "
1st digital art.
N. to innr head of FLEXR BREV. POLL.
N. to 2nd DORSL INT.
" " 1st PALMR "
" " DORSL "
N. to ADDR POLL.
SESAMOID bones
2nd digital art.
Interosseous arts

M. Cohn, ad naturam del.

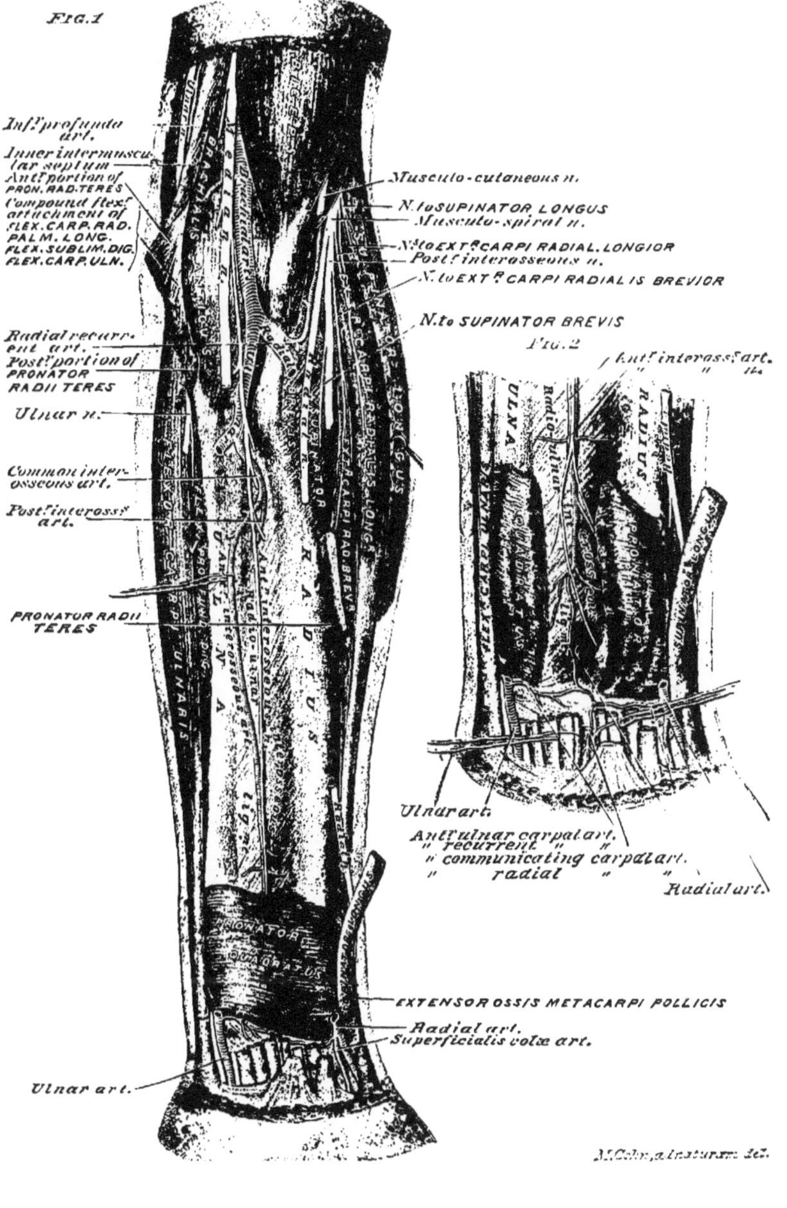

Fig. 1
Inf! profunda art.
Inner intermuscular septum
Ant! portion of PRON. RAD. TERES
Compound flex! attachment of FLEX. CARP. RAD. PALM. LONG. FLEX. SUBLIM. DIG. FLEX. CARP. ULN.
Radial recurrent art.
Post! portion of PRONATOR RADII TERES
Ulnar n.
Common interosseous art.
Post! inteross! art.
PRONATOR RADII TERES
Ulnar art.
Musculo-cutaneous n.
N. to SUPINATOR LONGUS
Musculo-spiral n.
N. to EXT. CARPI RADIAL. LONGIOR
Post! interosseous n.
N. to EXT. CARPI RADIALIS BREVIOR
N. to SUPINATOR BREVIS
EXTENSOR OSSIS METACARPI POLLICIS
Radial art.
Superficialis volæ art.
Fig. 2
Ant! inteross! art.
ULNA
RADIUS
Ulnar art.
Ant! ulnar carpal art.
" recurrent " "
" communicating carpal art.
" radial " "
Radial art.
M. Cohn, ad naturam del.

FIG. 2

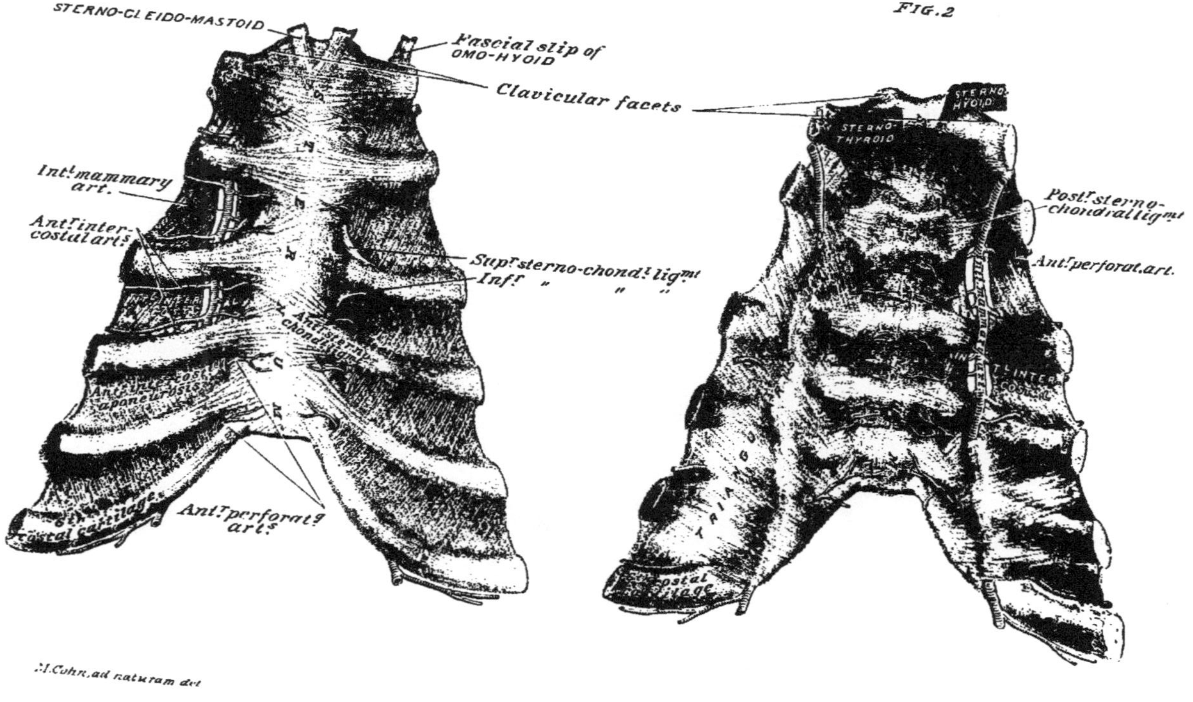

STERNO-CLEIDO-MASTOID

Fascial slip of OMO-HYOID

Clavicular facets

STERNO-HYOID

STERNO-THYROID

Intl. mammary art.

Antr. inter-costal arts.

Supr. sterno-chondl. ligmt.

Infr. " " "

Ant. sterno-chondl. ligmt.

Antr. inter-costal aponeurosis

6th costal cartilage

Antr. perforatg. arts.

Postr. sterno-chondral ligmt.

Antr. perforat. art.

TRIANGULARIS STERNI

Costal cartilage

M. Cohn, ad naturam del.

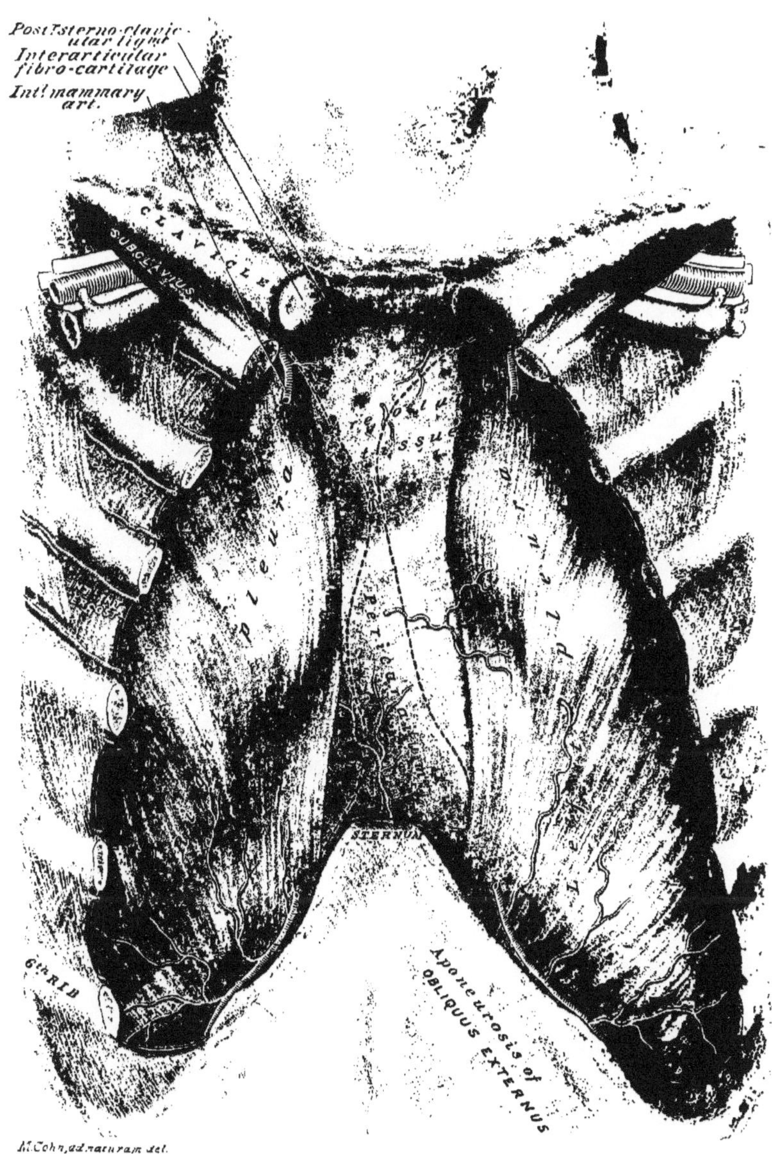
Post! sterno-clavicular ligmt
Interarticular fibro-cartilage
Int! mammary art.
CLAVICLE
SUBCLAVIUS
pleura
pleura
STERNUM
6th RIB
Aponeurosis of OBLIQUUS EXTERNUS

M. Cohn, ad naturam del.

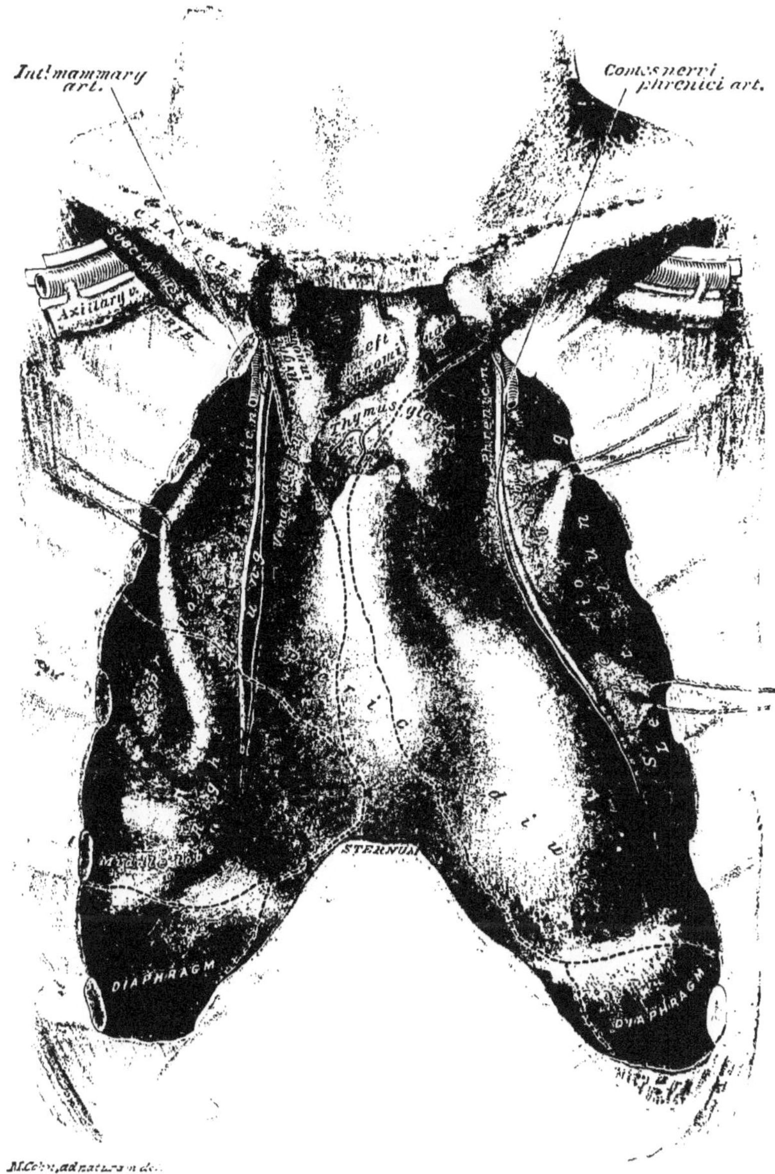
Int! mammary art.
Comes nervi phrenici art.
CLAVICLE
SUBCLAVIUS
Axillary v.
RIB
Thymus gland
STERNUM
Middle lobe
DIAPHRAGM
DIAPHRAGM

M.Cohn, ad naturam del.

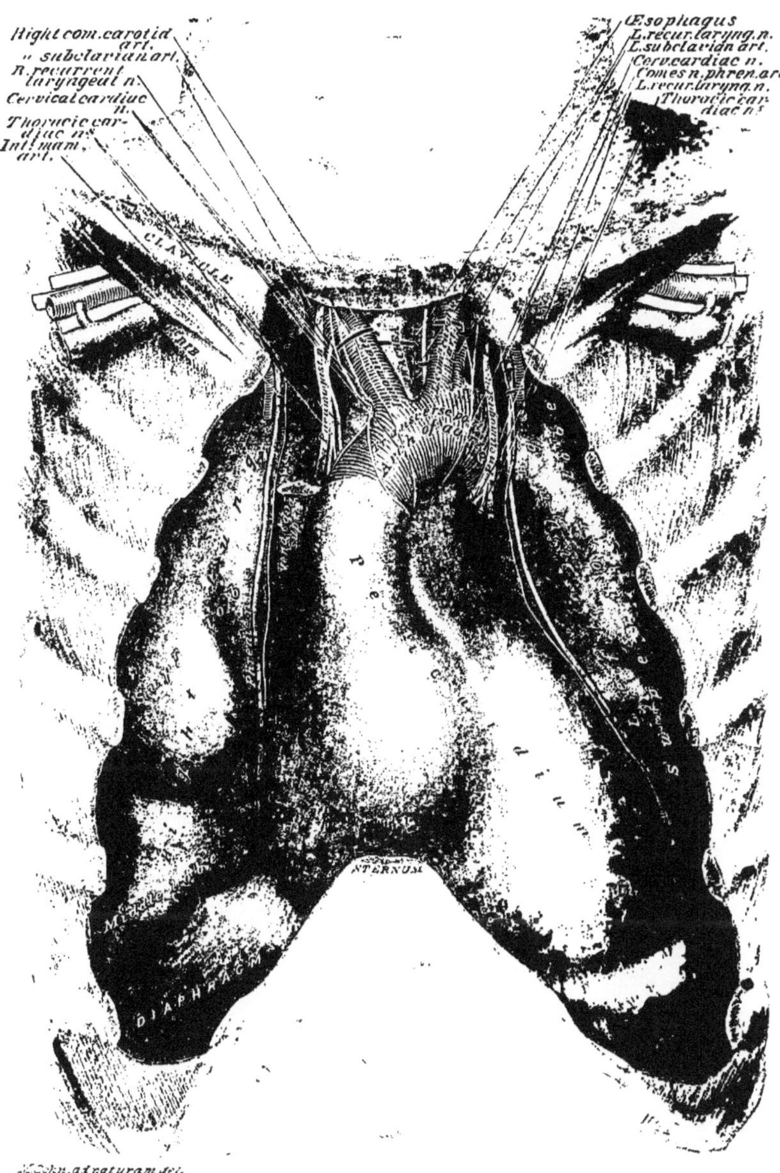
Right com.carotid art.
" subclavian art.
R. recurrent laryngeal n.
Cervical cardiac n.
Thoracic cardiac n's
Int. mam. art.
Œsophagus
L. recur. laryng. n.
L. subclavian art.
Cerv. cardiac n.
Comes n. phren. art.
L. recur. laryng. n.
Thoracic cardiac n's
CLAVICLE
STERNUM
DIAPHRAGM
Middle lobe
Pericardium

M. Cohn, ad naturam del.

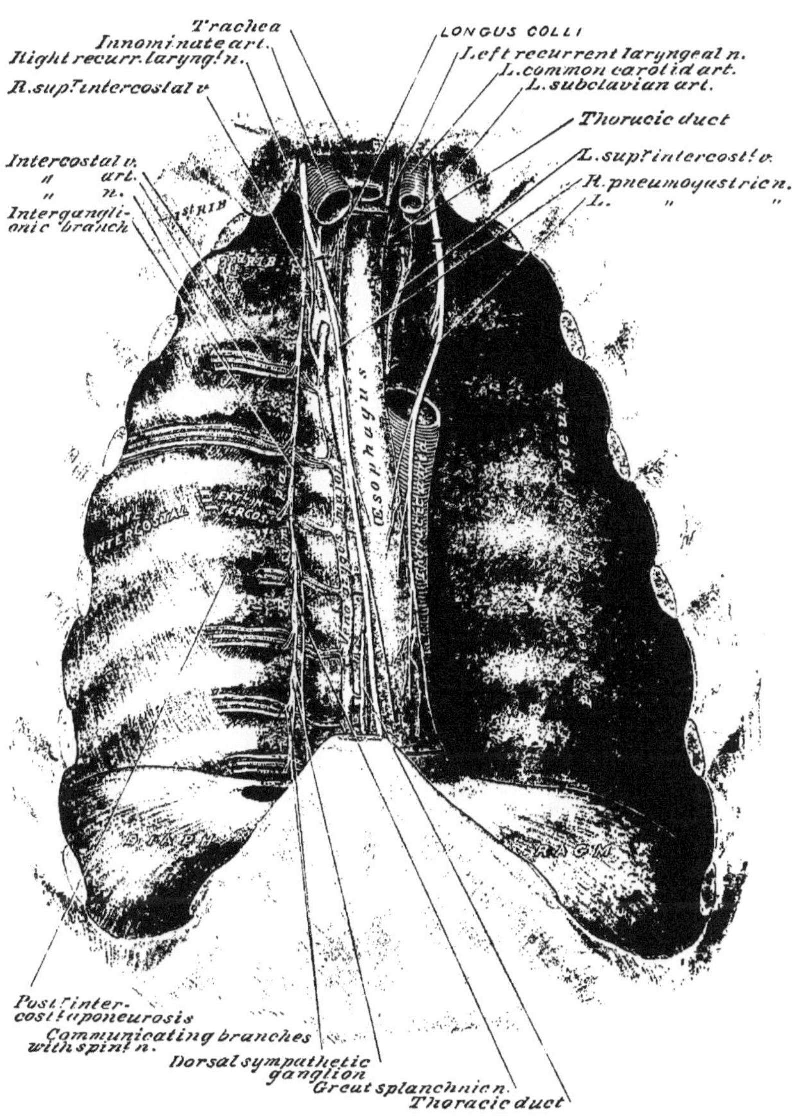
Trachea
Innominate art.
Right recurr. laryng! n.
R. sup! intercostal v.
LONGUS COLLI
Left recurrent laryngeal n.
L. common carotid art.
L. subclavian art.
Thoracic duct
L. sup! intercost! v.
R. pneumogastric n.
L. " "
Intercostal v.
" art.
" n.
Intergangli-
onic branch
Œsophagus
Post! inter-
cost! aponeurosis
Communicating branches
with spin! n.
Dorsal sympathetic
ganglion
Great splanchnic n.
Thoracic duct

M. Cohn, ad naturam del.

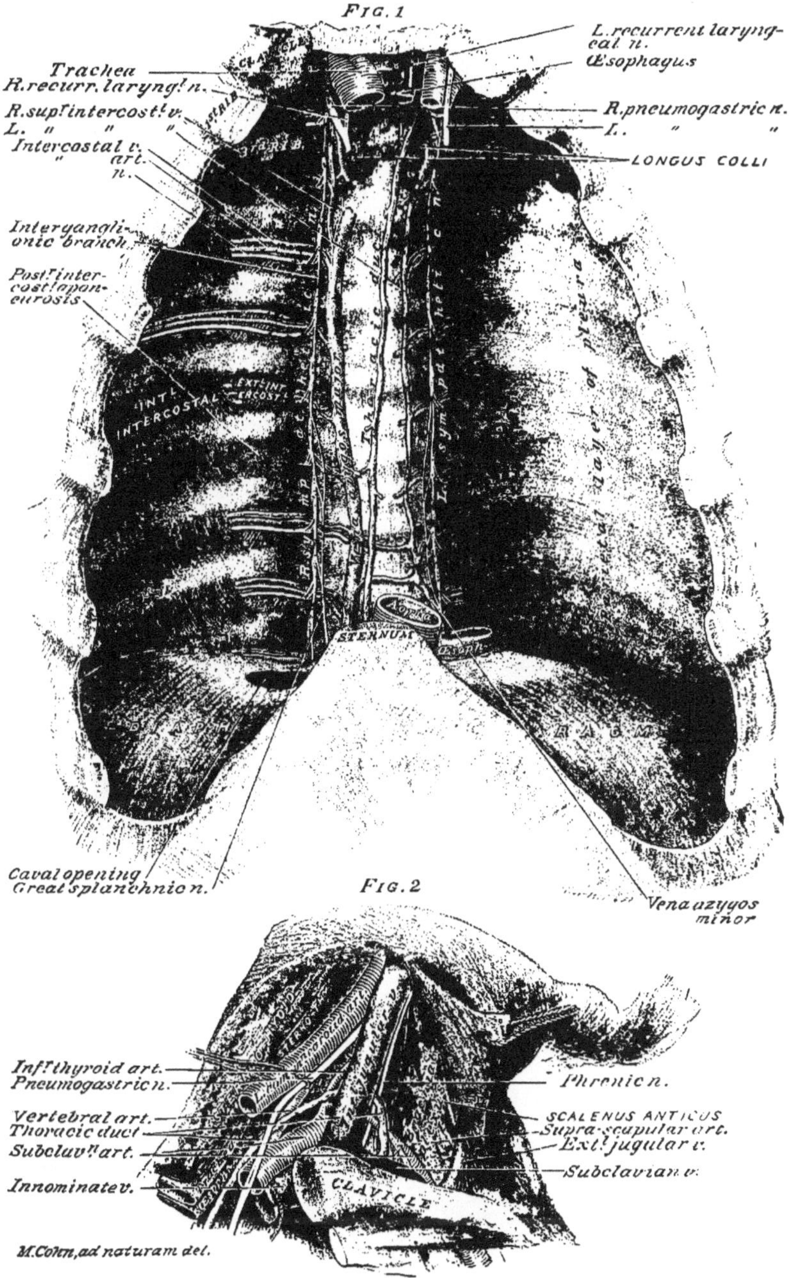
FIG. 1
L. recurrent laryngeal n.
Œsophagus
Trachea
R. recurr. laryng! n.
R. pneumogastric n.
R. supr intercost! v.
L. " " "
L. " "
Intercostal v.
" art.
n.
LONGUS COLLI
Interganglionic branch
Post! intercost! aponeurosis
INTL. INTERCOSTAL
EXTL. INTERCOSTAL
STERNUM
Aorta
Caval opening
Great splanchnic n.
Vena azygos minor
FIG. 2
Infr thyroid art.
Pneumogastric n.
Vertebral art.
Thoracic duct
Subclav! art.
Innominate v.
Phrenic n.
SCALENUS ANTICUS
Supra-scapular art.
Ext! jugular v.
Subclavian v.
CLAVICLE

M. Cohn, ad naturam del.

PLATE 141

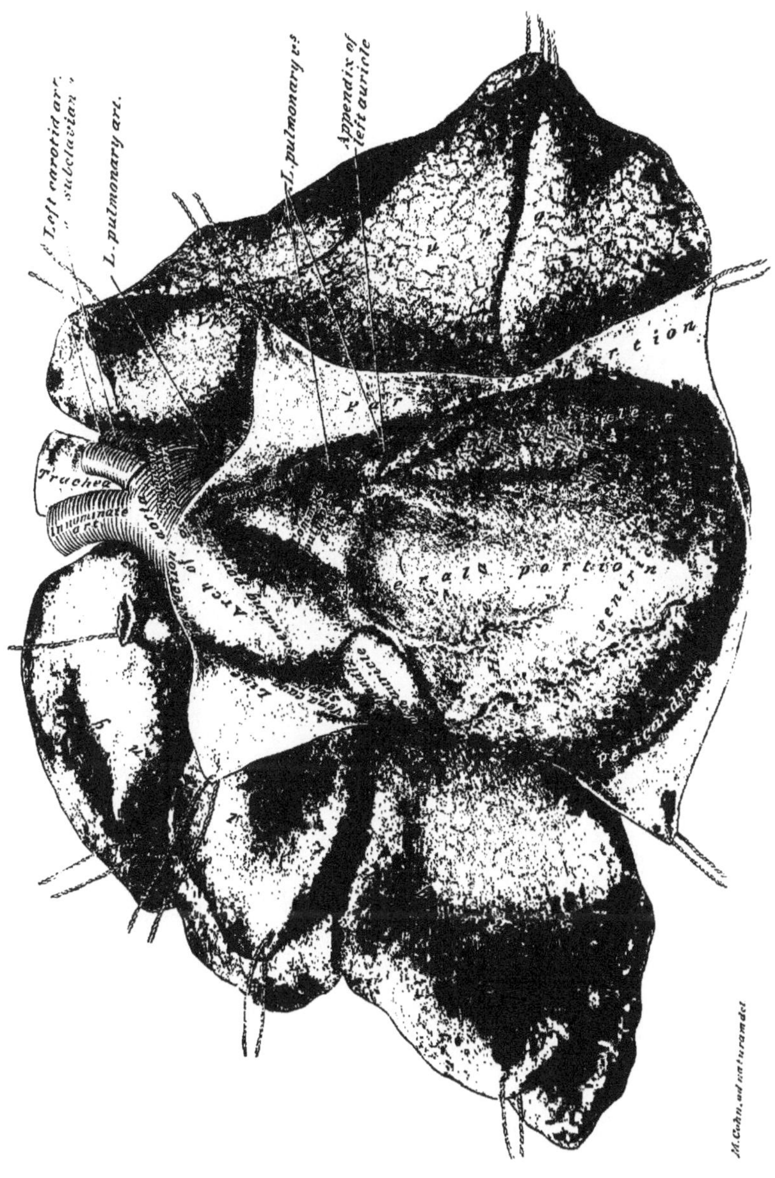

Left carotid art.
L. pulmonary art.
L. pulmonary v.s
Appendix of left auricle
Trachea
Innominate art.
Arch of aorta
Pericardium
M. Cohn, ad naturam del.

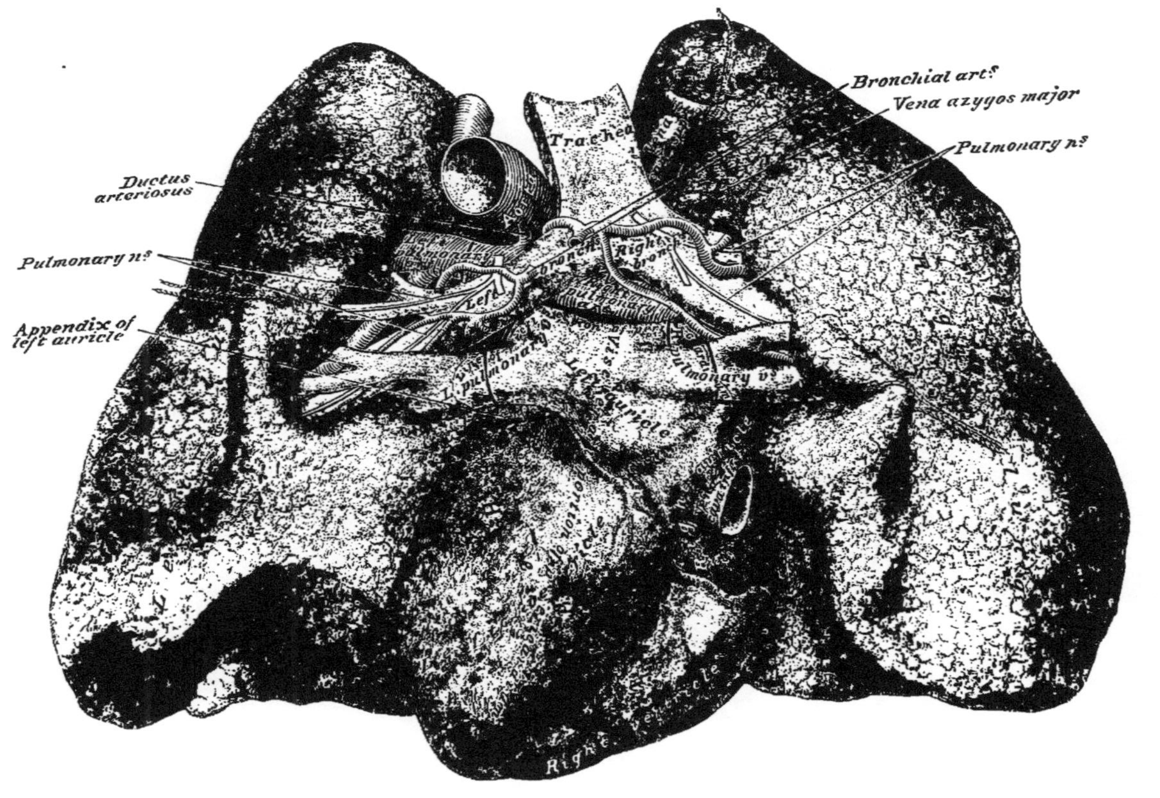
Bronchial art.s
Vena azygos major
Pulmonary n.s
Ductus arteriosus
Pulmonary n.s
Appendix of left auricle
Trachea
Left auricle
Pulmonary v.
Right ventricle

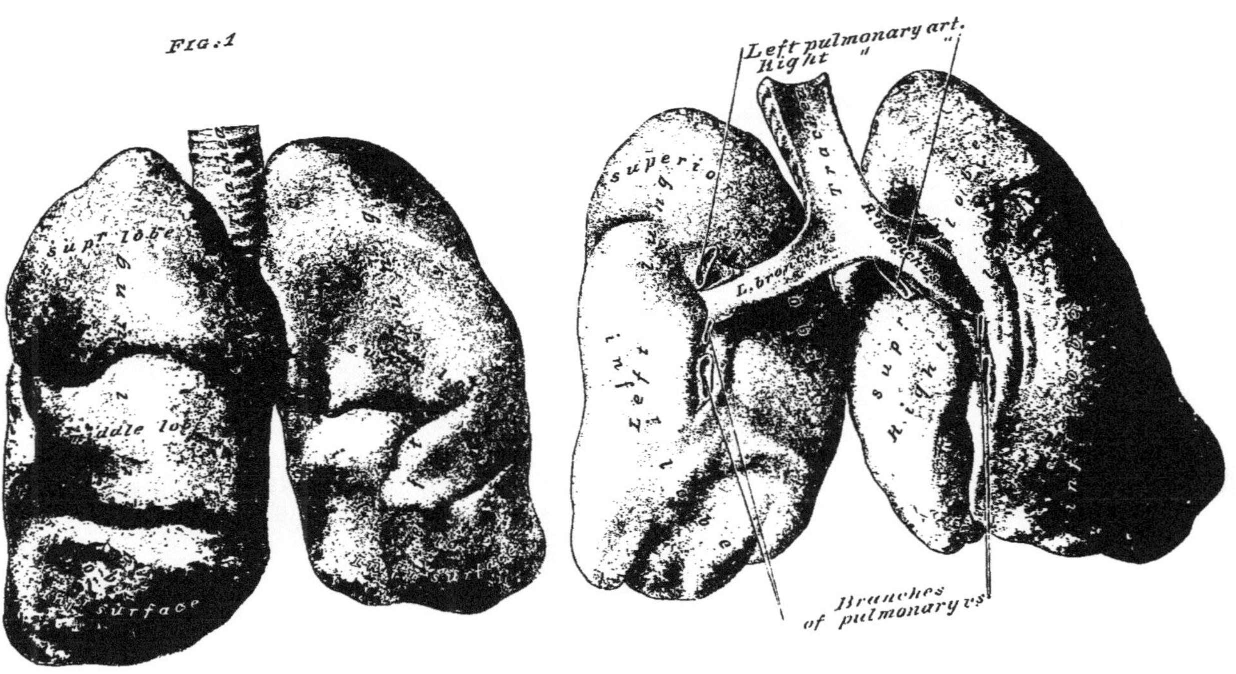
FIG. 1
Trachea
supr. lobe
middle lobe
surface
Left pulmonary art.
Right "   "
superior
L. bronchus
R. bronchus
Left infr.
supr. Right
Branches
of pulmonary v's

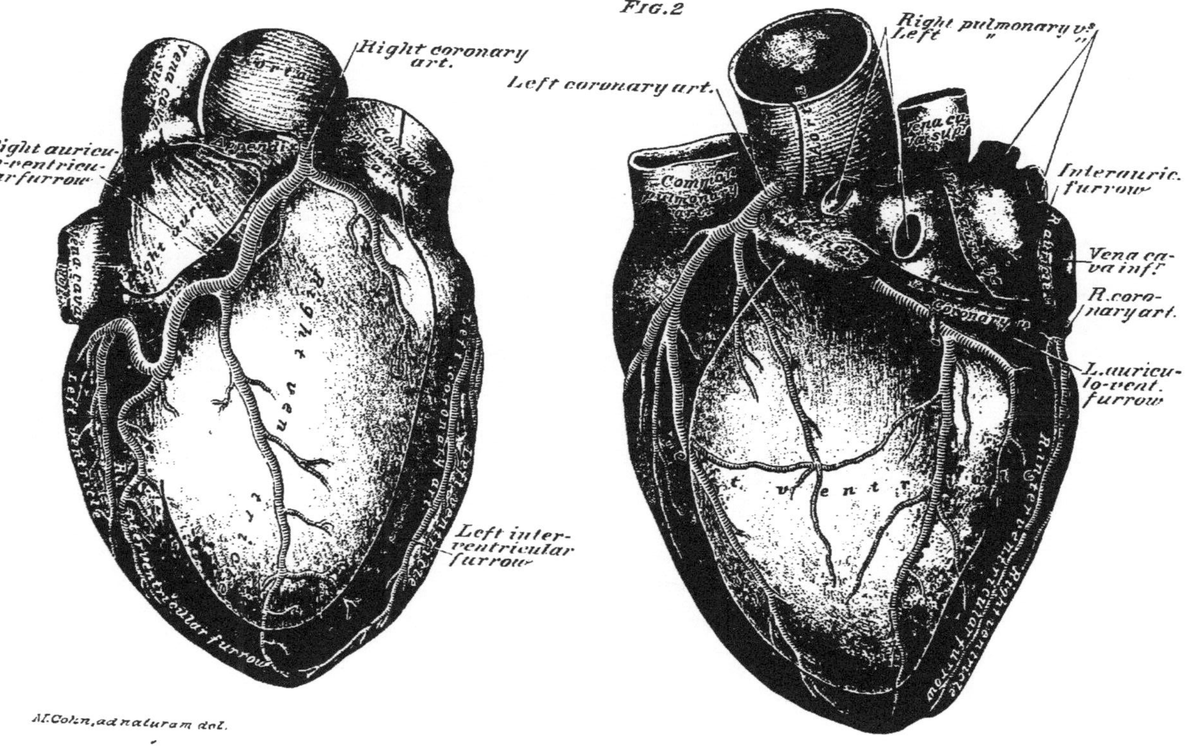
FIG. 2
Right coronary art.
Right auriculo-ventricular furrow
Vena cava sup.
Aorta
Appendix
Right auricle
Vena cava inf.
Common pulmonary art.
Right ventricle
Left ventricle
R. interventricular furrow
Left coronary art.
Left ventricle
Left inter-ventricular furrow
Left coronary art.
Right pulmonary v.s
Left "
Common pulmonary art.
Vena cava sup.
Interauric. furrow
R. auricle
Vena cava inf.
R. coronary art.
Coronary
L. auriculo-vent. furrow
R. interventricular furrow
Right ventricle
M. Cohn, ad naturam del.

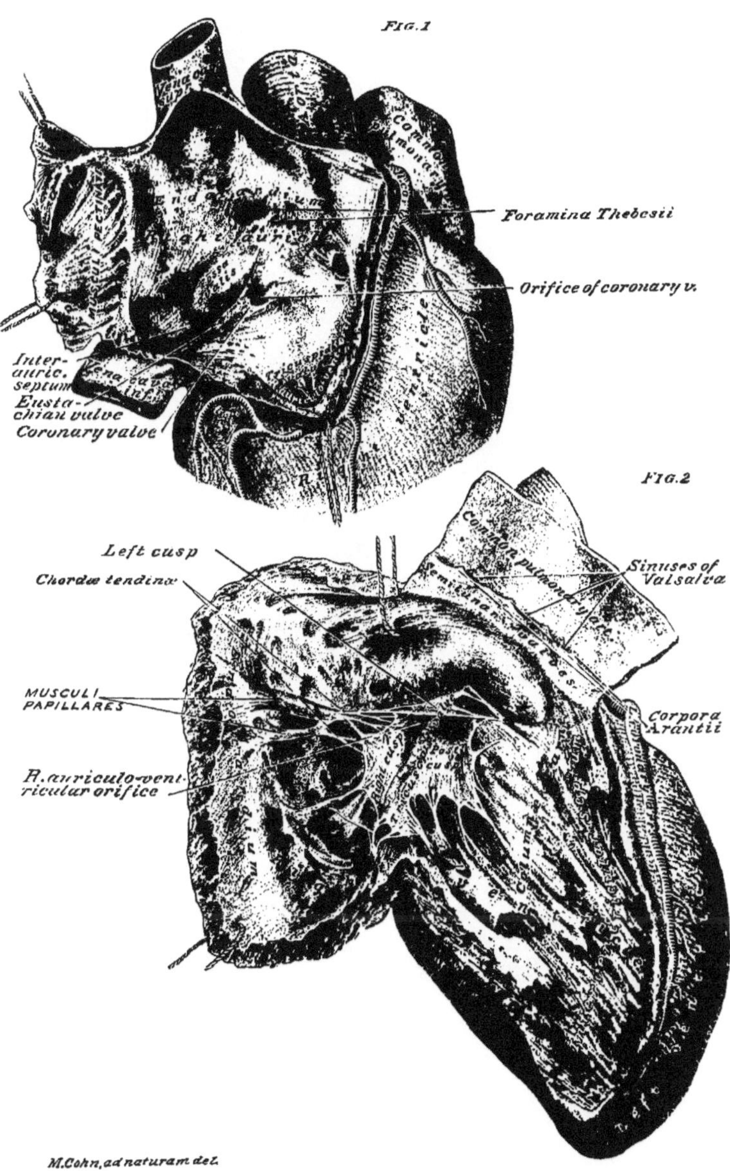
FIG. 1
Foramina Thebesii
Orifice of coronary v.
Inter-auric. septum
Eustachian valve
Coronary valve
FIG. 2
Left cusp
Chordæ tendinæ
Sinuses of Valsalva
MUSCULI PAPILLARES
Corpora Arantii
R. auriculo-ventricular orifice

M. Cohn, ad naturam del.

FIG. 1

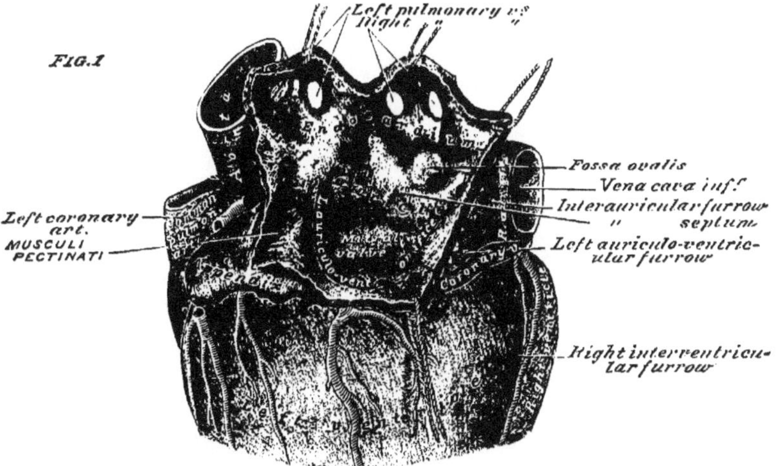
Left pulmonary v.s
Right " "
Fossa ovalis
Vena cava inf.r
Interauricular furrow
" septum
Left auriculo-ventricular furrow
Right interventricular furrow
Left coronary art.
MUSCULI PECTINATI
Mitral valve

FIG. 2

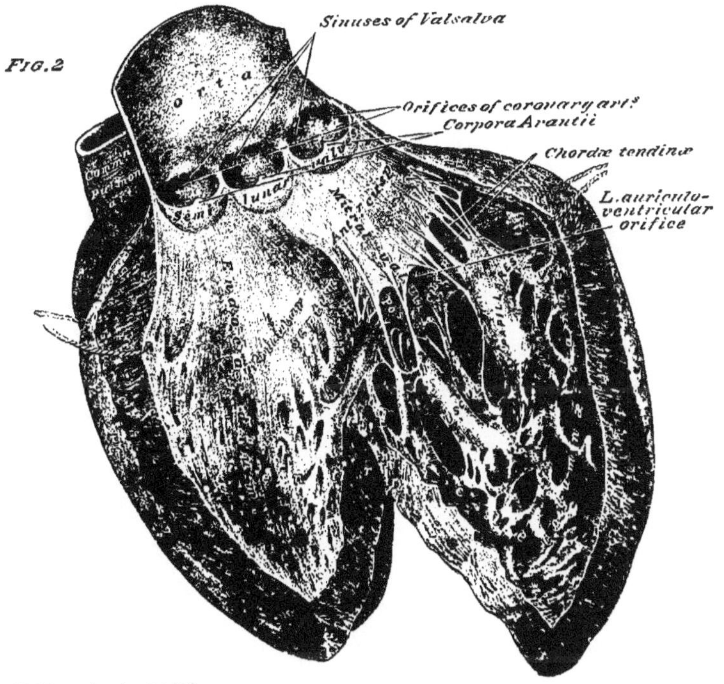
Sinuses of Valsalva
Orifices of coronary art.s
Corpora Arantii
Chordæ tendinæ
L. auriculo-ventricular orifice
orta
Semi lunar valv

M. Cohn, ad naturam del.

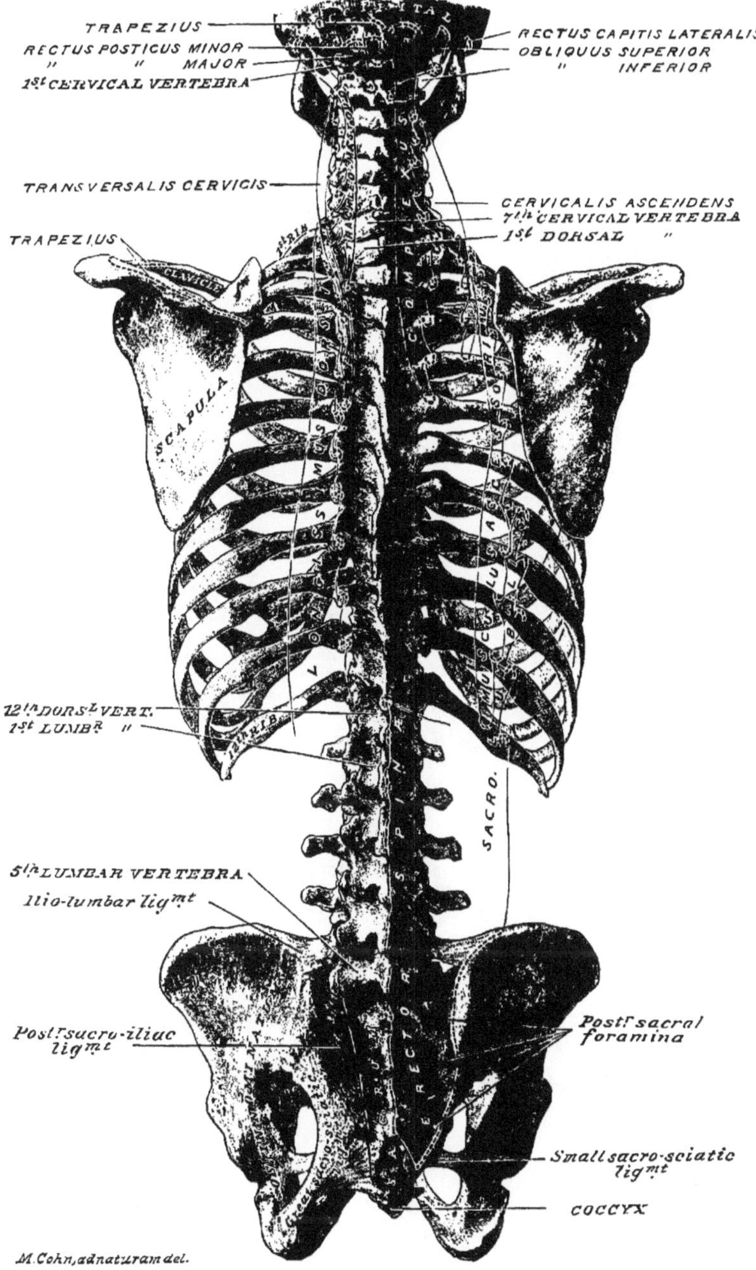
TRAPEZIUS
RECTUS POSTICUS MINOR
" " MAJOR
1st CERVICAL VERTEBRA
RECTUS CAPITIS LATERALIS
OBLIQUUS SUPERIOR
" INFERIOR
TRANSVERSALIS CERVICIS
CERVICALIS ASCENDENS
7th CERVICAL VERTEBRA
1st DORSAL "
TRAPEZIUS
SCAPULA
12th DORSL VERT.
1st LUMBR "
SACRO.
5th LUMBAR VERTEBRA
Ilio-lumbar ligmt
Postr sacro-iliac ligmt
Postr sacral foramina
Small sacro-sciatic ligmt
COCCYX

M. Cohn, ad naturam del.

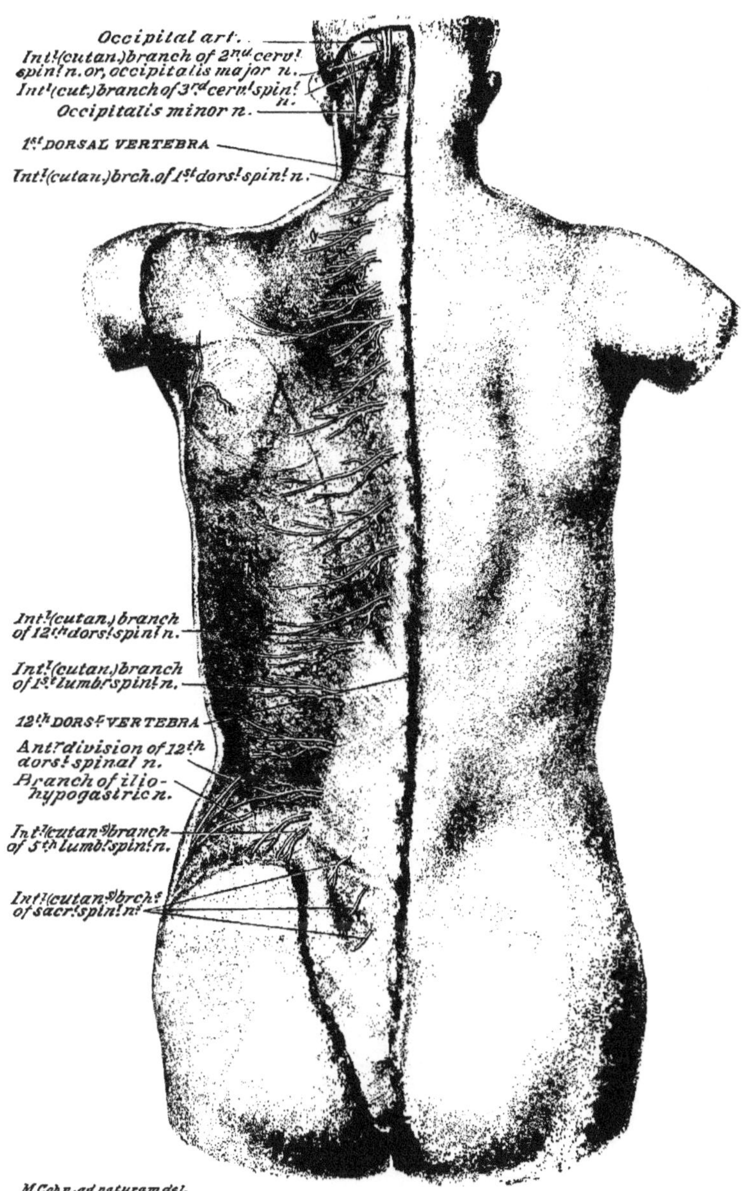
Occipital art.
Intl. (cutan.) branch of 2nd cervl. spinl. n. or, occipitalis major n.
Intl. (cut.) branch of 3rd cervl. spinl. n.
Occipitalis minor n.
1st DORSAL VERTEBRA
Intl. (cutan.) brch. of 1st dorsl. spinl. n.
Intl. (cutan.) branch of 12th dorsl. spinl. n.
Intl. (cutan.) branch of 1st lumbr. spinl. n.
12th DORSl. VERTEBRA
Antr. division of 12th dorsl. spinal n.
Branch of ilio-hypogastric n.
Intl. (cutan.) branch of 5th lumbr. spinl. n.
Intl. (cutan.) brchs. of sacr. spinl. ns.
M. Cohn, ad naturam del.

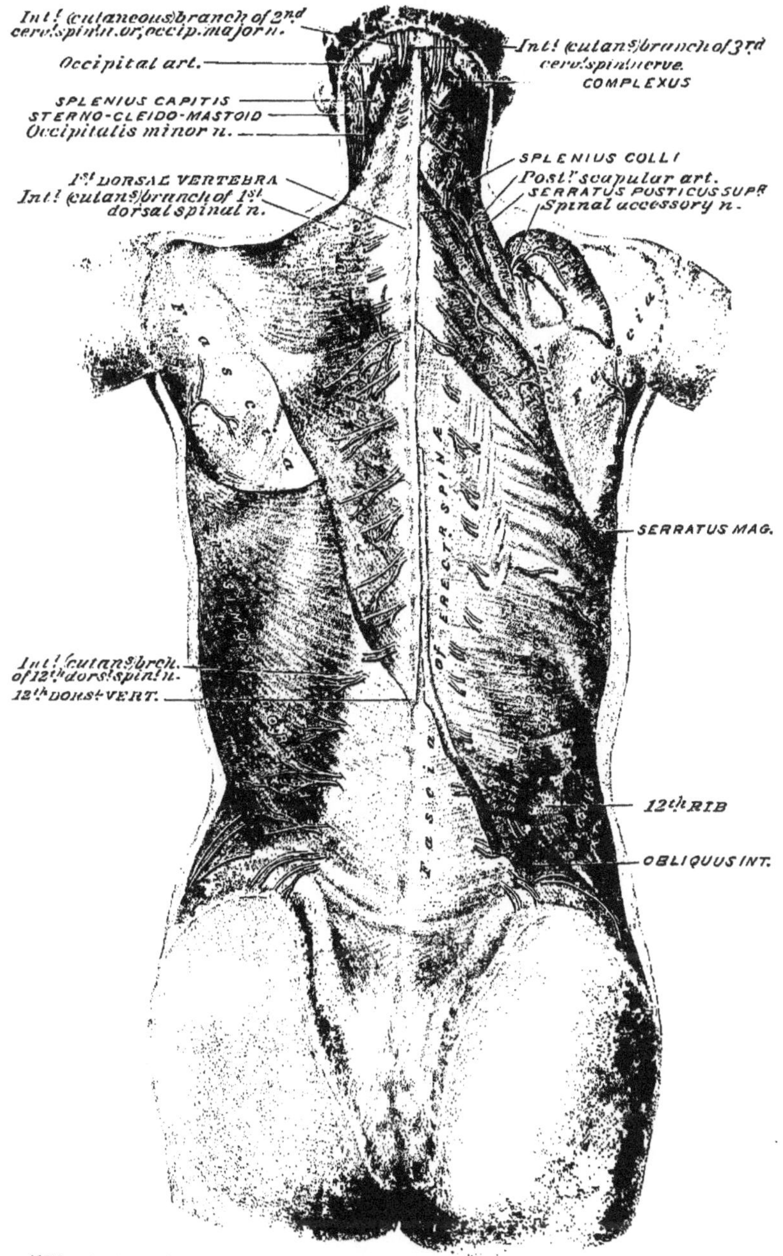
Int.l (cutaneous) branch of 2nd cerv.l spin.l n. or, occip. major n.
Occipital art.
SPLENIUS CAPITIS
STERNO-CLEIDO-MASTOID
Occipitalis minor n.
1st DORSAL VERTEBRA
Int.l (cutans) branch of 1st dorsal spinal n.
Int.l (cutans) branch of 3rd cerv.l spin.l nerve.
COMPLEXUS
SPLENIUS COLLI
Post.r scapular art.
SERRATUS POSTICUS SUPR
Spinal accessory n.
SERRATUS MAG.
Int.l (cutans) brch. of 12th dors.l spin.l n.
12th DORS.l VERT.
12th RIB
OBLIQUUS INT.
M. Cohn ad naturam del.

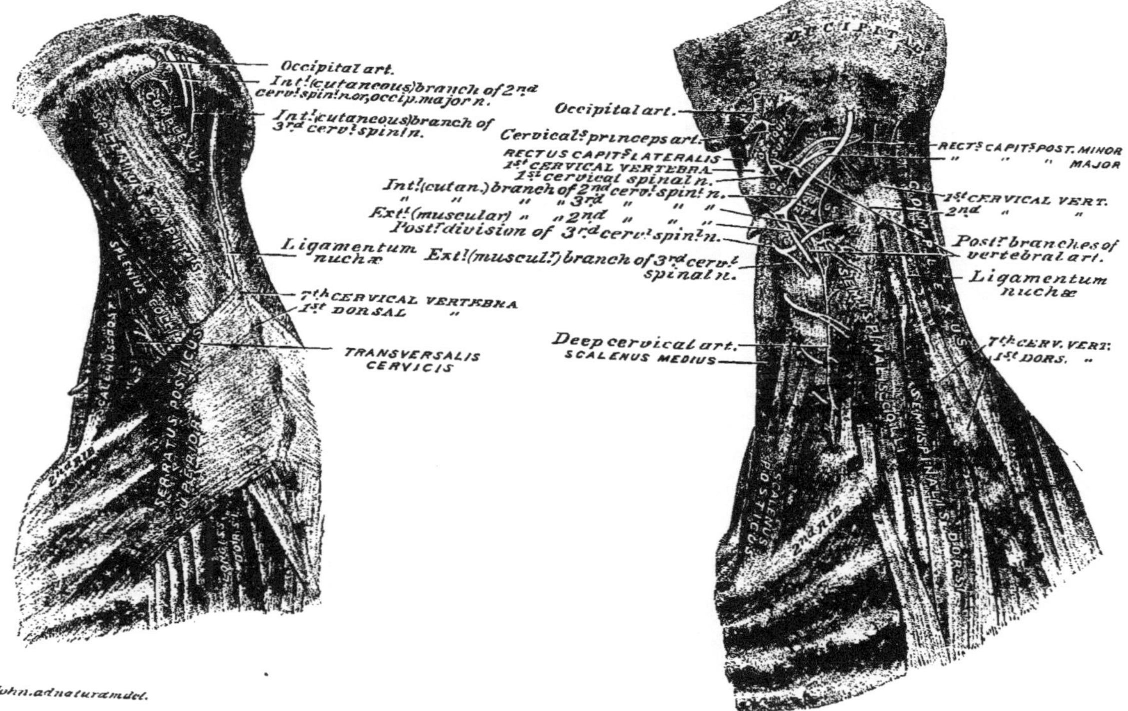
FIG. 1
Occipital art.
Int.l (cutaneous) branch of 2nd cerv.l spin.l n. or, occip. major n.
Int.l (cutaneous) branch of 3rd cerv.l spin.l n.
Ligamentum nuchæ
7th CERVICAL VERTEBRA
1st DORSAL "
TRANSVERSALIS CERVICIS
SPLENIUS CAPITIS
COMPLEXUS
SPLENIUS COLLI
SERRATUS POSTICUS SUPERIOR
SCALENUS POST.
LONGISS. DORSI
2nd RIB
FIG. 2
OCCIPITAL
Occipital art.
Cervical.s princeps art.
RECTUS CAPIT.S LATERALIS
1st CERVICAL VERTEBRA
1st cervical spinal n.
Int.l (cutan.) branch of 2nd cerv.l spin.l n.
" " " 3rd " " "
Ext.l (muscular) " " 2nd " " "
Post.r division of 3rd cerv.l spin.l n.
Ext.l (muscul.r) branch of 3rd cerv.l spinal n.
Deep cervical art.
SCALENUS MEDIUS
RECT.S CAPIT.S POST. MINOR
" " " MAJOR
1st CERVICAL VERT.
2nd " "
Post.r branches of vertebral art.
Ligamentum nuchæ
7th CERV. VERT.
1st DORS. "
COMPLEXUS
SEMISPINALIS COLLI
SEMISPINALIS DORSI
SCALENUS POSTICUS
2nd RIB
M.Cohn.ad natura m del.

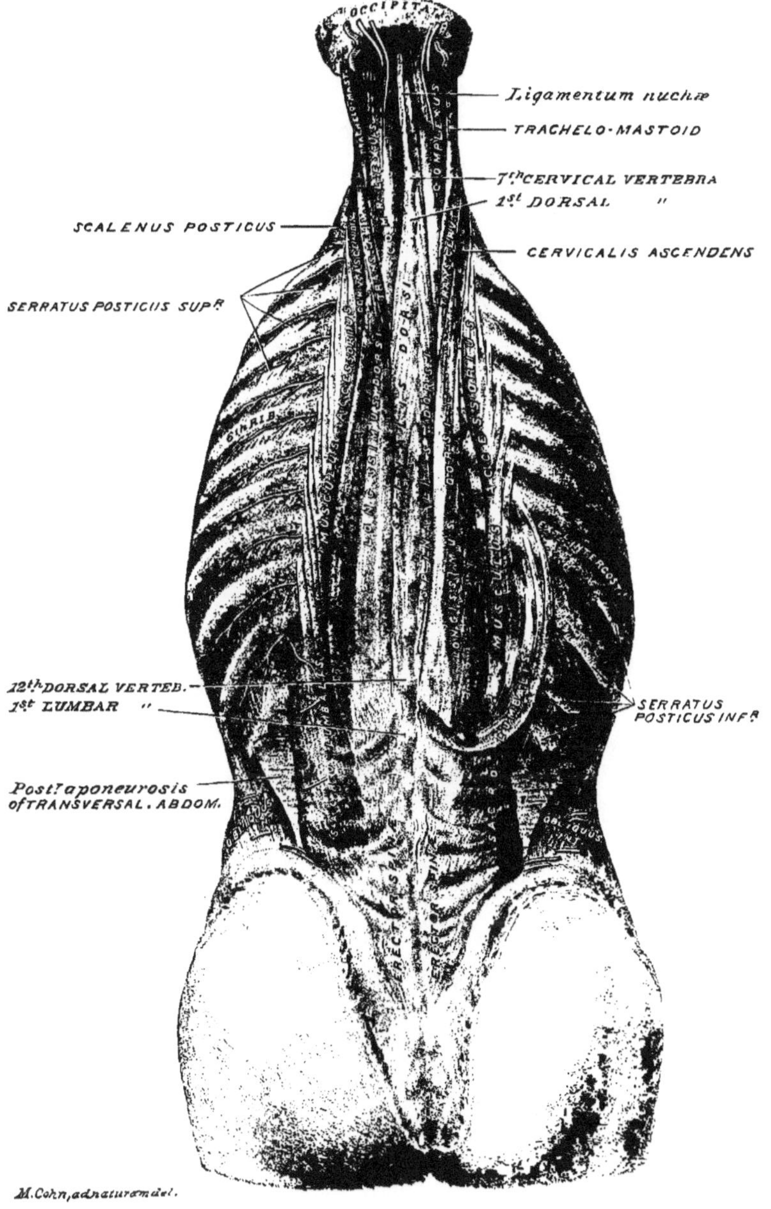
Ligamentum nuchæ
TRACHELO-MASTOID
7th CERVICAL VERTEBRA
1st DORSAL "
SCALENUS POSTICUS
CERVICALIS ASCENDENS
SERRATUS POSTICUS SUPR
12th DORSAL VERTEB.
1st LUMBAR "
SERRATUS POSTICUS INFR
Postr aponeurosis of TRANSVERSAL. ABDOM.
OCCIPITAL
COMPLEXUS
6th RIB
M.Cohn, ad naturam del.

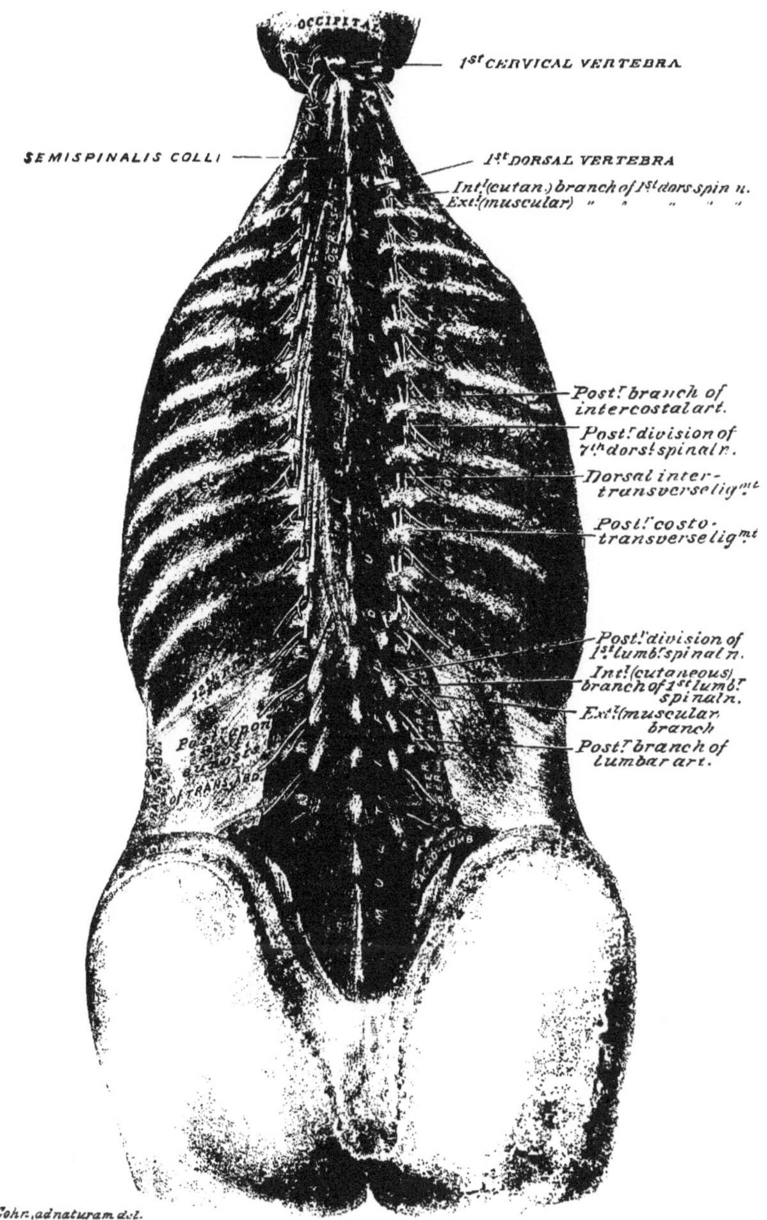
OCCIPITAL
1st CERVICAL VERTEBRA
SEMISPINALIS COLLI
1st DORSAL VERTEBRA
Intl (cutan.) branch of 1st dors spin n.
Extl (muscular) " " " "
Postr branch of intercostal art.
Postr division of 7th dorsl spinal n.
Dorsal inter-transverse ligmt
Postr costo-transverse ligmt
Postl division of 1st lumbr spinal n.
Intl (cutaneous) branch of 1st lumbr spinal n.
Extl (muscular branch
Postr branch of lumbar art.
SACRO-LUMB
M.Cohn, ad naturam del.

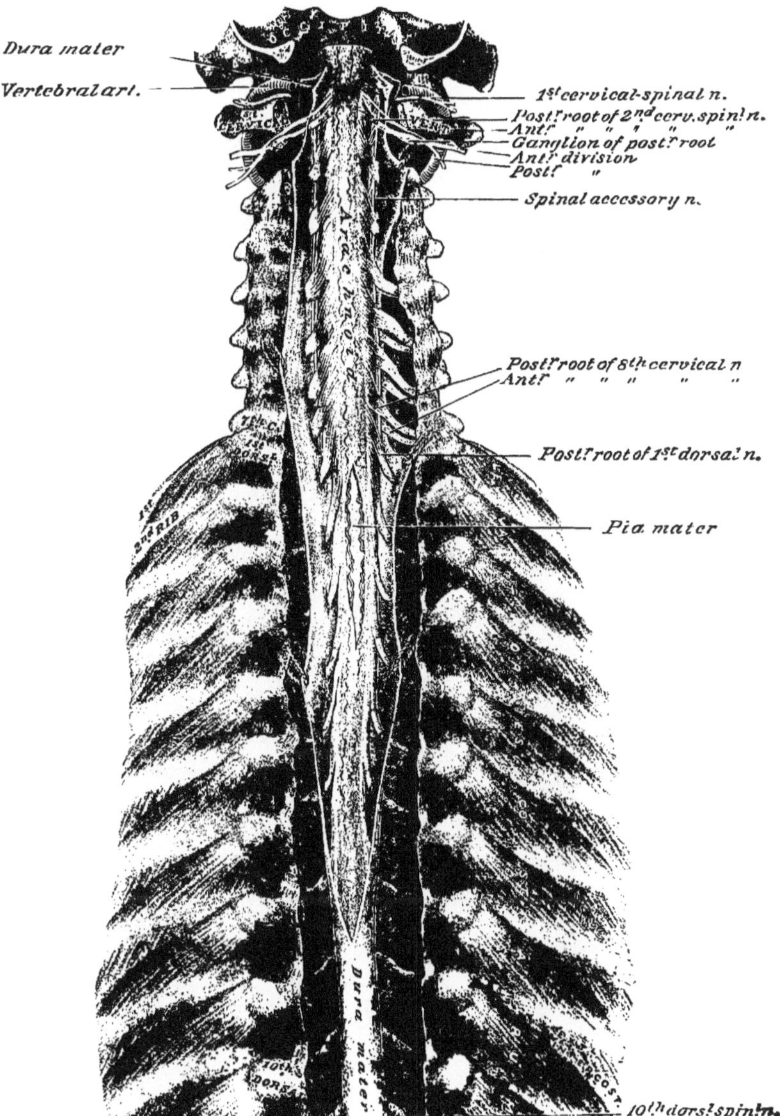
Dura mater
Vertebral art.
1st cervical-spinal n.
Postr root of 2nd cerv. spinl n.
Antr " " " " "
Ganglion of postr root
Antr division
Postr "
Spinal accessory n.
Postr root of 8th cervical n
Antr " " " " "
Postr root of 1st dorsal n.
Pia mater
10th dorsl spinl n.
Arachnoid
Dura mater

M. Cohn, ad naturam del

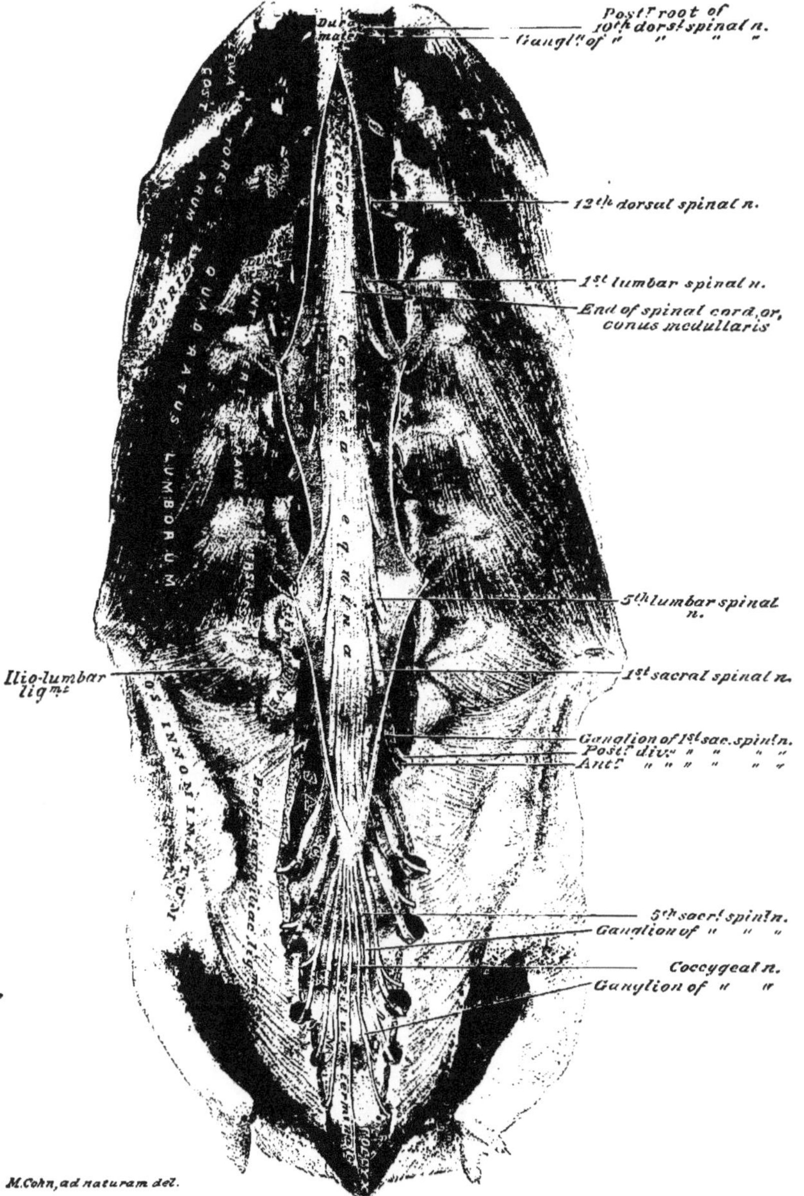
Post.r root of 10th dors.l spinal n.
Gangl.n of " " " "
Dura mater
Spinal cord
LEVATORES COSTARUM
12th RIB
QUADRATUS LUMBORUM
12th dorsal spinal n.
1st lumbar spinal n.
End of spinal cord, or, conus medullaris
Cauda equina
5th lumbar spinal n.
Ilio-lumbar ligmt
OS INNOMINATUM
Post. sacro-iliac lig.
1st sacral spinal n.
Ganglion of 1st sac. spin.l n.
Post.r div.n " " " " "
Ant.r " " " " " "
5th sacr.l spin.l n.
Ganglion of " " "
Coccygeal n.
Ganglion of " "
COCCYX

M.Cohn, ad naturam del.

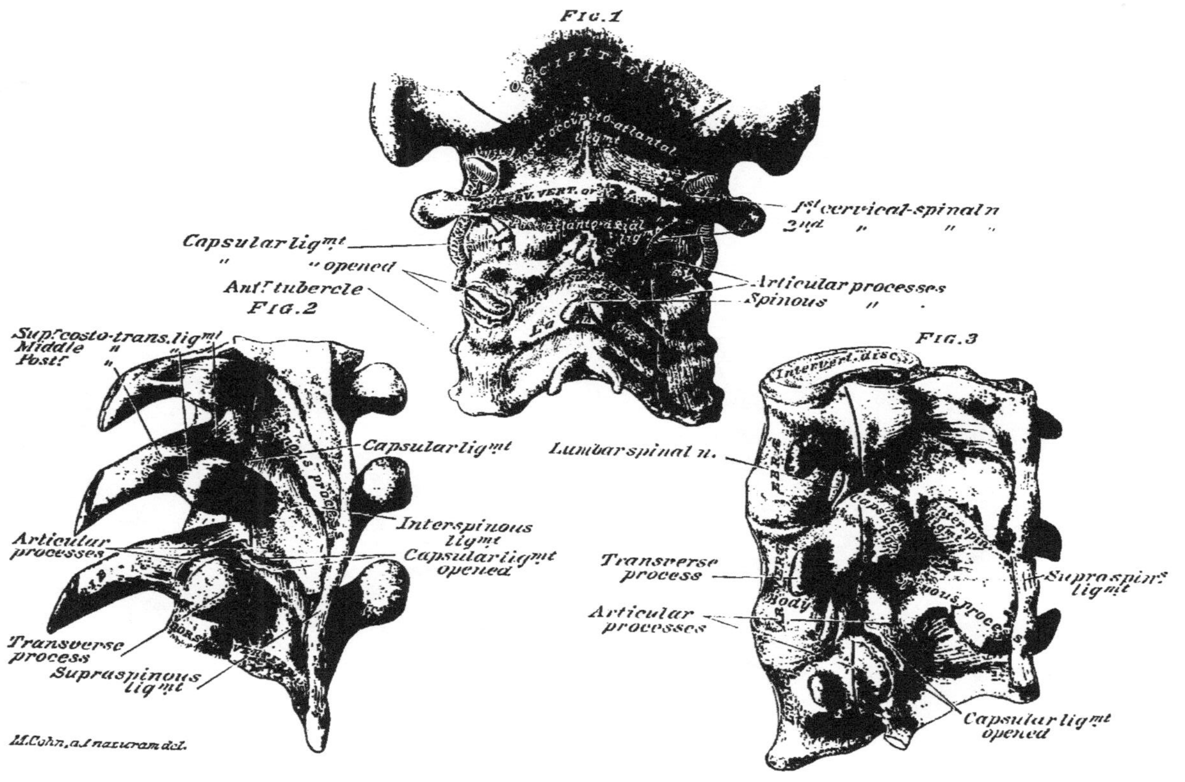
Fig. 1
Post. occipito-atlantal ligmt.
1st. cervical-spinal n.
2nd " "
Capsular ligmt.
" " opened
Antr. tubercle
Articular processes
Spinous "
Fig. 2
Supr. costo-trans. ligmt.
Middle "
Postr. "
Capsular ligmt.
Interspinous ligmt.
Articular processes
Capsular ligmt. opened
Transverse process
Supraspinous ligmt.
M. Cohn, ad naturam del.
Fig. 3
Intervert. disc.
Lumbar spinal n.
Transverse process
Body
Articular processes
Supraspinr. ligmt.
Capsular ligmt. opened

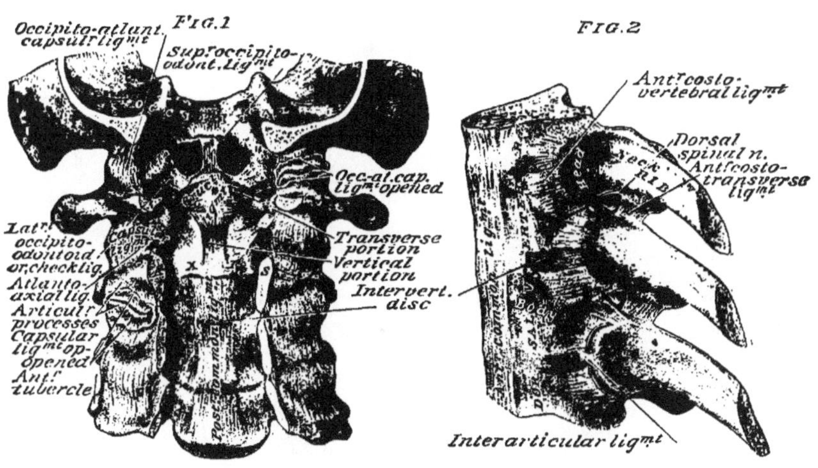
FIG. 1
Occipito-atlant. capsul.r ligmt
Sup.r occipito-odont. Ligmt
Occ-at. cap. ligmt opened
Lat.rl occipito-odontoid. or, check lig.
Transverse portion
Vertical portion
Atlanto-axial lig.
Articul.r processes
Intervert. disc
Capsular ligmt op-opened
Ant.r tubercle
FIG. 2
Ant.r costo-vertebral ligmt
Dorsal spinal n.
Ant.r costo-transverse ligmt
Interarticular ligmt

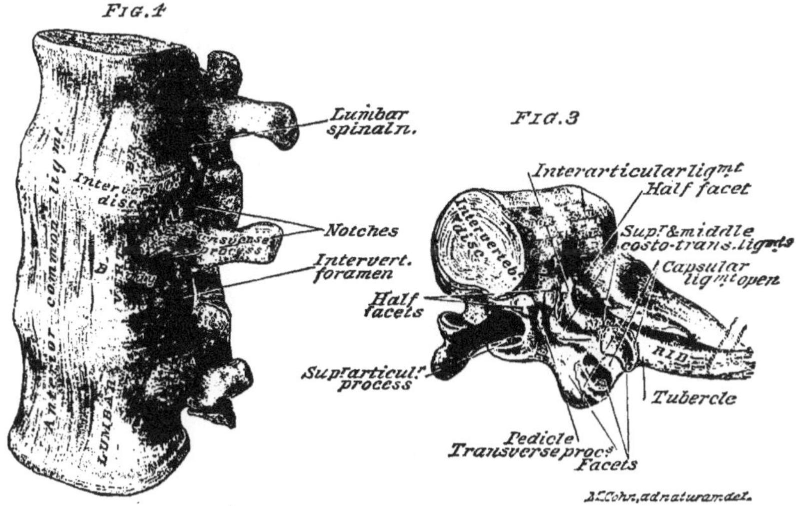
FIG. 4
Lumbar spinal n.
Notches
Intervert. foramen
FIG. 3
Interarticular ligmt
Half facet
Sup.r & middle costo-trans. ligmts
Capsular ligmt open
Half facets
Sup.r articul.r process
Tubercle
Pedicle
Transverse proc.s
Facets

M.L. Cohn, ad naturam del.

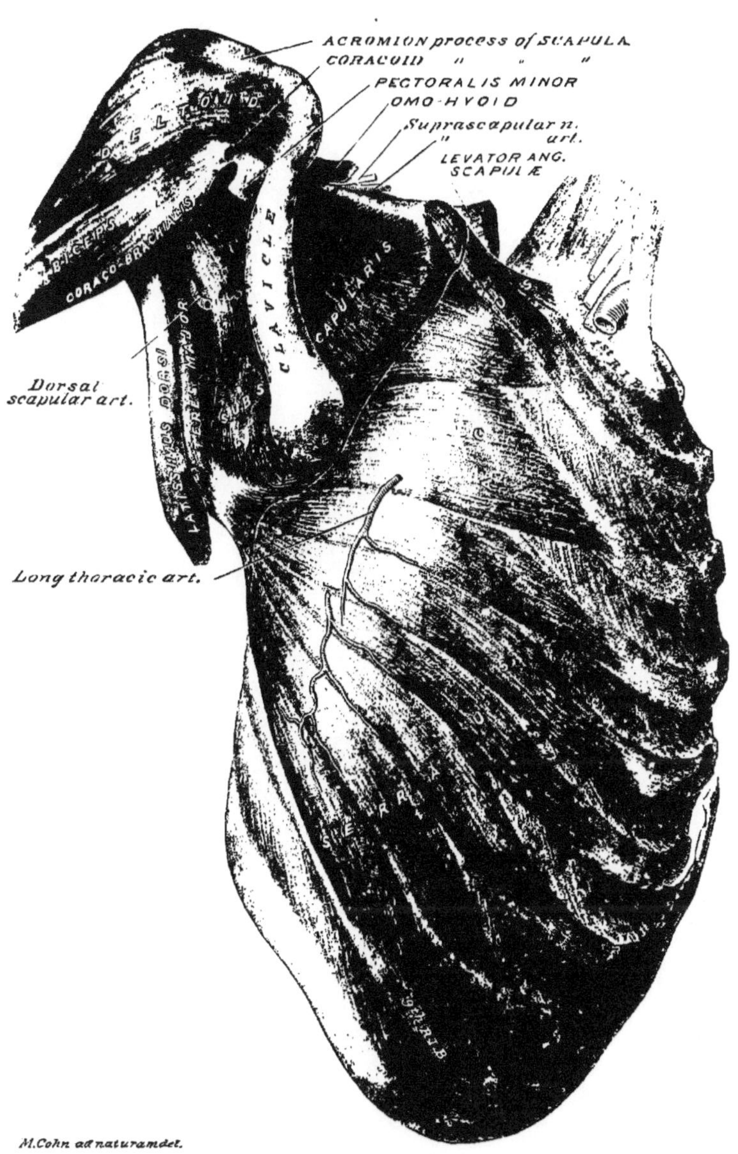
ACROMION process of SCAPULA
CORACOID " " "
PECTORALIS MINOR
OMO-HYOID
Suprascapular n.
" art.
LEVATOR ANG. SCAPULÆ
CLAVICLE
BICEPS
CORACO-BRACHIALIS
LATISSIMUS DORSI
TERES MAJOR
Dorsal scapular art.
Long thoracic art.
M.Cohn ad naturam del.

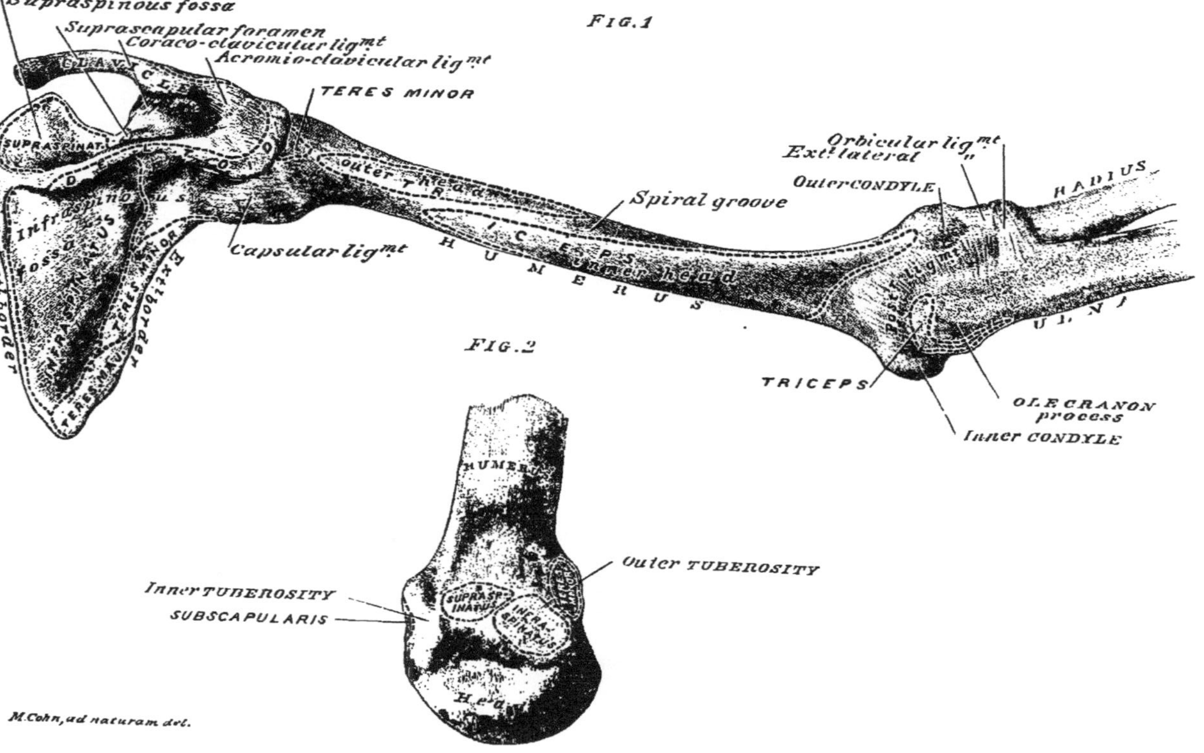
FIG.1
Supraspinous fossa
Suprascapular foramen
Coraco-clavicular ligmt
Acromio-clavicular ligmt
TERES MINOR
CLAVICLE
SUPRASPINAT.
Infraspinous fossa
INFRASPINATUS
TERES MINOR
TERES MAJ.
Intl border
Extl border
Capsular ligmt.
outer head
inner head
TRICEPS
HUMERUS
Spiral groove
Orbicular ligmt
Extl lateral
Outer CONDYLE
RADIUS
ULNA
Post. ligmt
TRICEPS
OLECRANON process
Inner CONDYLE
FIG.2
HUMERUS
Outer TUBEROSITY
Inner TUBEROSITY
SUBSCAPULARIS
SUPRASPINATUS
INFRASPINATUS
TERES MINOR
Head
M.Cohn, ad naturam del.

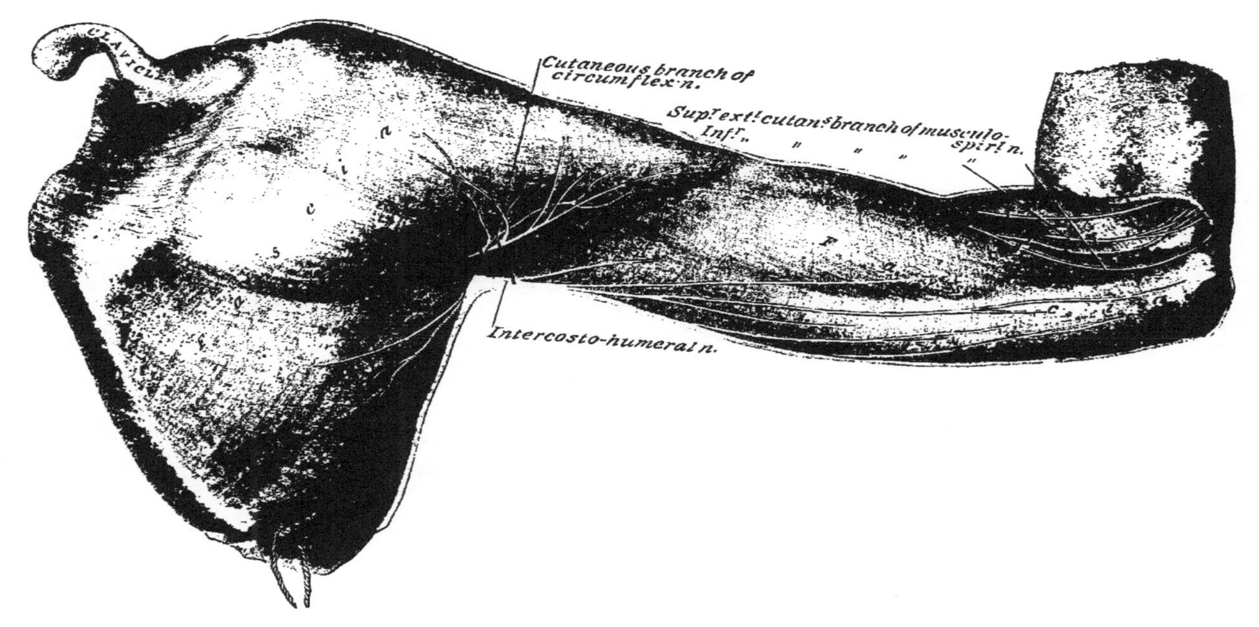
CLAVICLE
Cutaneous branch of circumflex n.
Supr. extl. cutans. branch of musculo-spiral n.
Infr. " " " "
Intercosto-humeral n.

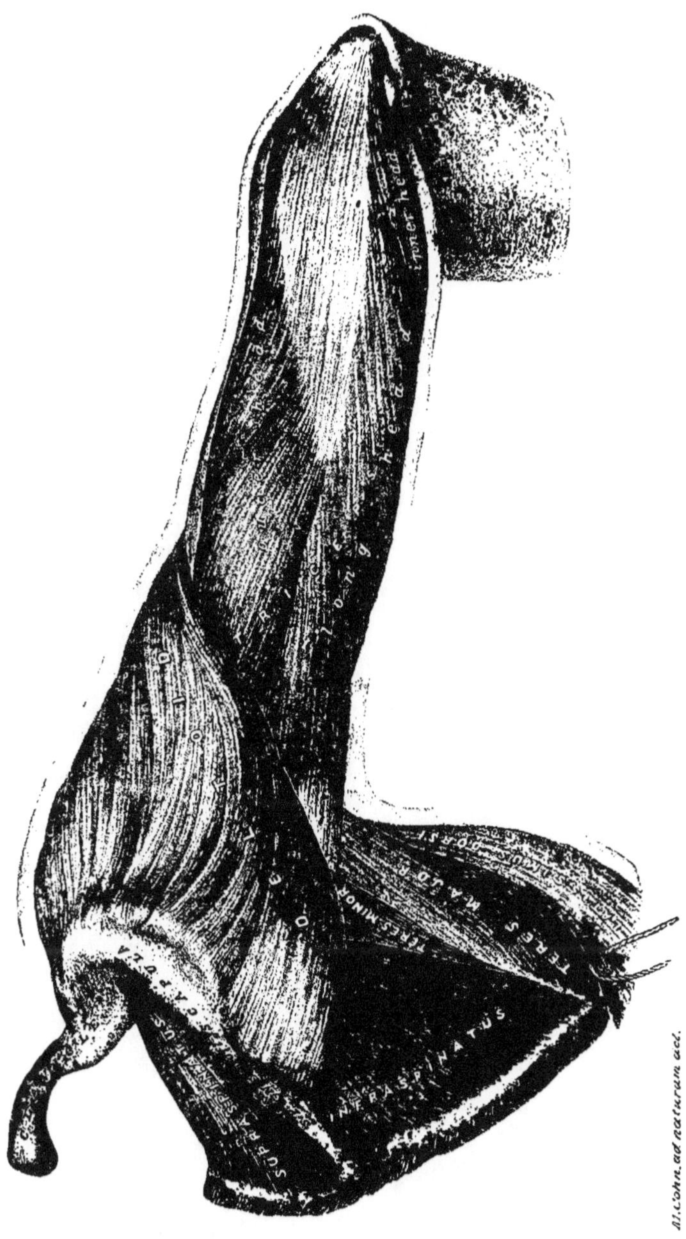
inner head
long head
TERES MINOR
TERES MAJOR
NFRASPINATUS
CAPSULA
M. Cohn ad naturam del.

PLATE 161

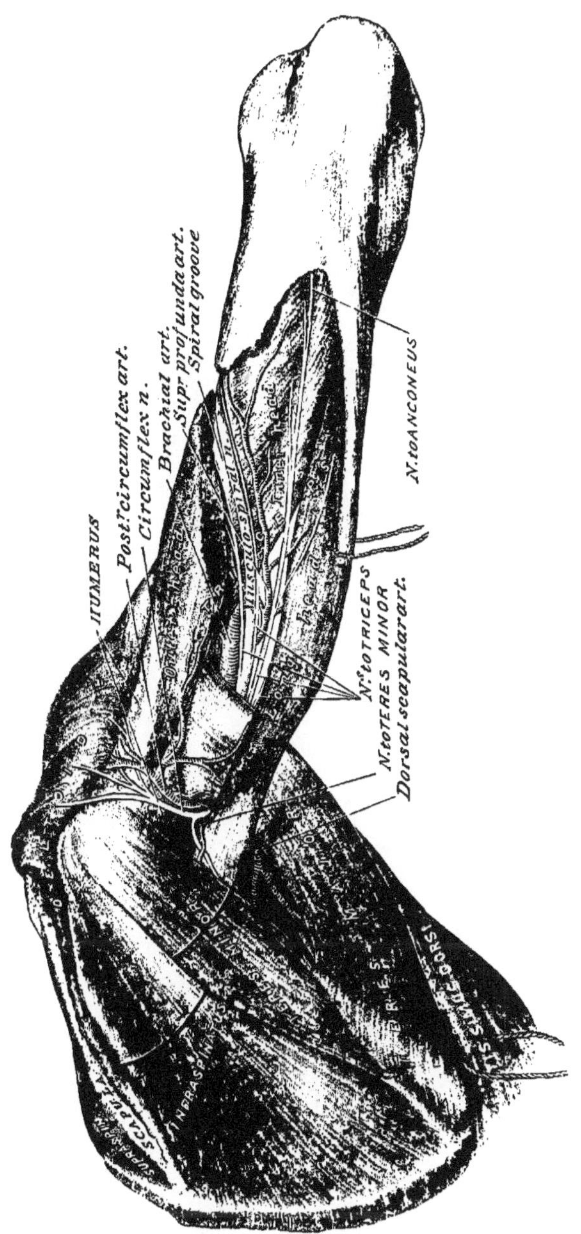
HUMERUS
Post.r circumflex art.
Circumflex n.
Brachial art.
Supr. profunda art.
Spiral groove
N. to ANCONEUS
N.rs to TRICEPS
N. to TERES MINOR
Dorsal scapular art.

M. Cohn, ad naturam del.

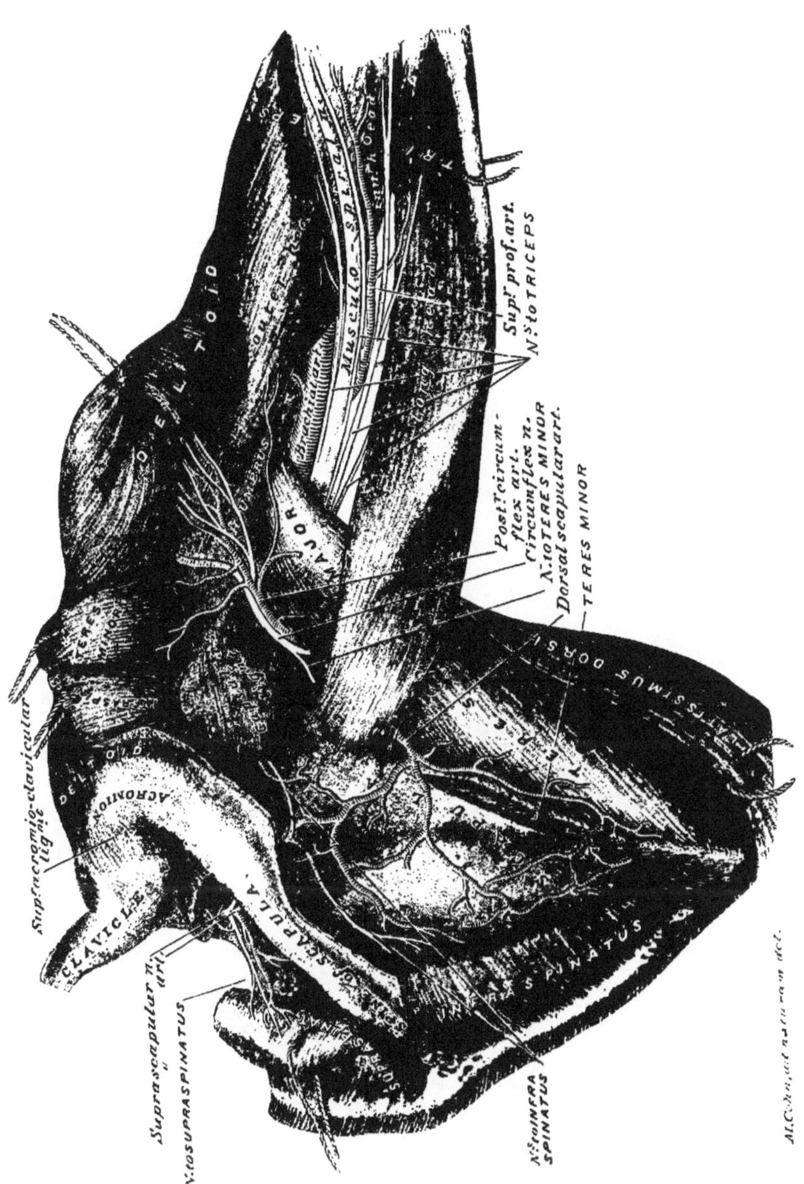
DELTOID
Musculo-spiral n.
Brachial art.
HUMERUS
MAJOR
TERES
Supr prof. art.
N.s to TRICEPS
Post. circum-
flex art.
Circumflex n.
N. to TERES MINOR
Dorsal scapular art.
TERES MINOR
LATISSIMUS DORSI
Supracromio-clavicular
lig.nt
ACROMION
CLAVICLE
SCAPULA
Suprascapular n.
art.
N. to SUPRASPINATUS
INFRASPINATUS
N. to INFRA
SPINATUS

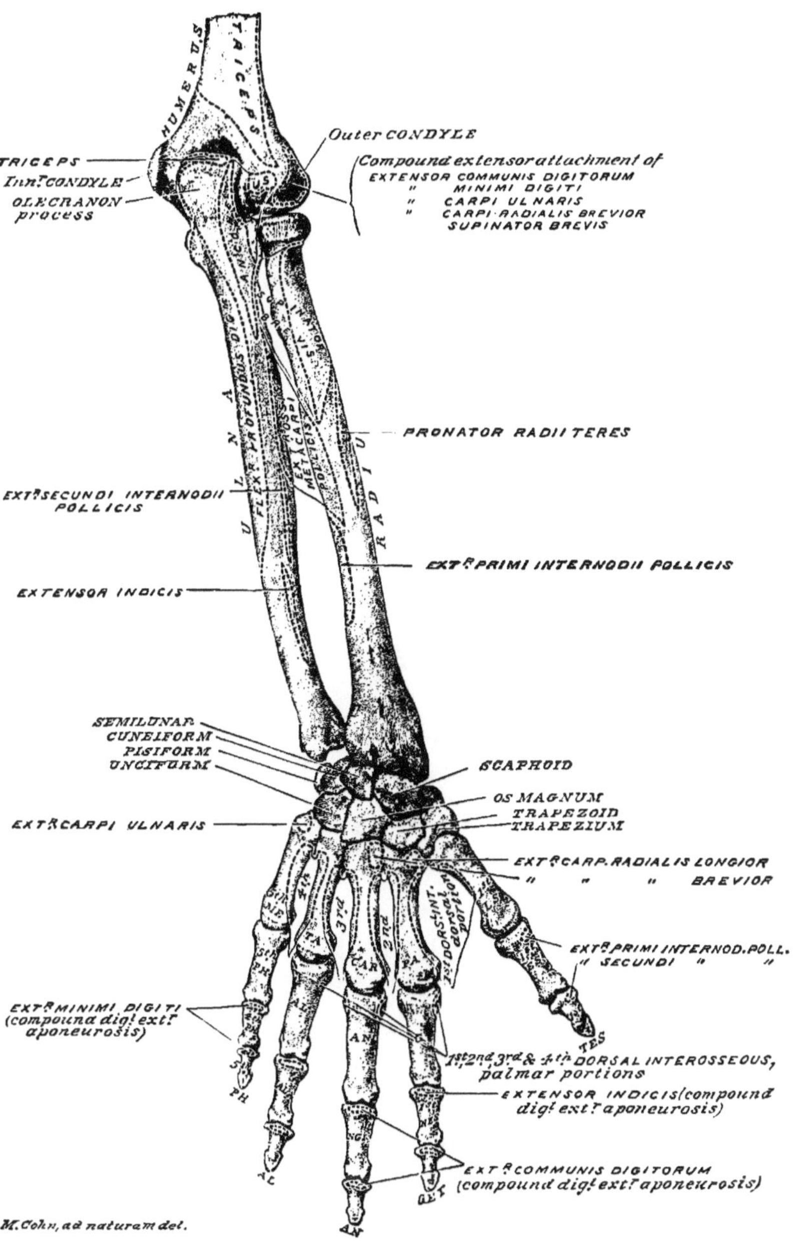
HUMERUS
TRICEPS
Outer CONDYLE
TRICEPS
Intn. CONDYLE
OLECRANON process
Compound extensor attachment of
EXTENSOR COMMUNIS DIGITORUM
" MINIMI DIGITI
" CARPI ULNARIS
" CARPI RADIALIS BREVIOR
SUPINATOR BREVIS
PRONATOR RADII TERES
EXTR. SECUNDI INTERNODII POLLICIS
EXTR. PRIMI INTERNODII POLLICIS
EXTENSOR INDICIS
ULNA
RADIUS
SEMILUNAR
CUNEIFORM
PISIFORM
UNCIFORM
SCAPHOID
OS MAGNUM
TRAPEZOID
TRAPEZIUM
EXTR. CARPI ULNARIS
EXTR. CARP. RADIALIS LONGIOR
" " " BREVIOR
EXTR. PRIMI INTERNOD. POLL.
" SECUNDI " "
EXTR. MINIMI DIGITI
(compound digl. extr. aponeurosis)
1st, 2nd, 3rd & 4th DORSAL INTEROSSEOUS, palmar portions
EXTENSOR INDICIS (compound digl. extr. aponeurosis)
EXTR. COMMUNIS DIGITORUM
(compound digl. extr. aponeurosis)
M. Cohn, ad naturam del.

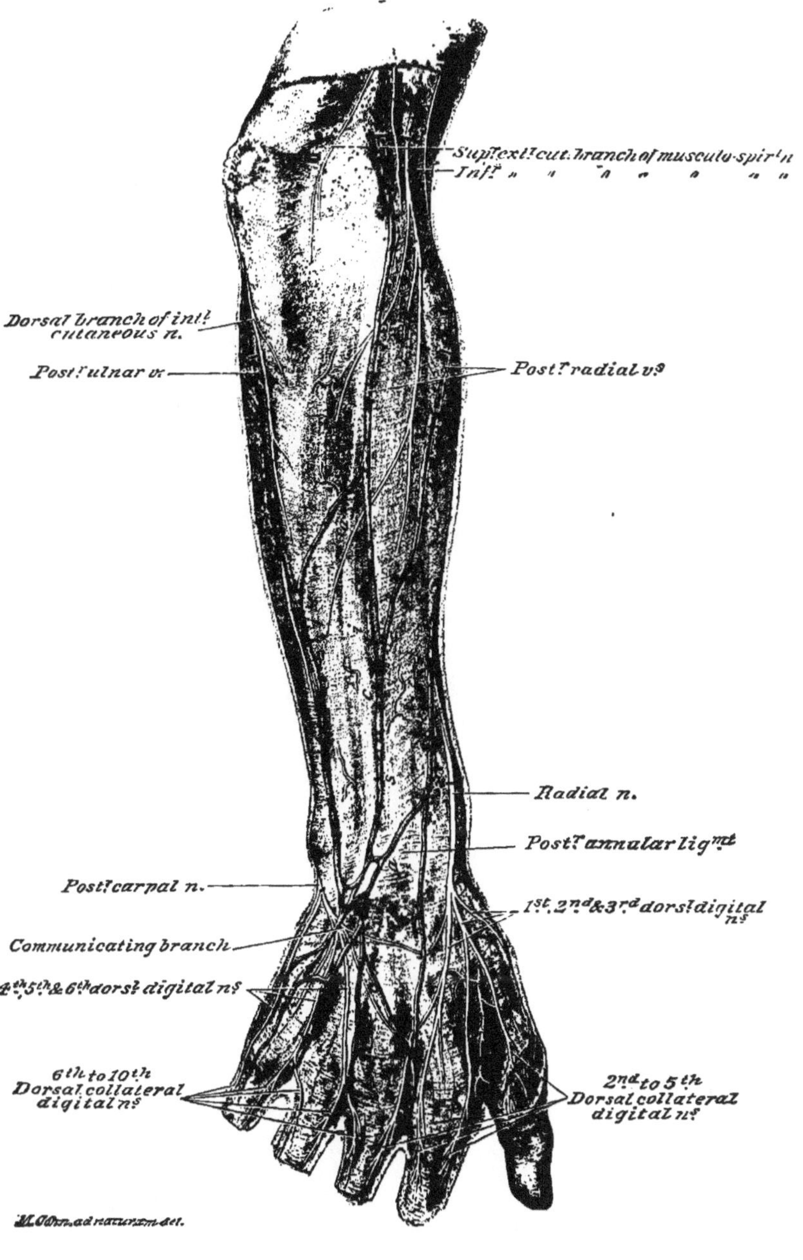

Supr extl cut. branch of musculo-spirl n
Infr " " " " " " "
Dorsal branch of intl cutaneous n.
Postr ulnar v.
Postr radial vs
Radial n.
Postr annular ligmt
Postr carpal n.
1st, 2nd & 3rd dorsl digital ns
Communicating branch
4th, 5th & 6th dorsl digital ns
6th to 10th Dorsal collateral digital ns
2nd to 5th Dorsal collateral digital ns
M. Odin. ad naturam del.

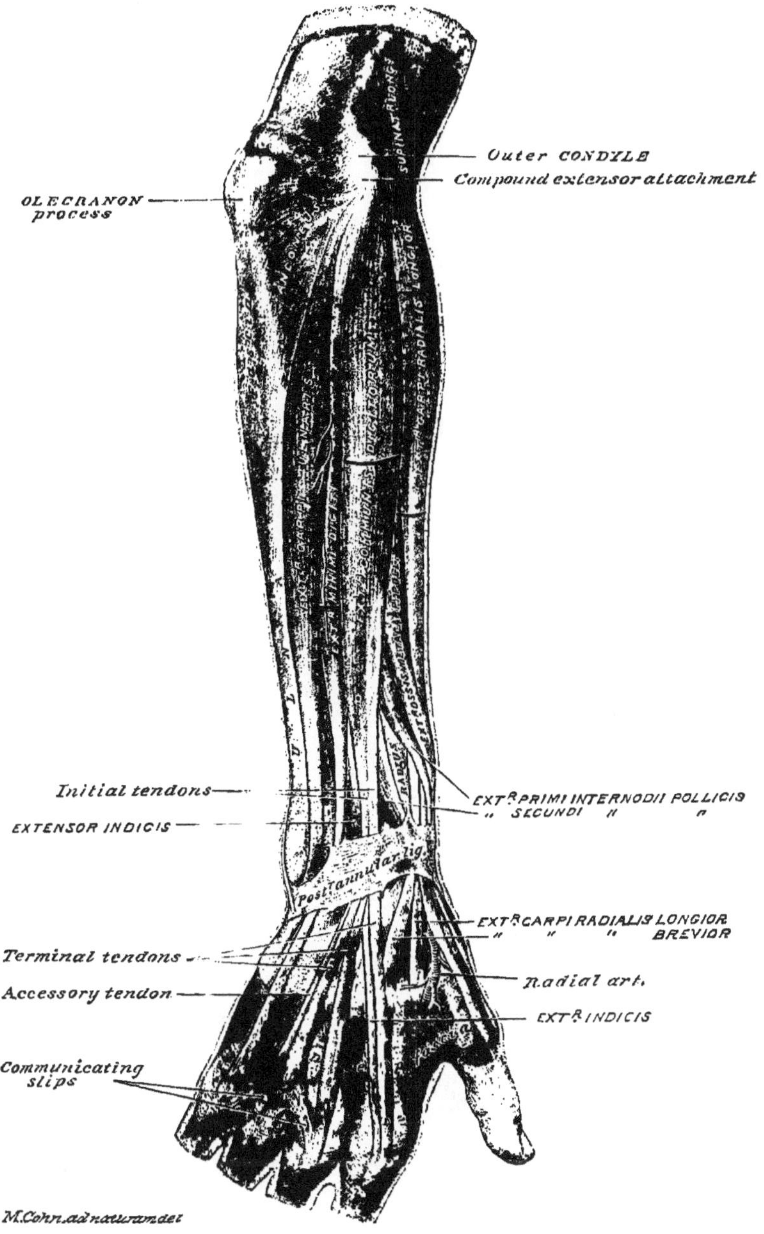
Outer CONDYLE
Compound extensor attachment
OLECRANON process
Initial tendons
EXTENSOR INDICIS
EXTR. PRIMI INTERNODII POLLICIS
" SECUNDI " "
Post. annular lig.
EXTR. CARPI RADIALIS LONGIOR
" " " BREVIOR
Terminal tendons
Accessory tendon
Radial art.
EXTR. INDICIS
Communicating slips
M.Cohn ad naturam del

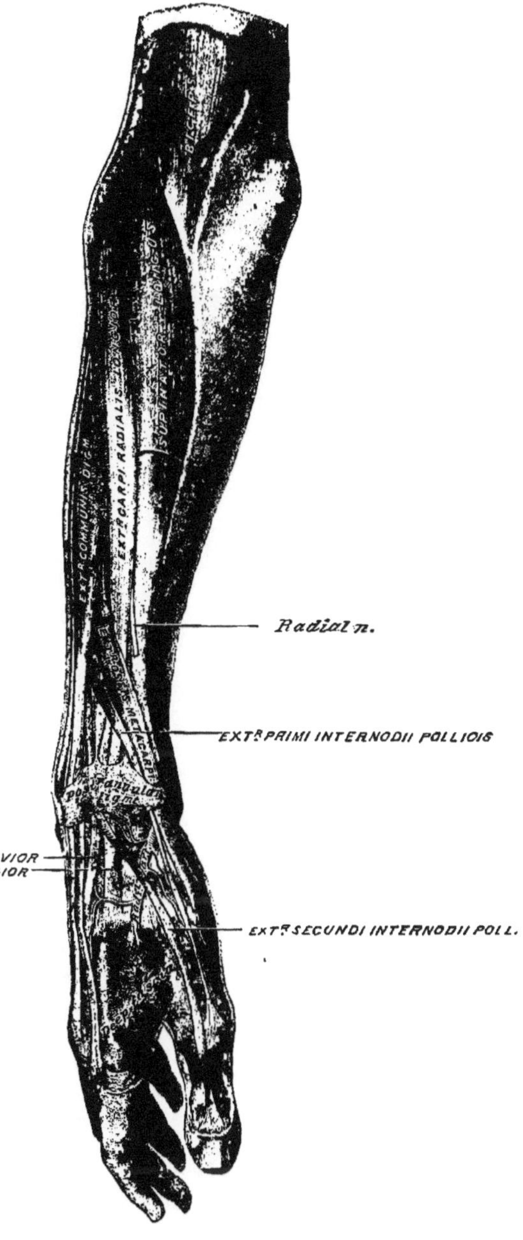
BICEPS
SUPINATOR LONGUS
EXTR CARPI RADIALIS LONGIOR
EXTR COMMUNIS DIGM
EXTR CARPI RADIALIS
Radial n.
EXTR PRIMI INTERNODII POLLICIS
Post. annular ligmt.
EXTR CARPI RADIALIS BREVIOR
" " " LONGIOR
EXTR SECUNDI INTERNODII POLL.
M. Cohn. ad naturam del.

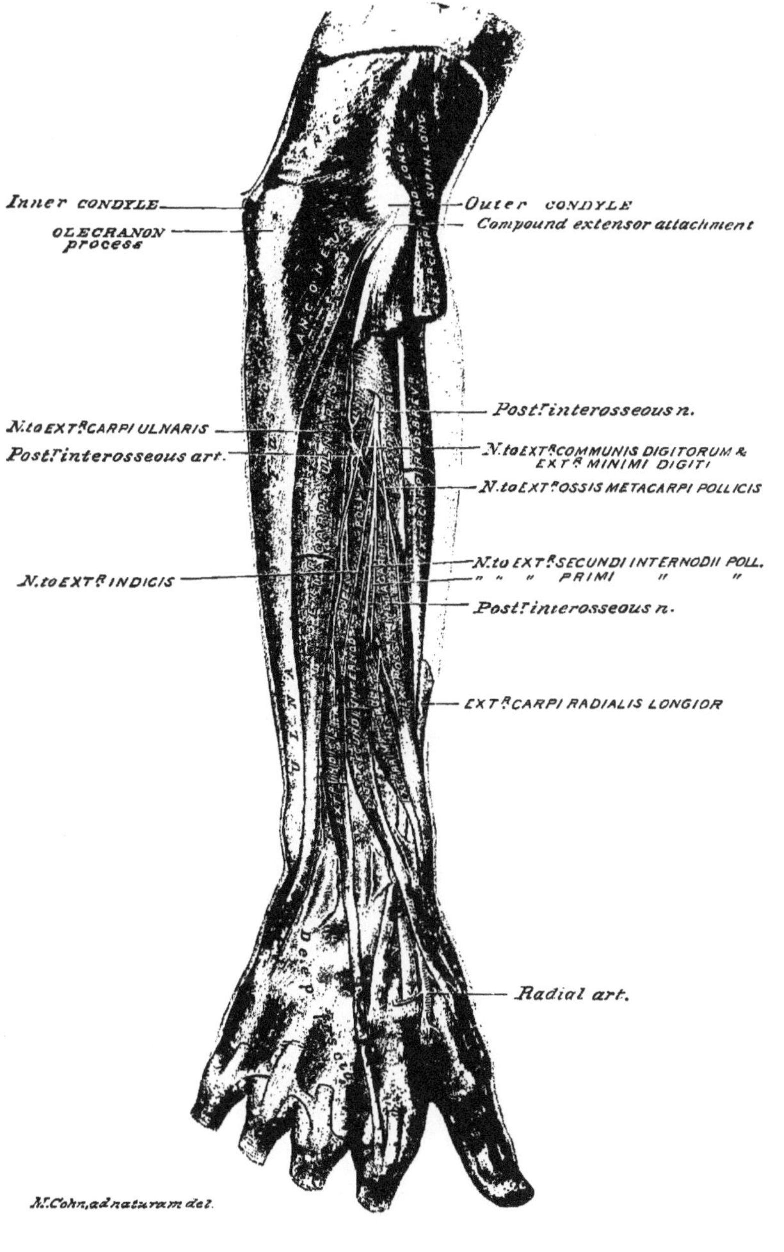
Inner CONDYLE
OLECRANON process
Outer CONDYLE
Compound extensor attachment
Post.r interosseous n.
N. to EXT.R CARPI ULNARIS
Post.r interosseous art.
N. to EXT.R COMMUNIS DIGITORUM & EXT.R MINIMI DIGITI
N. to EXT.R OSSIS METACARPI POLLICIS
N. to EXT.R SECUNDI INTERNODII POLL.
" " " PRIMI " "
N. to EXT.R INDICIS
Post.r interosseous n.
EXT.R CARPI RADIALIS LONGIOR
Radial art.
N. Cohn, ad naturam del.

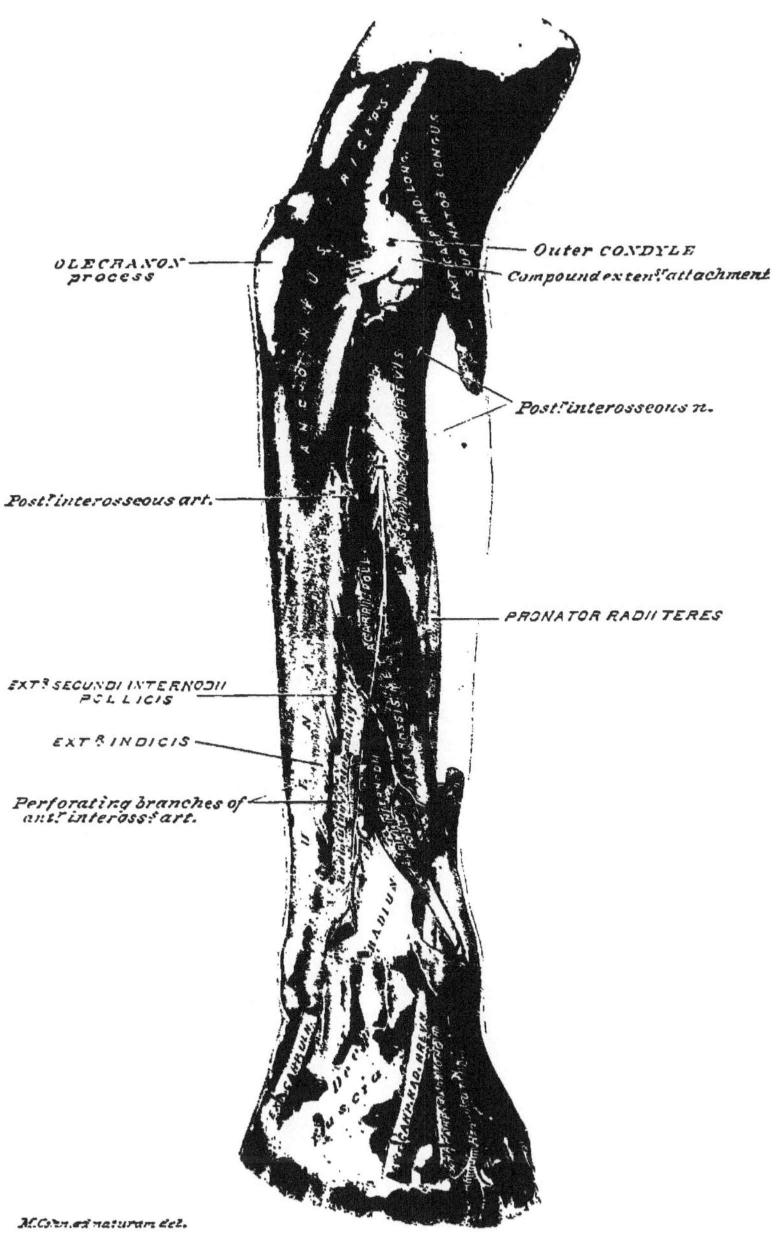
OLECRANON process
Outer CONDYLE
Compound extens.r attachment
Post.r interosseous n.
Post.r interosseous art.
PRONATOR RADII TERES
EXT.R SECUNDI INTERNODII POLLICIS
EXT.R INDICIS
Perforating branches of ant.r inteross.s art.
EXT. CARP. RAD. LONG.
SUPINATOR LONGUS
ANCONEUS
RADIUS
FASCIA

M.Cohn.ad naturam del.

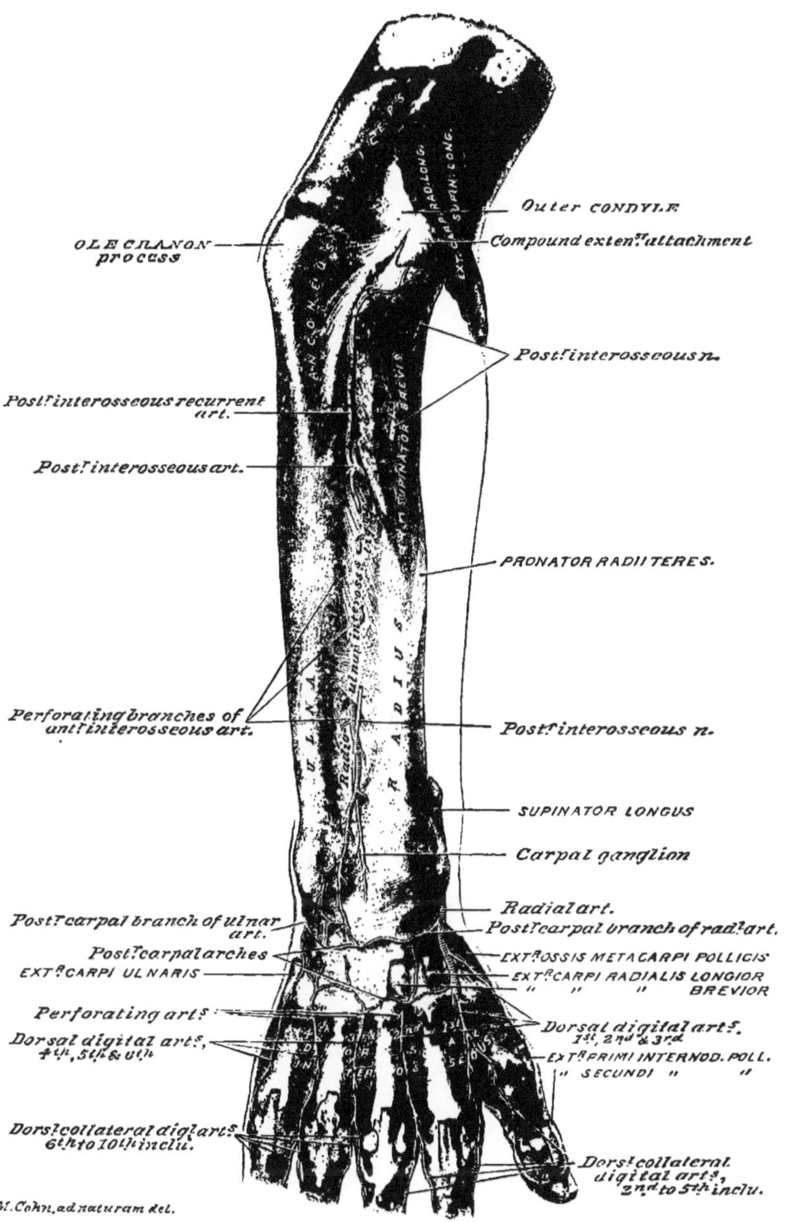
Outer CONDYLE
OLECRANON process
Compound extenr. attachment
Postr. interosseous n.
Postr. interosseous recurrent art.
Postr. interosseous art.
PRONATOR RADII TERES.
Perforating branches of antr. interosseous art.
Postr. interosseous n.
SUPINATOR LONGUS
Carpal ganglion
Radial art.
Postr. carpal branch of ulnar art.
Postr. carpal branch of radl. art.
Postr. carpal arches
EXTR. OSSIS METACARPI POLLICIS
EXTR. CARPI ULNARIS
EXTR. CARPI RADIALIS LONGIOR
" " " BREVIOR
Perforating artS.
Dorsal digital artS. 1st, 2nd & 3rd
Dorsal digital artS., 4th, 5th & 6th
EXTR. PRIMI INTERNOD. POLL.
" SECUNDI " "
Dorsl. collateral digl. artS. 6th to 10th inclu.
Dorsl. collateral digital artS., 2nd to 5th inclu.
ANCONEUS
RAD. LONG.
SUPIN. LONG.
SUPINATOR BREVIS
ULNA
RADIUS
M. Cohn, ad naturam del.

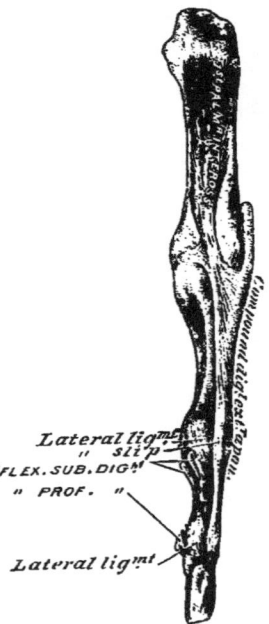

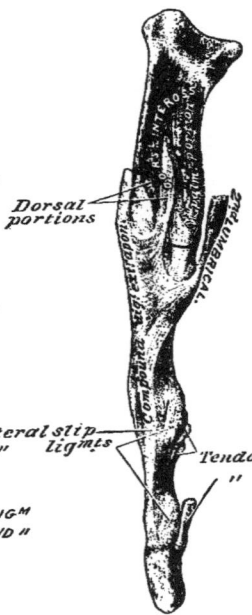

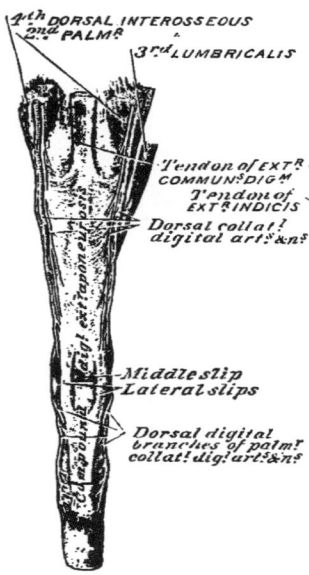

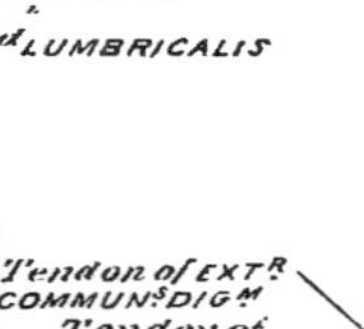
FIG. 1
4th DORSAL INTEROSSEOUS
2nd PALMR
3rd LUMBRICALIS
Tendon of EXTR COMMUNS DIGM
Tendon of EXTR INDICIS
Dorsal collatl digital artS &nS
Compound digl extr aponeurosis
Middle slip
Lateral slips
Dorsal digital branches of palmr collatl diglartS &nS
FIG. 2
DORS INTEROSSEU dorsl portion
1st LUMBRICALIS
Palmr portion
Compound digl extr aponeu.
Lateral ligmt
" slip
Tendon of FLEX. SUB. DIGM
" " " PROFUND "
Lateral ligmt
FIG. 3
Dorsal portions
2nd LUMBRICAL.
Compound digl extr apon.
Lateral slip
" ligmts
Tendons of FLEX. SUB. DIGM
" " " PROF. "
FIG. 4
1st PALMR INTEROSS
Compound diglextr apon.
Lateral ligmt
" slip
Lateral ligmt

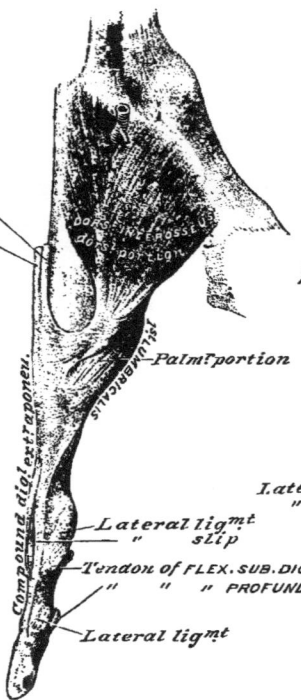

M. Cohn, ad naturam del.

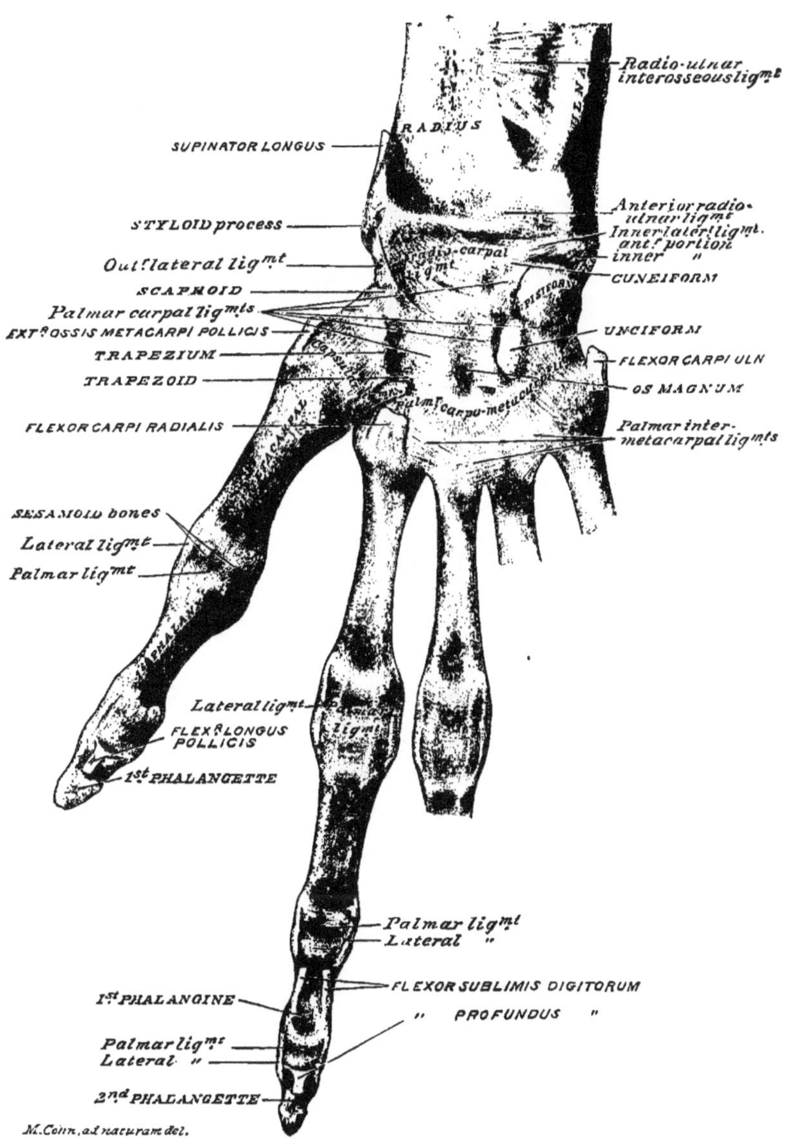
RADIUS
ULNA
Radio-ulnar interosseous ligmt
SUPINATOR LONGUS
Anterior radio-ulnar ligmt
STYLOID process
Inner lateral ligmt. ant. portion
inner "
Out lateral ligmt
Radio-carpal ligmt
CUNEIFORM
SCAPHOID
PISIFORM
Palmar carpal ligmts
EXT. OSSIS METACARPI POLLICIS
UNCIFORM
TRAPEZIUM
FLEXOR CARPI ULN
TRAPEZOID
OS MAGNUM
Palmr carpo-metacarpal
FLEXOR CARPI RADIALIS
Palmar inter-metacarpal ligmts
SESAMOID bones
Lateral ligmt
Palmar ligmt
Lateral ligmt
Palmar ligmt
FLEXR LONGUS POLLICIS
1st PHALANGETTE
Palmar ligmt
Lateral "
FLEXOR SUBLIMIS DIGITORUM
1st PHALANGINE
" PROFUNDUS "
Palmar ligmt
Lateral "
2nd PHALANGETTE

M. Cohn, ad naturam del.

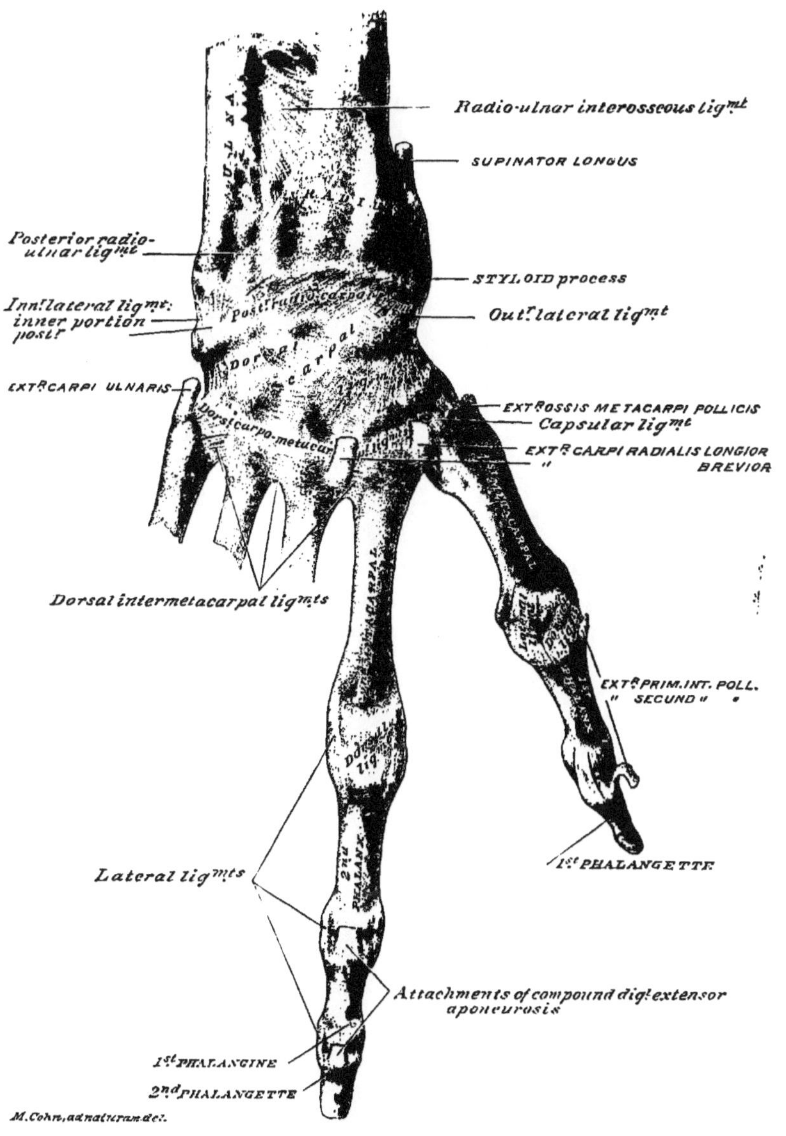
Radio-ulnar interosseous ligmt
SUPINATOR LONGUS
Posterior radio-ulnar ligmt
STYLOID process
Innr lateral ligmt: inner portion
postr
Outr lateral ligmt
EXTR CARPI ULNARIS
EXTR OSSIS METACARPI POLLICIS
Capsular ligmt
EXTR CARPI RADIALIS LONGIOR
" BREVIOR
Dorsal intermetacarpal ligmts
EXTR PRIM. INT. POLL.
" SECUND "
1st PHALANGETTE
Lateral ligmts
Attachments of compound digl extensor aponeurosis
1st PHALANGINE
2nd PHALANGETTE
M. Cohn, ad naturam del.

PLATE 173

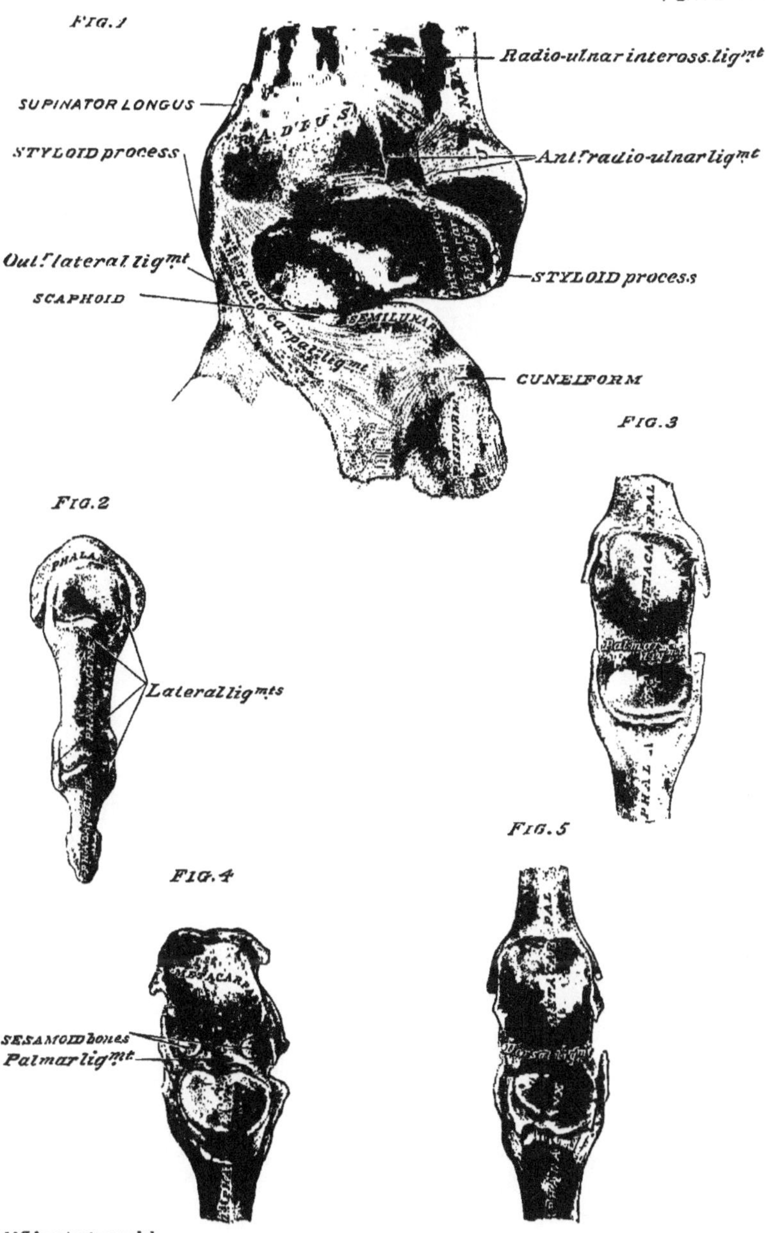

FIG. 1
Radio-ulnar inteross. ligmt
SUPINATOR LONGUS
STYLOID process
RADIUS
Ant. radio-ulnar ligmt
Out. lateral ligmt
STYLOID process
SCAPHOID
SEMILUNAR
CUNEIFORM
FIG. 3
FIG. 2
Lateral ligmts
Palmar ligmt
FIG. 5
FIG. 4
SESAMOID bones
Palmar ligmt
Dorsal ligmt

M. Cohn, ad naturam del.

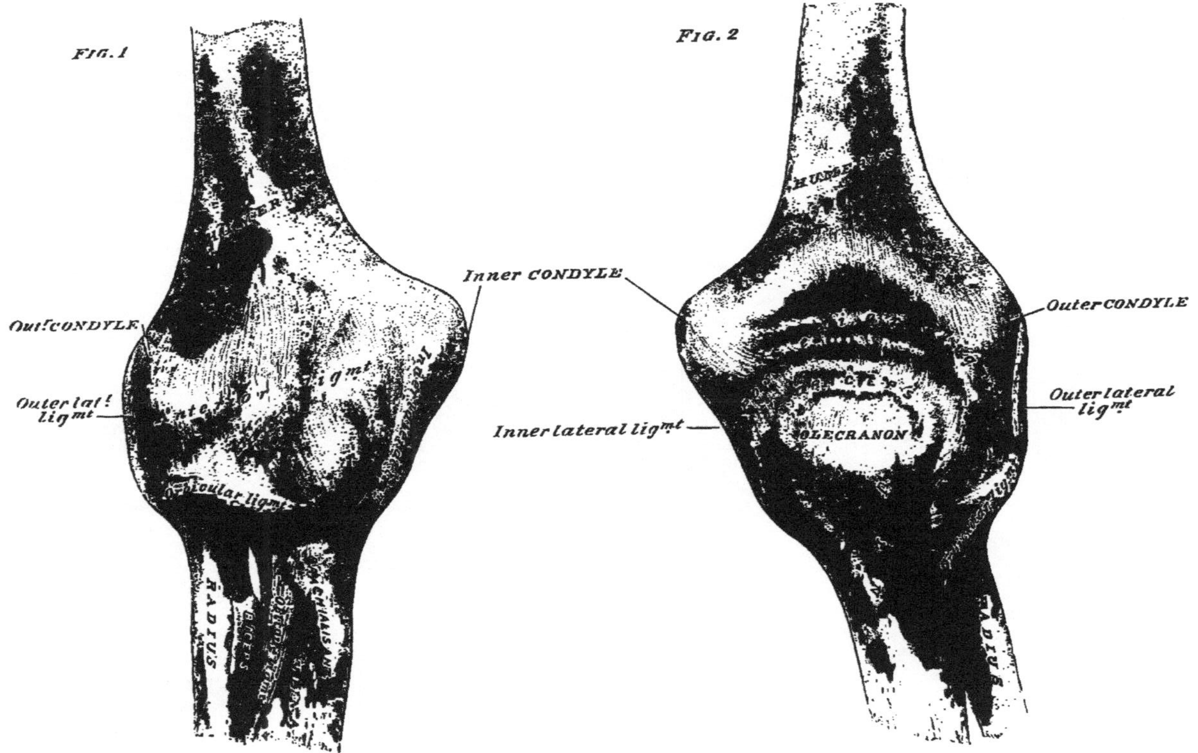
Fig. 1
Out.r CONDYLE
Outer lat.l ligmt
Inner CONDYLE
RADIUS
Fig. 2
Outer CONDYLE
Outer lateral ligmt
Inner lateral ligmt
OLECRANON

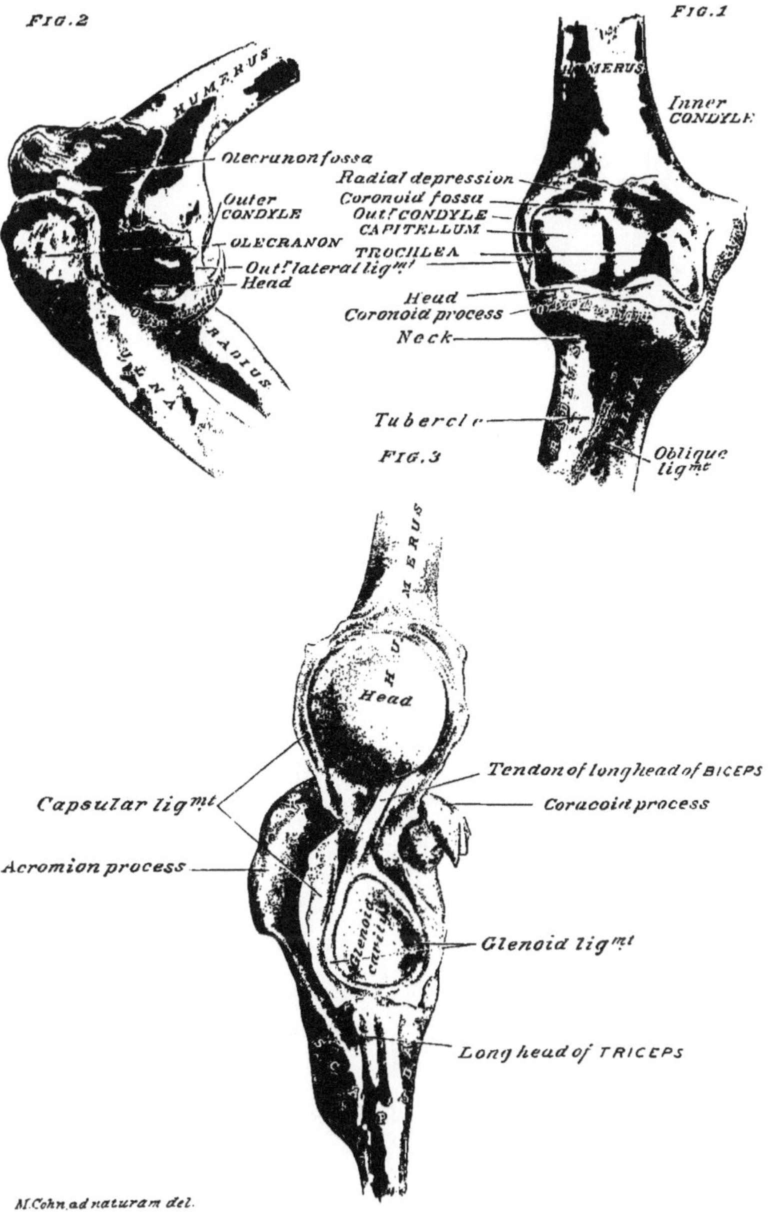
FIG. 2
HUMERUS
Olecranon fossa
Outer CONDYLE
OLECRANON
Out! lateral lig mt
Head
RADIUS
ULNA
FIG. 1
HUMERUS
Inner CONDYLE
Radial depression
Coronoid fossa
Out! CONDYLE
CAPITELLUM
TROCHLEA
Head
Coronoid process
Neck
Tubercle
ULNA
Oblique lig mt
FIG. 3
HUMERUS
Head
Tendon of long head of BICEPS
Capsular lig mt
Coracoid process
Acromion process
Glenoid cavity
Glenoid lig mt
Long head of TRICEPS
SCAPULA

M. Cohn ad naturam del.

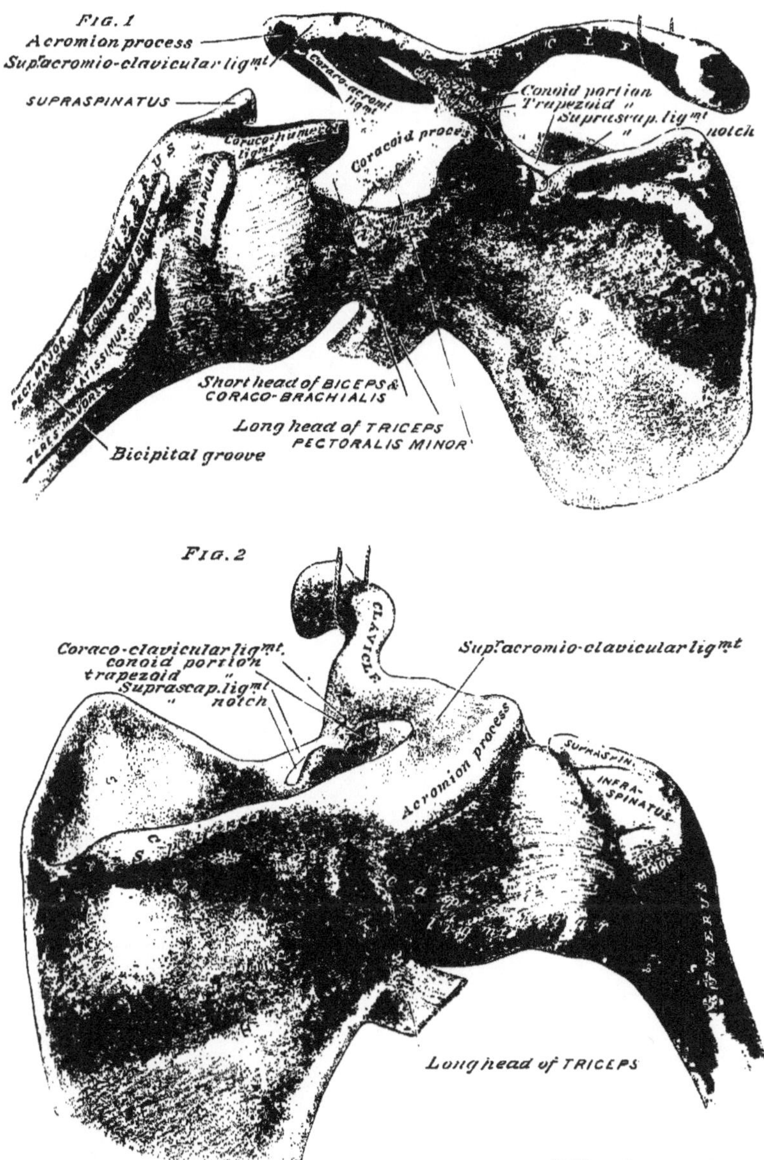
FIG. 1
Acromion process
Supr. acromio-clavicular ligmt.
SUPRASPINATUS
Coraco-acroml. ligmt.
Coraco-humeral ligmt.
Coracoid process
Conoid portion
Trapezoid "
Suprascap. ligmt.
" notch
SUBSCAPULARIS
Long head of BICEPS
LATISSIMUS DORSI
PECT. MAJOR
TERES MAJOR
Short head of BICEPS & CORACO-BRACHIALIS
Long head of TRICEPS
PECTORALIS MINOR
Bicipital groove
FIG. 2
CLAVICLE
Coraco-clavicular ligmt.
conoid portion
trapezoid "
Suprascap. ligmt.
" notch
Supr. acromio-clavicular ligmt.
Acromion process
SUPRASPIN.
INFRA-SPINATUS
MINOR
HUMERUS
Long head of TRICEPS
M. Cohn, ad naturam del.

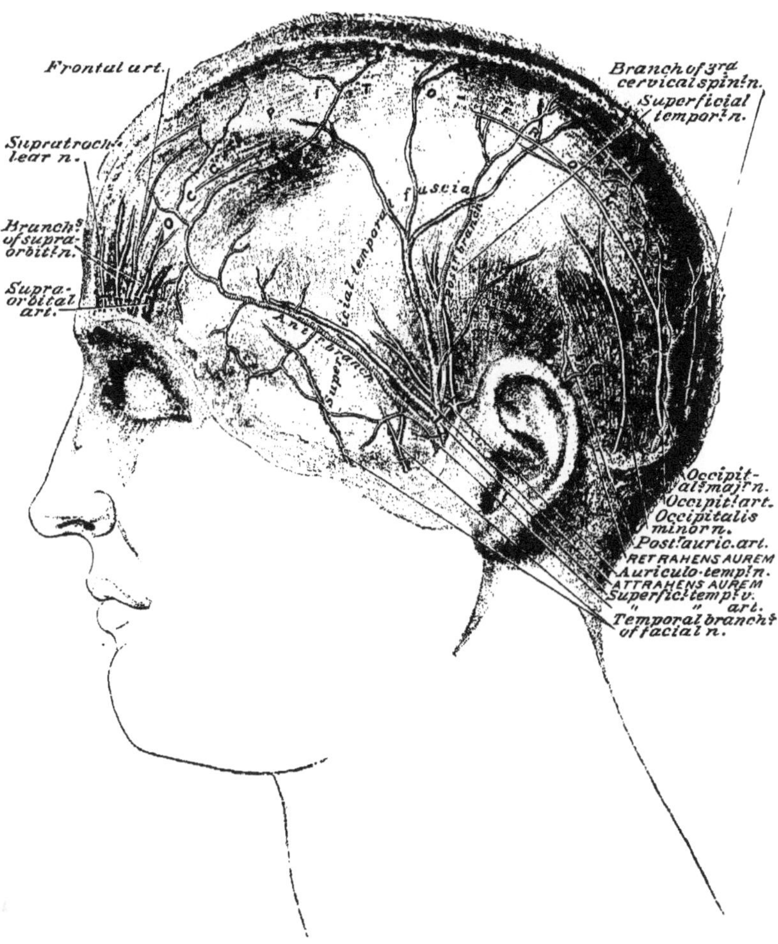
Frontal art.
Supratrochlear n.
Branch^s of supra-orbit^l n.
Supra-orbital art.
Branch of 3rd cervical spin^l n.
Superficial tempor^l n.
fascia
Superficial temporal
Post^r branch
Ant^r branch
Superficial branch
Occipit-al^s maj^r n.
Occipit^l art.
Occipitalis minor n.
Post^r auric. art.
RETRAHENS AUREM
Auriculo-temp^l n.
ATTRAHENS AUREM
Superfic^l temp^l v.
" " art.
Temporal branch^s of facial n.

M.Cohn, ad naturam del.

FIG. 1

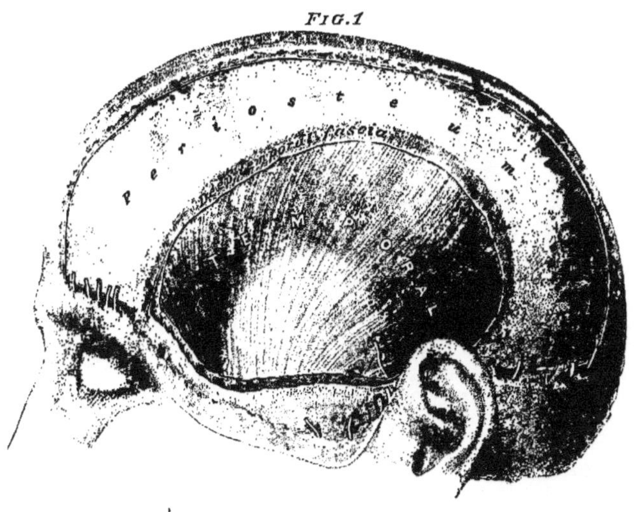
Periosteum
Deep temporal fascia
TEMPORAL

FIG. 2

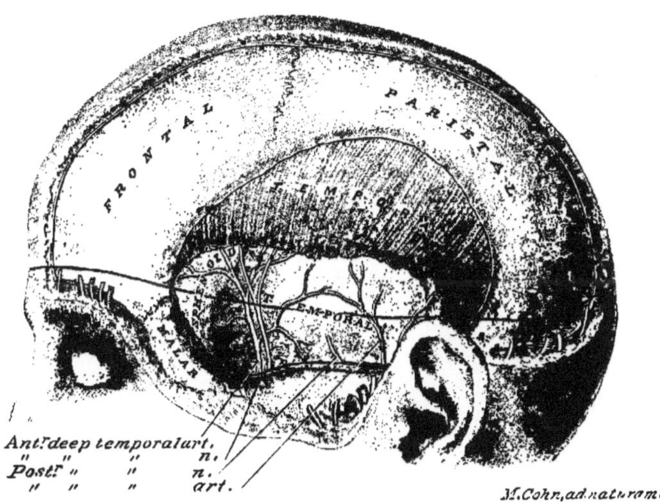
FRONTAL
PARIETAL
TEMPORAL
MALAR
TEMPORAL
Ant.r deep temporal art.
" " " n.
Post.r " " n.
" " " art.

M. Cohn, ad naturam del.

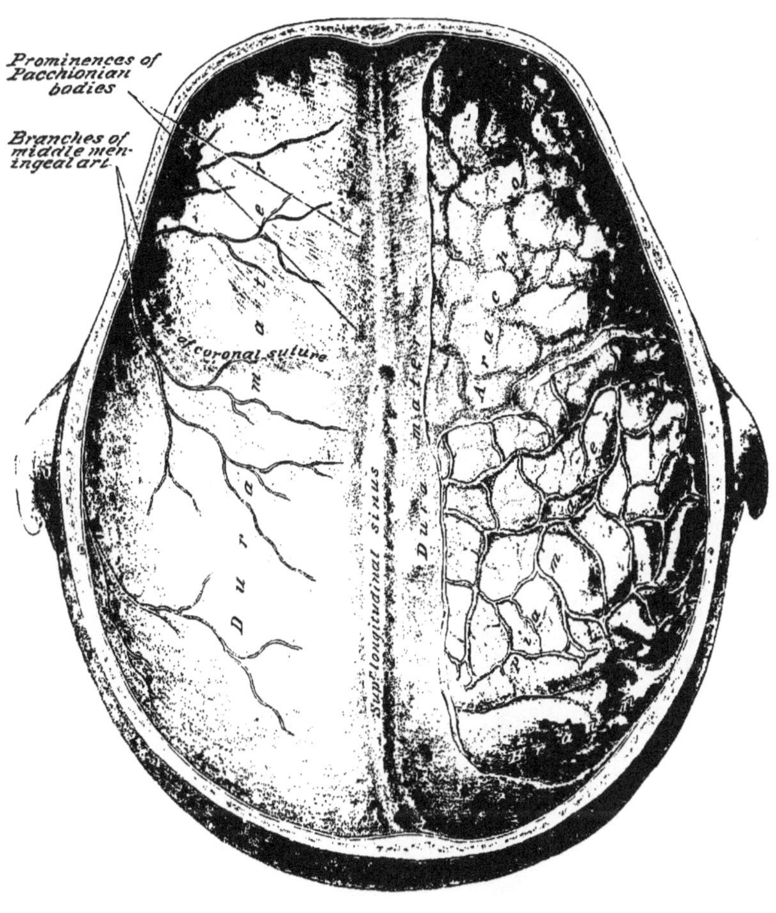
Prominences of Pacchionian bodies
Branches of middle meningeal art.
of coronal suture
Dura mater
Sup. longitudinal sinus
Dura mater
Brain

M. Cohn, ad naturam del.

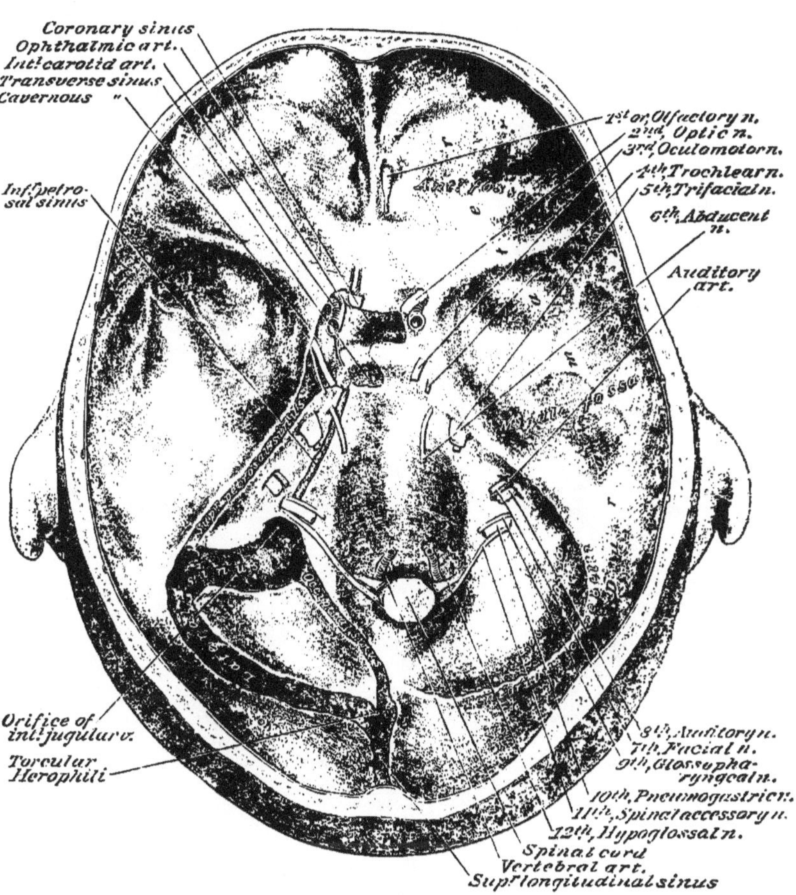
Coronary sinus
Ophthalmic art.
Intl. carotid art.
Transverse sinus
Cavernous "
Infr. petrosal sinus
1st or, Olfactory n.
2nd, Optic n.
3rd, Oculomotor n.
4th, Trochlear n.
5th, Trifacial n.
6th, Abducent n.
Auditory art.
Orifice of intl. jugular v.
Torcular Herophili
8th, Auditory n.
7th, Facial n.
9th, Glossopharyngeal n.
10th, Pneumogastric n.
11th, Spinal accessory n.
12th, Hypoglossal n.
Spinal cord
Vertebral art.
Supr. longitudinal sinus

M. Cohn ad naturam del.

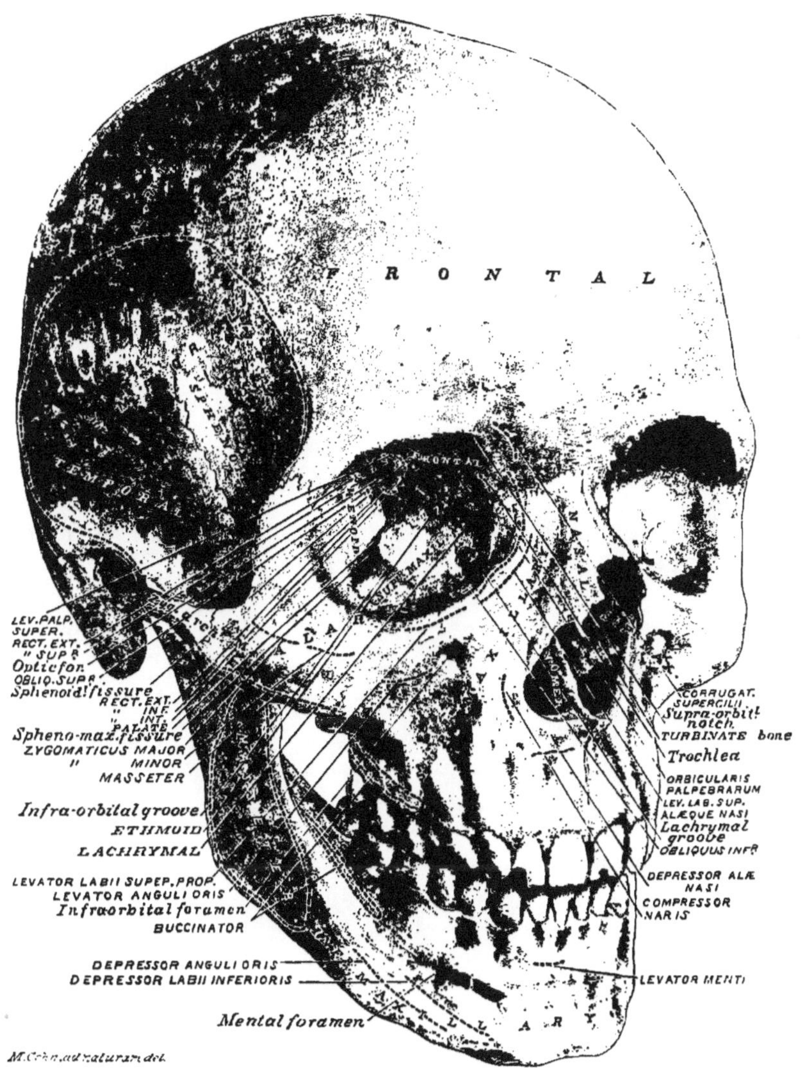
FRONTAL
TEMPORAL
NASAL
LEV. PALP. SUPER.
RECT. EXT.
" SUP R.
Optic for.
OBLIQ. SUP R.
Sphenoid! fissure
RECT. EXT.
" INF.
" INT.
PALATE
Spheno-max. fissure
ZYGOMATICUS MAJOR
" MINOR
MASSETER
Infra-orbital groove
ETHMOID
LACHRYMAL
LEVATOR LABII SUPEP. PROP.
LEVATOR ANGULI ORIS
Infraorbital foramen
BUCCINATOR
DEPRESSOR ANGULI ORIS
DEPRESSOR LABII INFERIORIS
Mental foramen
CORRUGAT. SUPERCILII
Supra-orbit! notch.
TURBINATE bone
Trochlea
ORBICULARIS PALPEBRARUM
LEV. LAB. SUP. ALÆQUE NASI
Lachrymal groove
OBLIQUUS INF R.
DEPRESSOR ALÆ NASI
COMPRESSOR NARIS
LEVATOR MENTI

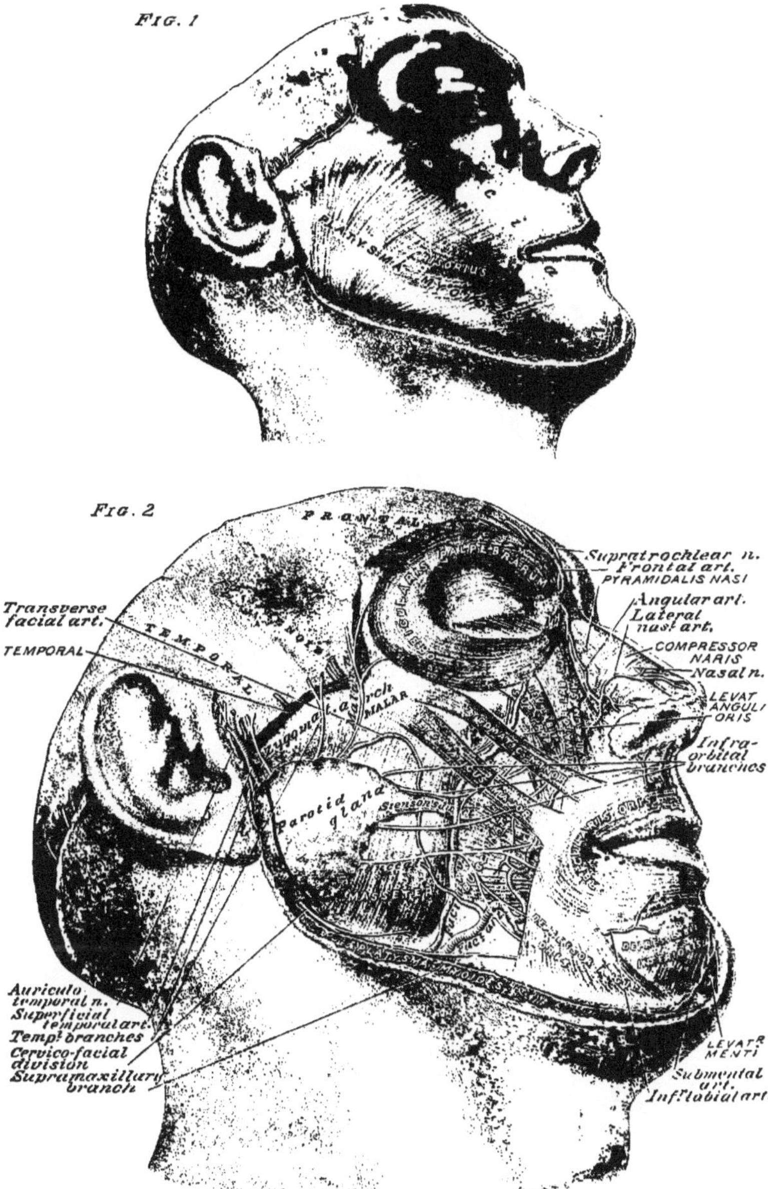
FIG. 1
PLATYSMA MYOIDES
RISORIUS
FIG. 2
FRONTAL
Supratrochlear n.
Frontal art.
PYRAMIDALIS NASI
Angular art.
Lateral nasl art.
COMPRESSOR NARIS
Nasal n.
LEVAT ANGULI ORIS
Infra-orbital branches
Transverse facial art.
TEMPORAL
TEMPORAL
MALAR
Zygomat. arch
Parotid gland
Stensons d.
MASSETER
Auriculo temporal n.
Superficial temporal art.
Templ branches
Cervico-facial division
Supramaxillary branch
LEVATR MENTI
Submental art.
Infrl labial art.

M.Cohn, ad naturam del.

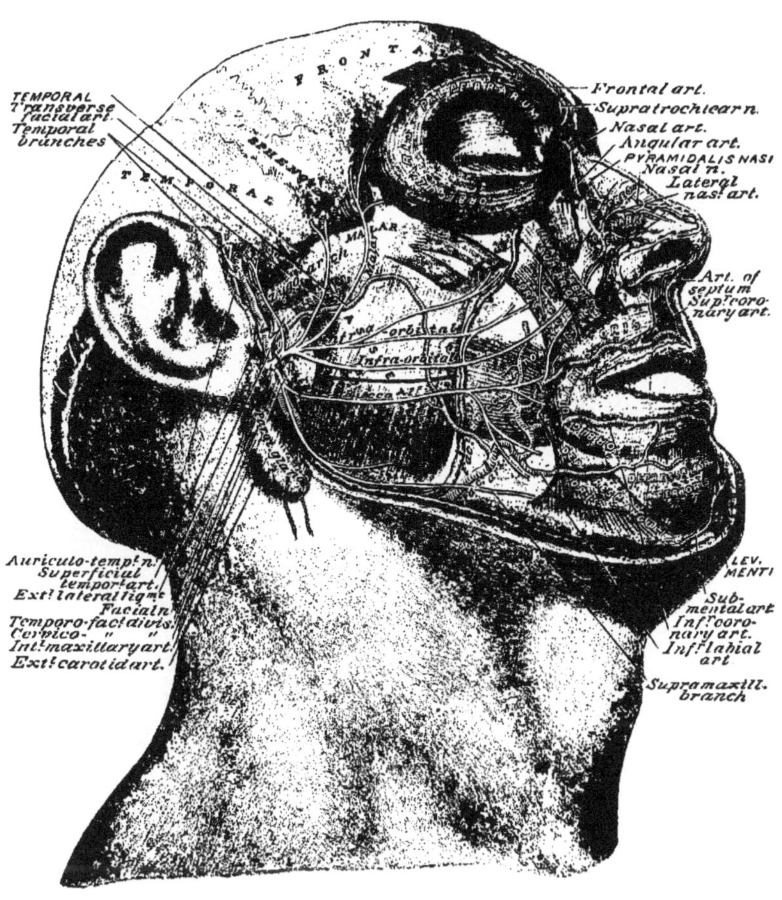
TEMPORAL
Transverse facial art.
Temporal branches
FRONTAL
TEMPORAL
MALAR
Infra-orbital
Frontal art.
Supratrochlear n.
Nasal art.
Angular art.
PYRAMIDALIS NASI
Nasal n.
Lateral nasl art.
Art. of septum
Supr coronary art.
Auriculo-templ n.
Superficial tempor art.
Extl lateral ligmt
Facial n
Temporo-facl divis.
Cervico- " "
Intl maxillary art.
Extl carotid art.
LEV. MENTI
Sub-mental art
Infr coronary art.
Infr labial art
Supramaxill. branch

M. Cohn, ad naturam del.

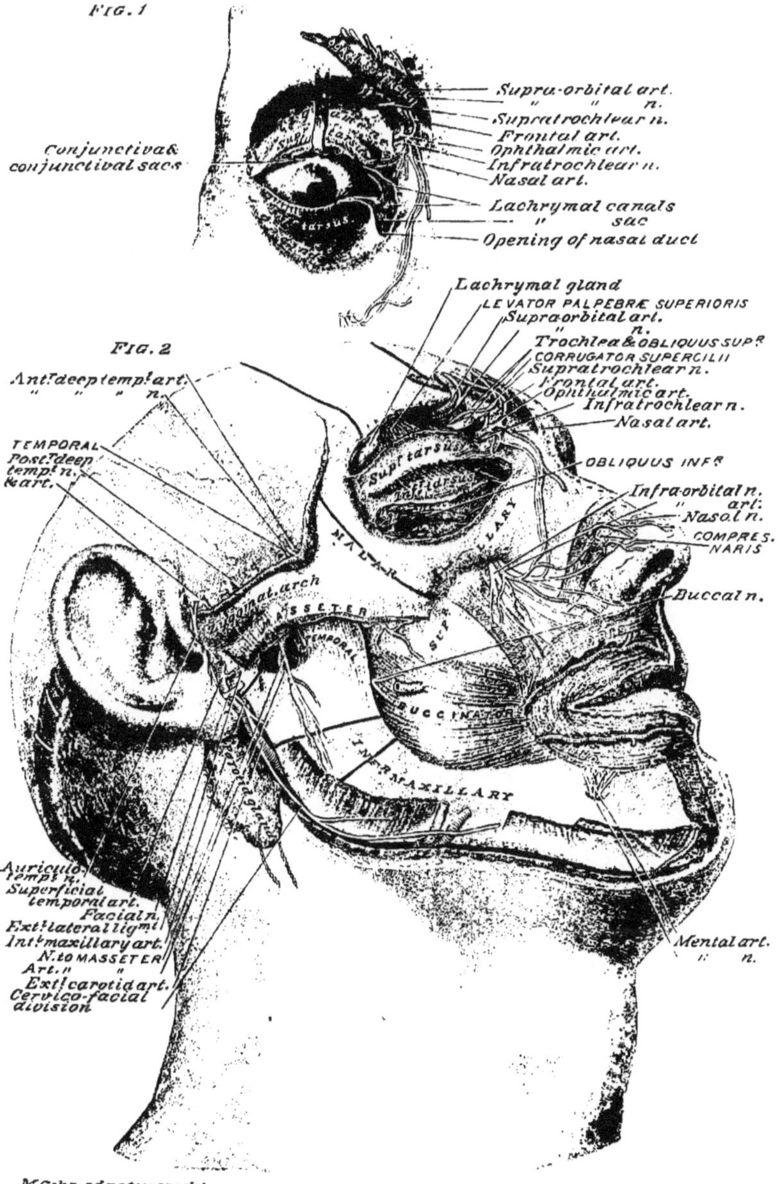
FIG. 1
Supra-orbital art.
" " n.
Supratrochlear n.
Frontal art.
Ophthalmic art.
Infratrochlear n.
Nasal art.
Lachrymal canals
" sac
Opening of nasal duct
Conjunctiva & conjunctival sacs
tarsus.
FIG. 2
Lachrymal gland
LEVATOR PALPEBRÆ SUPERIORIS
Supra-orbital art.
" n.
Trochlea & OBLIQUUS SUP.R
CORRUGATOR SUPERCILII
Supratrochlear n.
Frontal art.
Ophthalmic art.
Infratrochlear n.
Nasal art.
OBLIQUUS INF.R
Infra-orbital n.
" art.
Nasal n.
COMPRES. NARIS
Buccal n.
Ant.r deep temp.l art.
" " " n.
TEMPORAL
Post.r deep temp.l n. & art.
Sup.r tarsus
MALAR
Zygomat. arch
MASSETER
TEMPORAL
BUCCINATOR
INF. MAXILLARY
Auriculo-temp.l n.
Superficial temporal art.
Facial n.
Ext.l lateral lig.mt
Int.l maxillary art.
N. to MASSETER
Art. " "
Ext.l carotid art.
Cervico-facial division
Mental art.
" n.

M. Cohn, ad naturam del.

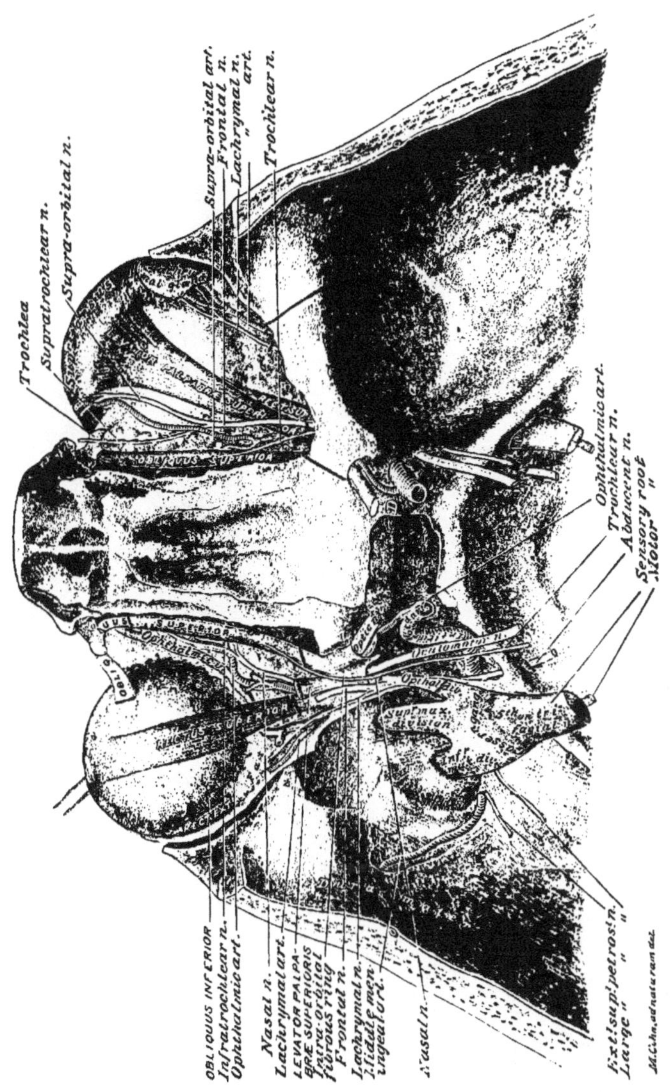
Trochlea
Supratrochlear n.
Supra-orbital n.
Supra-orbital art.
Frontal n.
Lachrymal n.
" art.
Trochlear n.
Ophthalmic art.
Trochlear n.
Abducent n.
Sensory root
Motor "
OBLIQUUS INFERIOR
Infratrochlear n.
Ophthalmic art.
Nasal n.
Lachrymal art.
LEVATOR PALPA-
BRÆ SUPERIORIS
Intra-orbital
fibrous ring
Frontal n.
Lachrymal n.
Middle men-
ingeal art.
Nasal n.
Ext. sup. petros. n.
Large " "
OBLIQUUS SUPERIOR
RECTUS SUPERIOR
Optic n.
Oculomotor n.
Sup. max. division
Gasserian ganglion

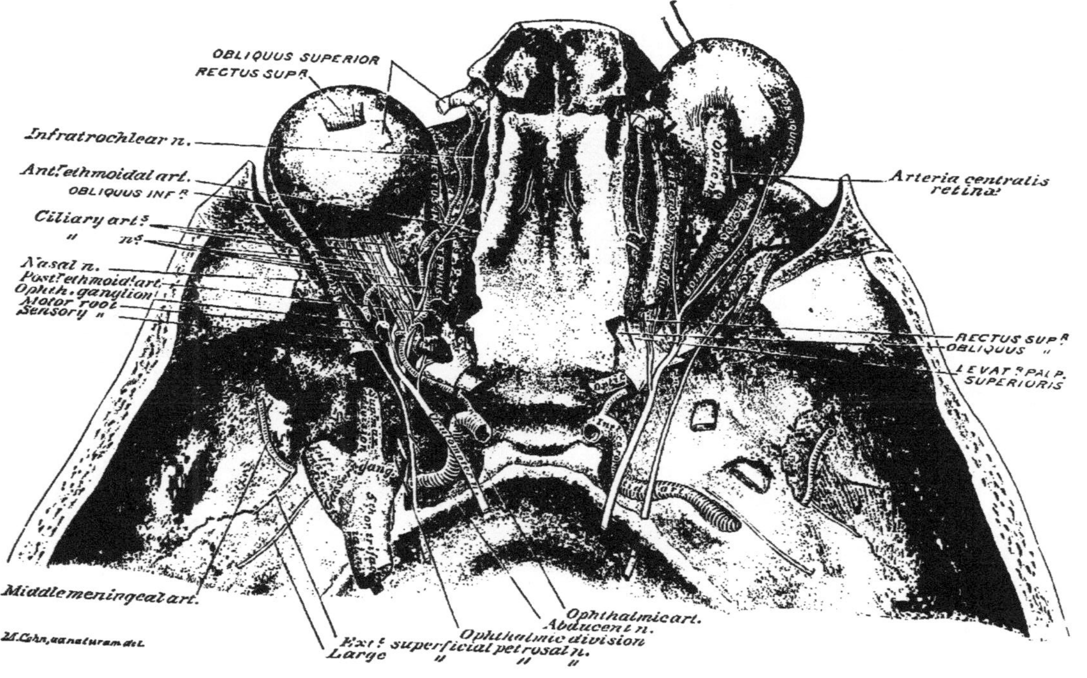
OBLIQUUS SUPERIOR
RECTUS SUPR.
Infratrochlear n.
Antr. ethmoidal art.
OBLIQUUS INFR.
Ciliary artS.
" nS.
Nasal n.
Postr. ethmoidal art.
Ophth. ganglion
Motor root
Sensory "
Middle meningeal art.
M. Cohn, ad naturam del.
Arteria centralis retinæ
RECTUS SUPR.
OBLIQUUS "
LEVATR. PALP. SUPERIORIS
Ophthalmic art.
Abducent n.
Ophthalmic division
Extl. superficial petrosal n.
Large " " "
RECTUS INTERNUS
RECTUS INFERIOR
Optic n.
5th or trifacial

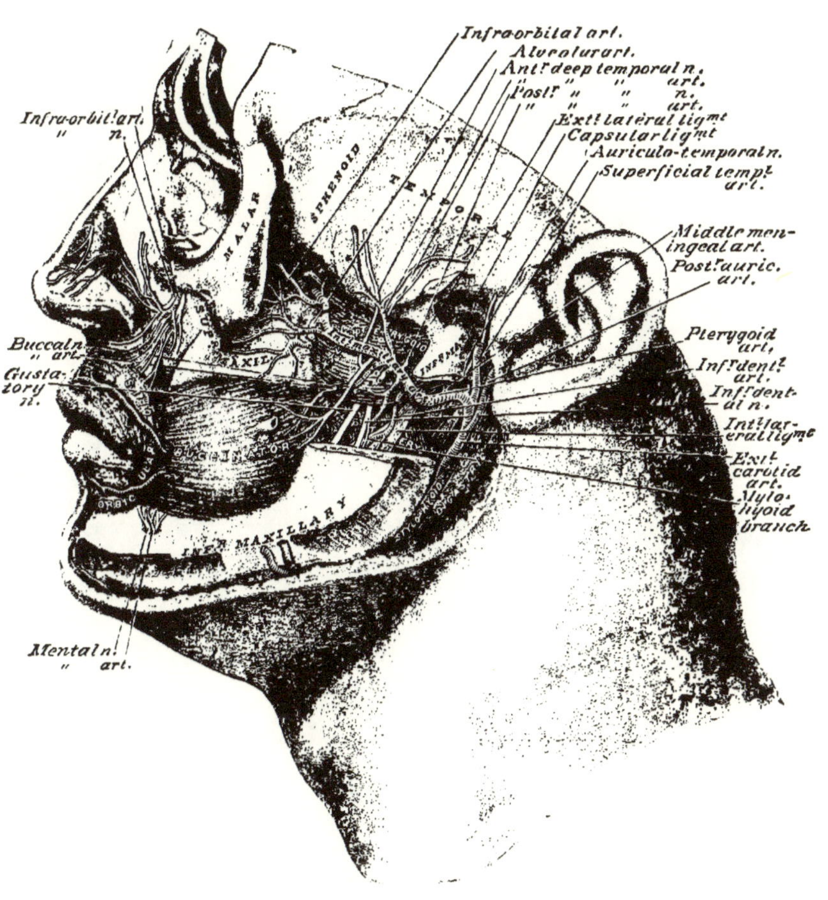
Infra-orbital art.
Alveolar art.
Antr deep temporal n.
" " " art.
Postr " " n.
" " " art.
Extl lateral ligmt
Capsular ligmt
Auriculo-temporal n.
Superficial templ art.
Middle meningeal art.
Postr auric. art.
Pterygoid art.
Infr dentl art.
Infr dental n.
Intl lateral ligmt
Extl carotid art.
Mylo-hyoid branch
Infra-orbitl art.
" n.
Buccal n.
" art.
Gustatory n.
Mental n.
" art.
MALAR
SPHENOID
TEMPORAL
BUCCINATOR
INFR MAXILLARY

M. Cohn, ad naturam del.

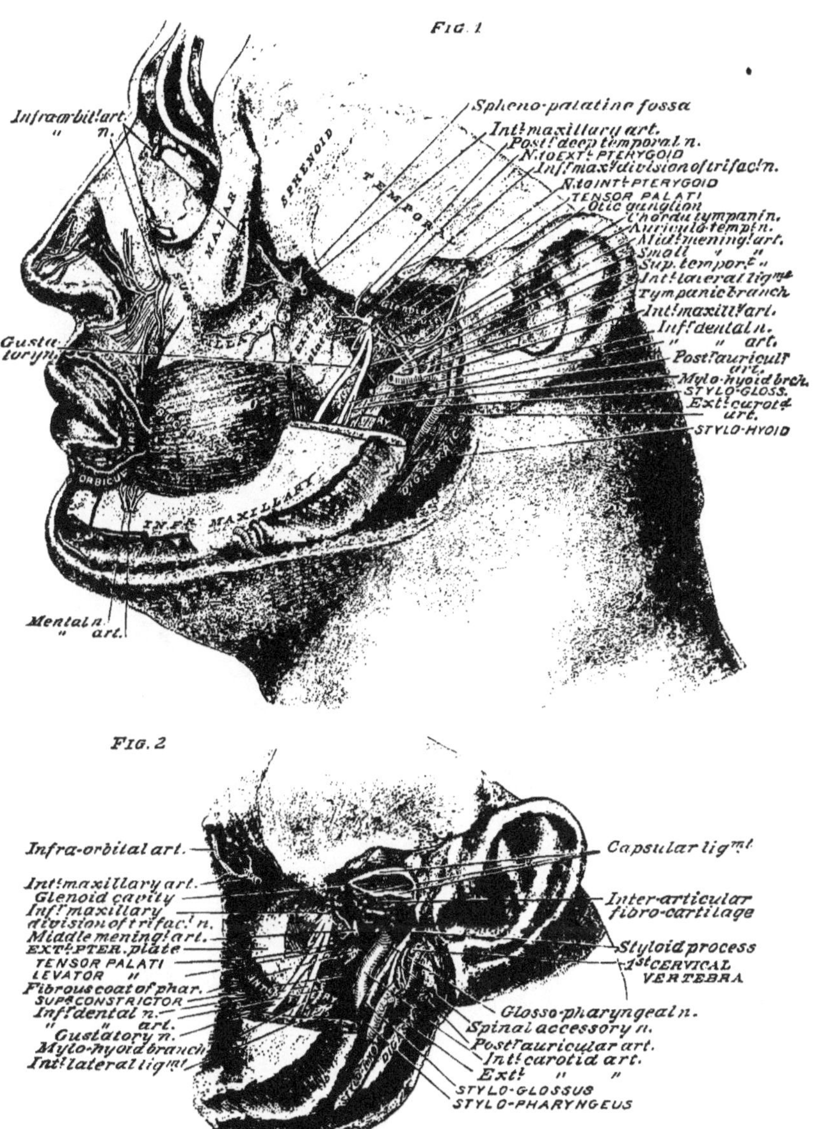
FIG. 1
Infraorbit! art.
" n.
Spheno-palatine fossa
Int! maxillary art.
Post! deep temporal n.
N. to EXT! PTERYGOID
Inf! max! division of trifac! n.
N. to INT! PTERYGOID
TENSOR PALATI
Otic ganglion
Chorda tympani n.
Auriculo-temp! n.
Mid! mening! art.
Small " "
Sup. tempor! "
Int! lateral ligmt
Tympanic branch
Int! maxill! art.
Inf! dental n.
" " art.
Post! auricul! art.
Mylo-hyoid brch.
STYLO-GLOSS.
Ext! carotd art.
STYLO-HYOID
Gustatory n.
SPHENOID
TEMPORAL
MALAR
INF! MAXILLARY
ORBICU
DIGASTRIC
Mental n.
" art.
FIG. 2
Infra-orbital art.
Int! maxillary art.
Glenoid cavity
Inf! maxillary division of trifac! n.
Middle mening! art.
EXT! PTER. plate
TENSOR PALATI
LEVATOR "
Fibrous coat of phar.
SUP! CONSTRICTOR
Inf! dental n.
" " art.
Gustatory n.
Mylo-hyoid branch
Int! lateral ligmt
Capsular ligmt
Inter-articular fibro-cartilage
Styloid process
1st CERVICAL VERTEBRA
Glosso-pharyngeal n.
Spinal accessory n.
Post! auricular art.
Int! carotid art.
Ext! " "
STYLO-GLOSSUS
STYLO-PHARYNGEUS
M. Cohn, ad naturam del.

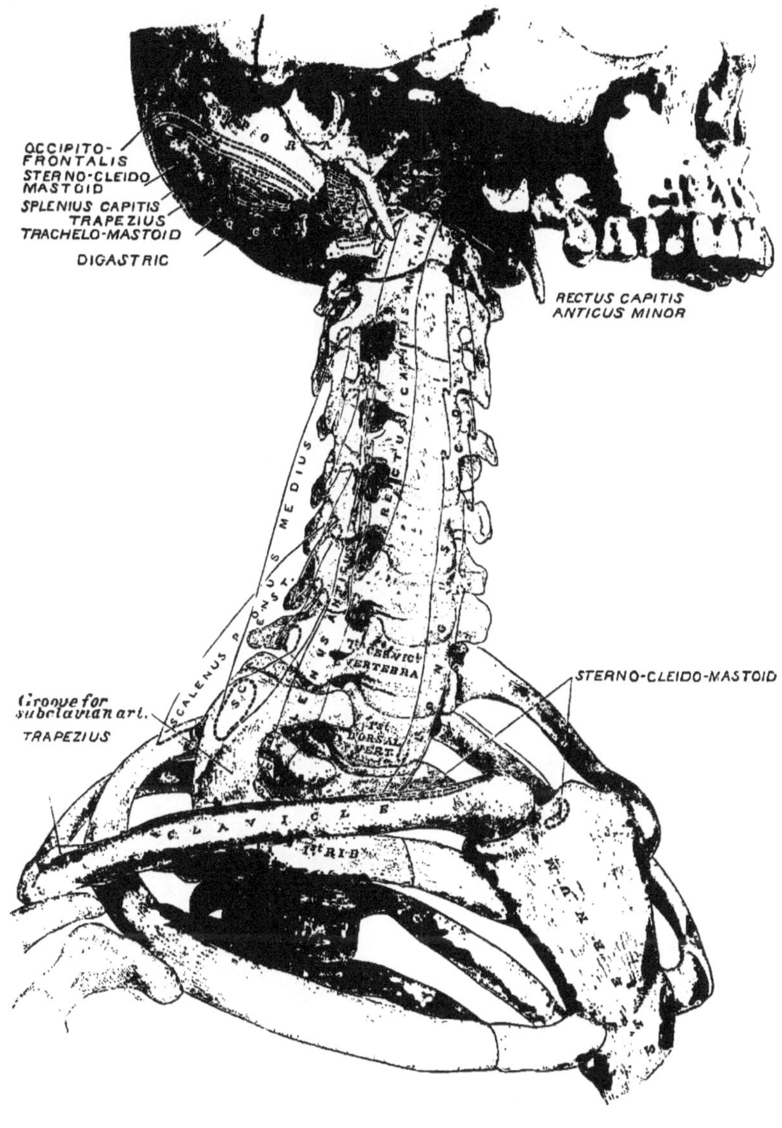
OCCIPITO-FRONTALIS
STERNO-CLEIDO MASTOID
SPLENIUS CAPITIS
TRAPEZIUS
TRACHELO-MASTOID
DIGASTRIC
RECTUS CAPITIS ANTICUS MINOR
SCALENUS MEDIUS
RECTUS CAPITIS ANT. MA
LONGUS COLLI
7th CERVIC VERTEBRA
First DORSAL VERT.
STERNO-CLEIDO-MASTOID
Groove for subclavian art.
TRAPEZIUS
CLAVICLE
1st RIB

M. Cohn. ad naturam del.

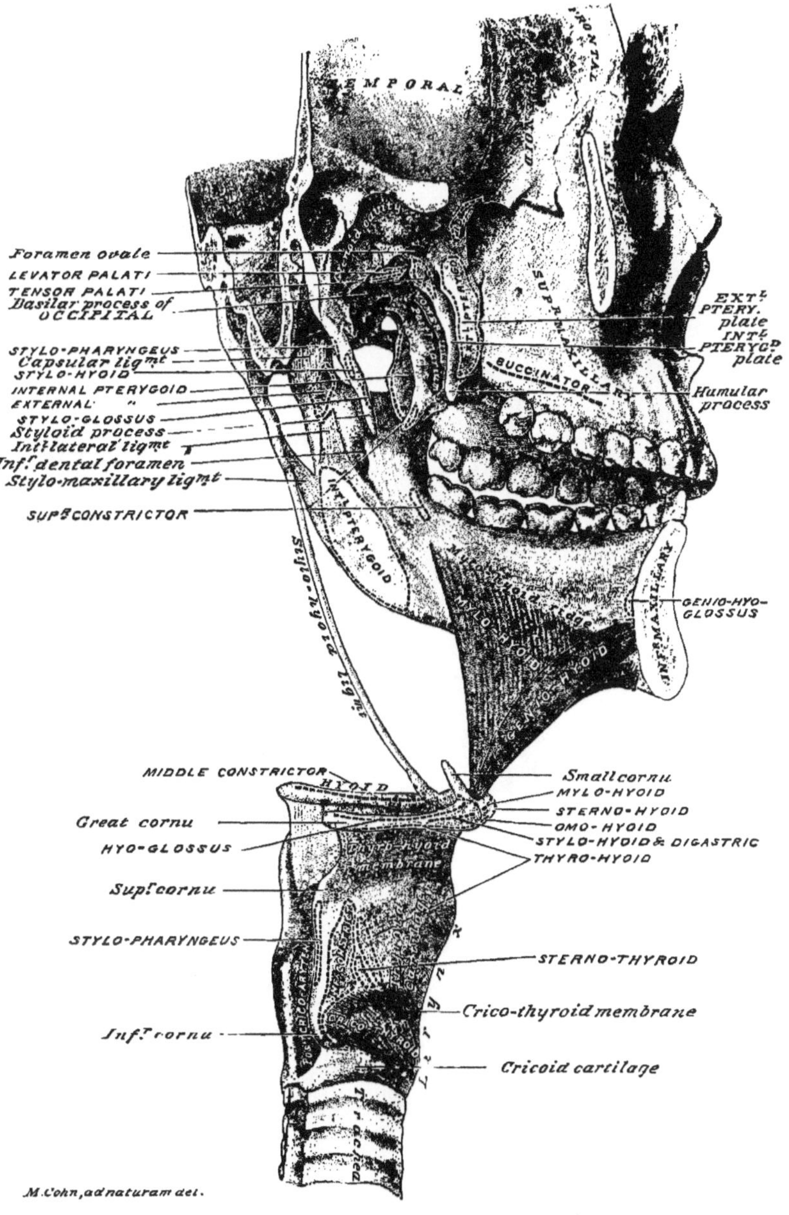
TEMPORAL
FRONTAL
MALAR
SUPR. MAXILLARY
Foramen ovale
LEVATOR PALATI
TENSOR PALATI
Basilar process of OCCIPITAL
STYLO-PHARYNGEUS
Capsular ligmt
STYLO-HYOID
INTERNAL PTERYGOID
EXTERNAL "
STYLO-GLOSSUS
Styloid process
Intl lateral ligmt
Infr dental foramen
Stylo-maxillary ligmt
SUPR CONSTRICTOR
EXTL PTERY. plate
INTL PTERYGD plate
BUCCINATOR
Humular process
INTL PTERYGOID
Stylo-hyoid ligmt
Mylo-hyoid ridge
MYLO-HYOID
GENIO-HYOID
INFR MAXILLARY
GENIO-HYO-GLOSSUS
MIDDLE CONSTRICTOR
HYOID
Small cornu
MYLO-HYOID
STERNO-HYOID
OMO-HYOID
STYLO-HYOID & DIGASTRIC
THYRO-HYOID
Great cornu
HYO-GLOSSUS
Thyro-hyoid membrane
Supr cornu
STYLO-PHARYNGEUS
STERNO-THYROID
Crico-thyroid membrane
Infr cornu
Cricoid cartilage
CRICO-THYROID
Trachea

M. Cohn, ad naturam del.

M. Cohn, ad naturam del.

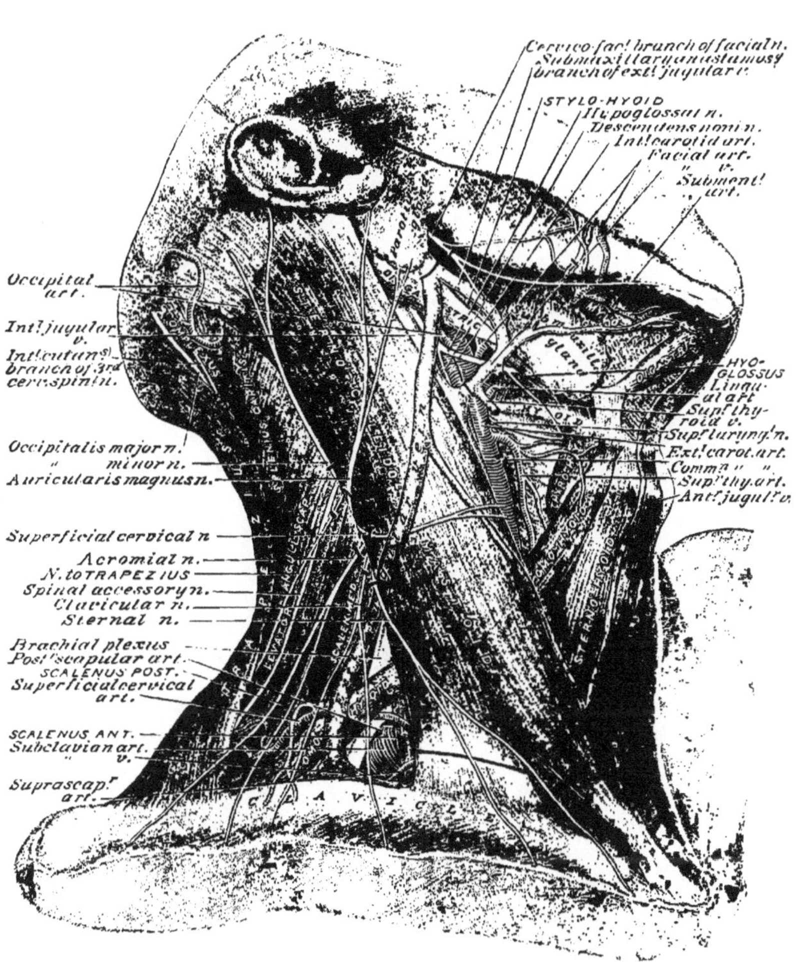
Cervico-fac! branch of facial n.
Submaxillary anastamos!
branch of ext! jugular v.
STYLO-HYOID
Hypoglossal n.
Descendens noni n.
Int! carotid art.
Facial art.
" v.
Subment!
art.
Occipital
art.
Int! jugular
v.
Int! cutan's
branch of 3rd
cerv. spin! n.
HYO-
GLOSSUS
Lingu-
al art
Sup! thy-
roid v.
Sup! laryng! n.
Ext! carot. art.
Comm " "
Sup! thy. art.
Ant! jugul! v.
Occipitalis major n.
" minor n.
Auricularis magnus n.
Superficial cervical n
Acromial n.
N. to TRAPEZIUS
Spinal accessory n.
Clavicular n.
Sternal n.
Brachial plexus
Post! scapular art.
SCALENUS POST.
Superficial cervical
art.
SCALENUS ANT.
Subclavian art.
" v.
Suprascap!
art.
CLAVICLE

N. Cohn, ad naturam del.

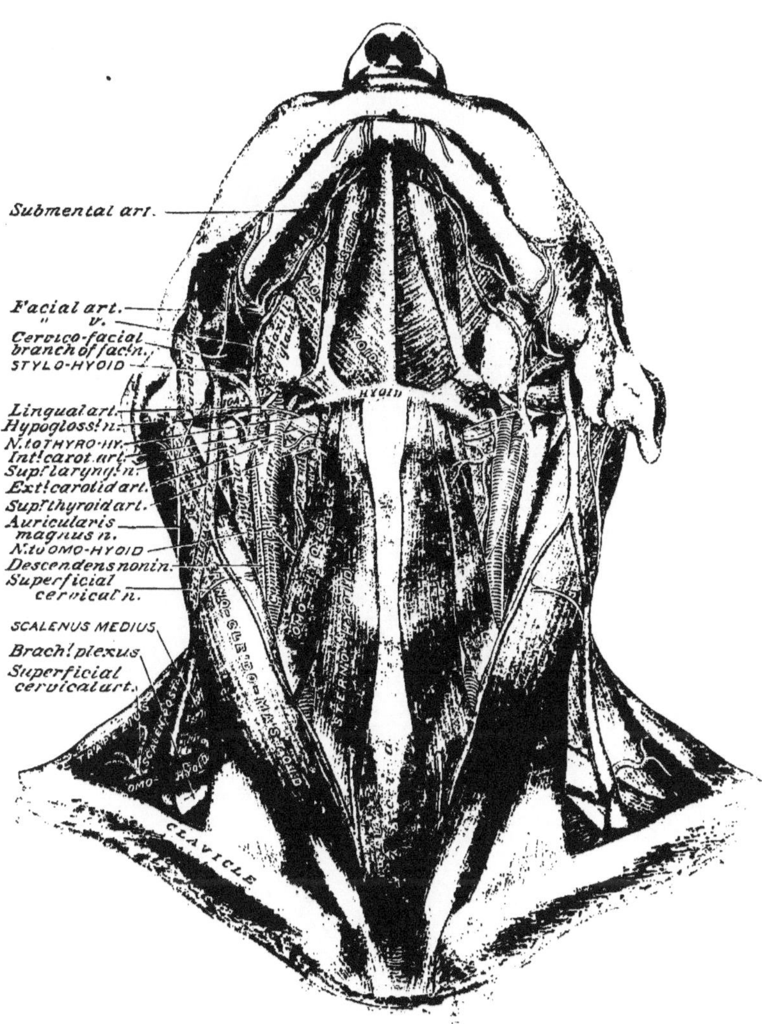
Submental art.
Facial art.
" v.
Cervico-facial branch of fac! n.
STYLO-HYOID
Lingual art.
Hypogloss! n.
N. to THYRO-HY.
Int! carot. art.
Sup! laryng! n.
Ext! carotid art.
Sup! thyroid art.
Auricularis magnus n.
N. to OMO-HYOID
Descendens noni n.
Superficial cervical n.
SCALENUS MEDIUS
Brach! plexus
Superficial cervical art.
HYOID
CLAVICLE

M.Cohn. ad naturam del.

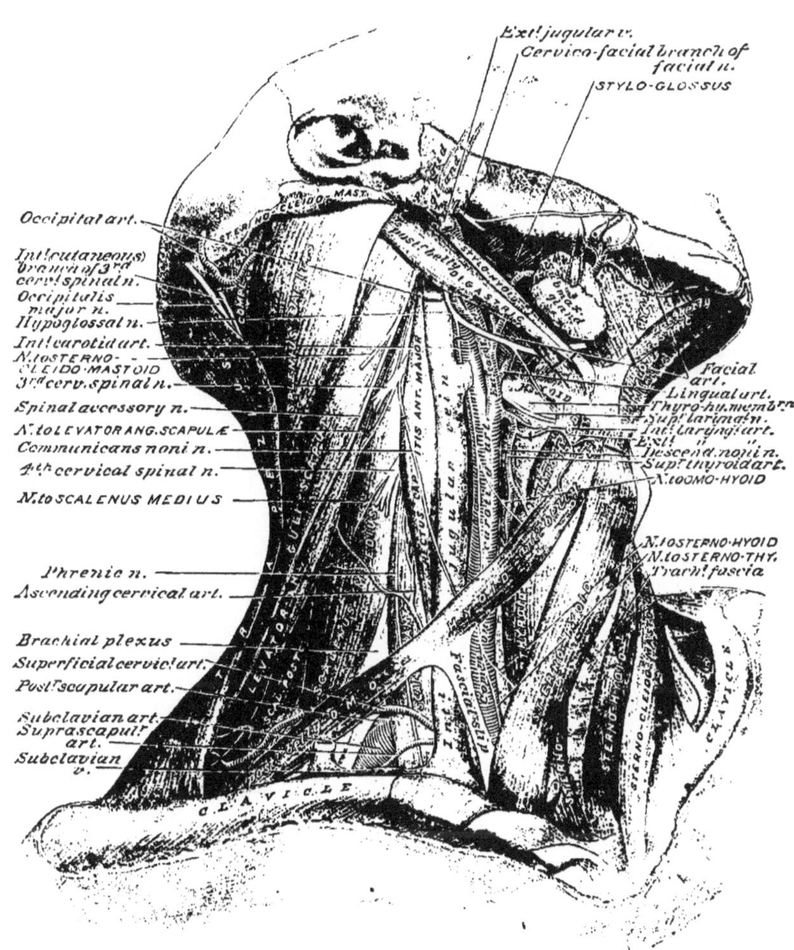
Ext! jugular v.
Cervico-facial branch of facial n.
STYLO-GLOSSUS
Occipital art.
Int! cutaneous) branch of 3rd cerv! spinal n.
Occipitalis major n.
Hypoglossal n.
Int! carotid art.
N. to STERNO-CLEIDO-MASTOID
3rd cerv. spinal n.
Spinal accessory n.
N. to LEVATOR ANG. SCAPULÆ
Communicans noni n.
4th cervical spinal n.
N. to SCALENUS MEDIUS
Phrenic n.
Ascending cervical art.
Brachial plexus
Superficial cervic! art.
Post! scapular art.
Subclavian art.
Suprascapul! art.
Subclavian v.
Facial art.
Lingual art.
Thyro-hy. memb!
Sup! laryng! n.
Int! laryng! art.
Ext! " "
Descend. noni n.
Sup! thyroid art.
N. to OMO-HYOID
N. to STERNO-HYOID
N. to STERNO-THY.
Trach! fascia
CLAVICLE
CLAVICLE

M. Cohn ad natur. del.

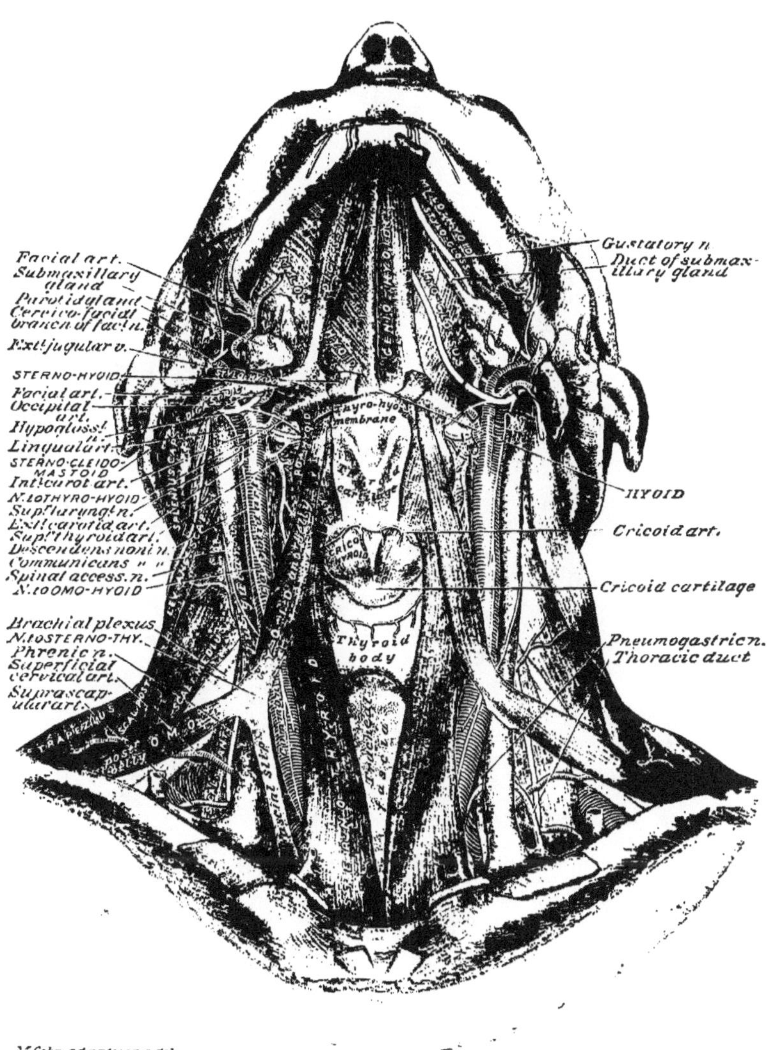
Facial art.
Submaxillary gland
Parotid gland
Cervico-facial branch of fac. n.
Extl jugular v.
STERNO-HYOID
Facial art.
Occipital art.
Hypogloss! n.
Lingual art.
STERNO-CLEIDO-MASTOID
Intl carot art.
N. to THYRO-HYOID
Supl laryngl n.
Extl carotid art.
Supl thyroid art.
Descendens noni n.
Communicans " "
Spinal access. n.
N. to OMO-HYOID
Brachial plexus
N. to STERNO-THY.
Phrenic n.
Superficial cervical art.
Suprascapular art.
Gustatory n.
Duct of submaxillary gland
HYOID
Cricoid art.
Cricoid cartilage
Pneumogastric n.
Thoracic duct
Thyro-hyoid membrane
Thyroid body

M. Cohn, ad naturam del.

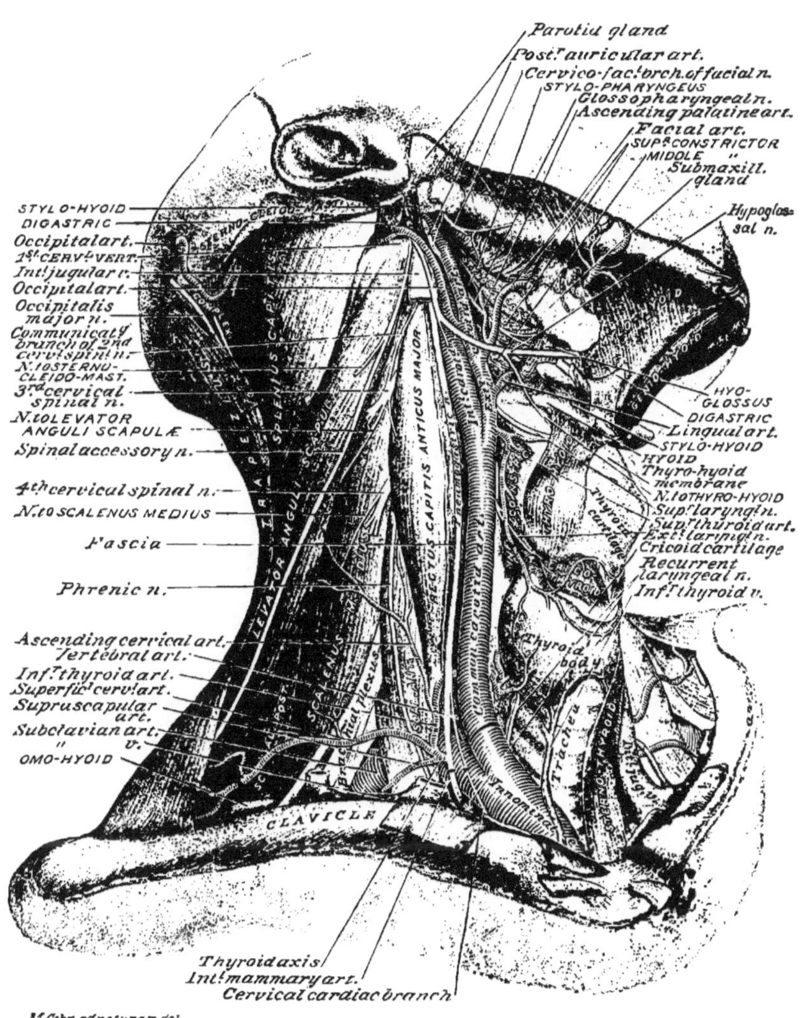
Parotid gland
Post.r auricular art.
Cervico-fac.l brch. of facial n.
STYLO-PHARYNGEUS
Glossopharyngeal n.
Ascending palatine art.
Facial art.
SUP.r CONSTRICTOR
MIDDLE "
Submaxill. gland
Hypoglossal n.
STYLO-HYOID
DIGASTRIC
Occipital art.
1st CERV.l VERT.
Int.l jugular v.
Occipital art.
Occipitalis major n.
Communicat.g branch of 2nd cervi. spin.l n.
N. to STERNO-CLEIDO-MAST.
3rd cervical spinal n.
N. to LEVATOR ANGULI SCAPULÆ
Spinal accessory n.
4th cervical spinal n.
N. to SCALENUS MEDIUS
Fascia
Phrenic n.
Ascending cervical art.
Vertebral art.
Inf.r thyroid art.
Superfic.l cervi.l art.
Suprascapular art.
Subclavian art.
" v.
OMO-HYOID
HYO-GLOSSUS
DIGASTRIC
Lingual art.
STYLO-HYOID
HYOID
Thyro-hyoid membrane
N. to THYRO-HYOID
Sup.r laryng.l n.
Sup.r thyroid art.
Ext.l laryng.l n.
Cricoid cartilage
Recurrent laryngeal n.
Inf.r thyroid v.
TRAPEZIUS
LEVATOR ANGULI SCAPULÆ
SPLENIUS CAPITIS
RECTUS CAPITIS ANTICUS MAJOR
Common carotid art.
Thyroid cartilage
Thyroid body
Trachea
CLAVICLE
Innominate
Thyroid axis
Int.l mammary art.
Cervical cardiac branch

M. Cohn ad naturam del.

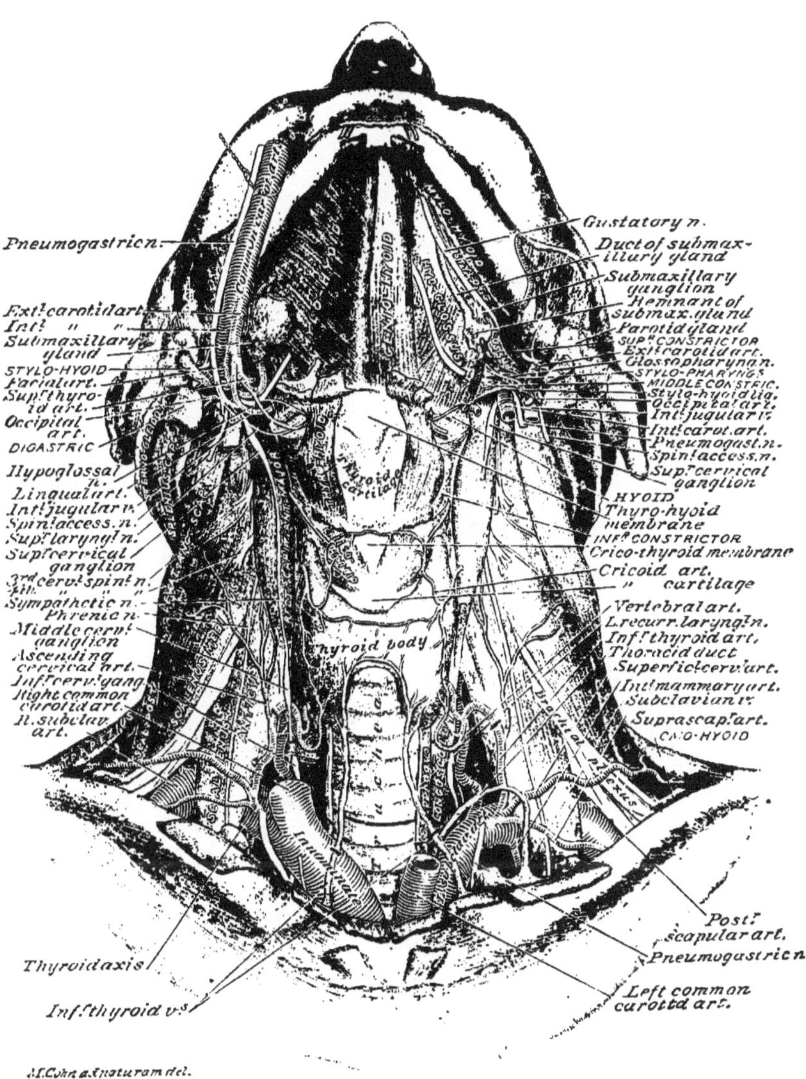
Pneumogastric n.
Extl carotid art.
Intl " "
Submaxillary gland
STYLO-HYOID
Facial art.
Supl thyroid art.
Occipital art.
DIGASTRIC
Hypoglossal n.
Lingual art.
Intl jugular v.
Spinl access. n.
Supl laryngl n.
Supl cervical ganglion
3rd cervl spinl n.
4th " " "
Sympathetic n.
Phrenic n.
Middle cervl ganglion
Ascending cervical art.
Infl cervl gang.
Right common carotid art.
R. subclav. art.
Gustatory n.
Duct of submaxillary gland
Submaxillary ganglion
Remnant of submax. gland
Parotid gland
SUPR CONSTRICTOR
Extl carotid art.
Glossopharyng. n.
STYLO-PHARYNG.
MIDDLE CONSTRIC.
Stylo-hyoid lig.
Occipital art.
Intl jugular v.
Intl carot. art.
Pneumogast. n.
Spinl access. n.
Supr cervical ganglion
HYOID
Thyro-hyoid membrane
INFR CONSTRICTOR
Crico-thyroid membrane
Cricoid art.
" cartilage
Vertebral art.
L. recurr. laryngl n.
Infr thyroid art.
Thoracic duct
Superficl cervl art.
Intl mammary art.
Subclavian v.
Suprascapr art.
OMO-HYOID
Postr scapular art.
Pneumogastric n.
Left common carotid art.
Thyroid axis
Infr thyroid vs
GENIO-HYOID
MYLO-HYOID
Thyroid cartilage
Thyroid body
Trachea

M. Cohn ad naturam del.

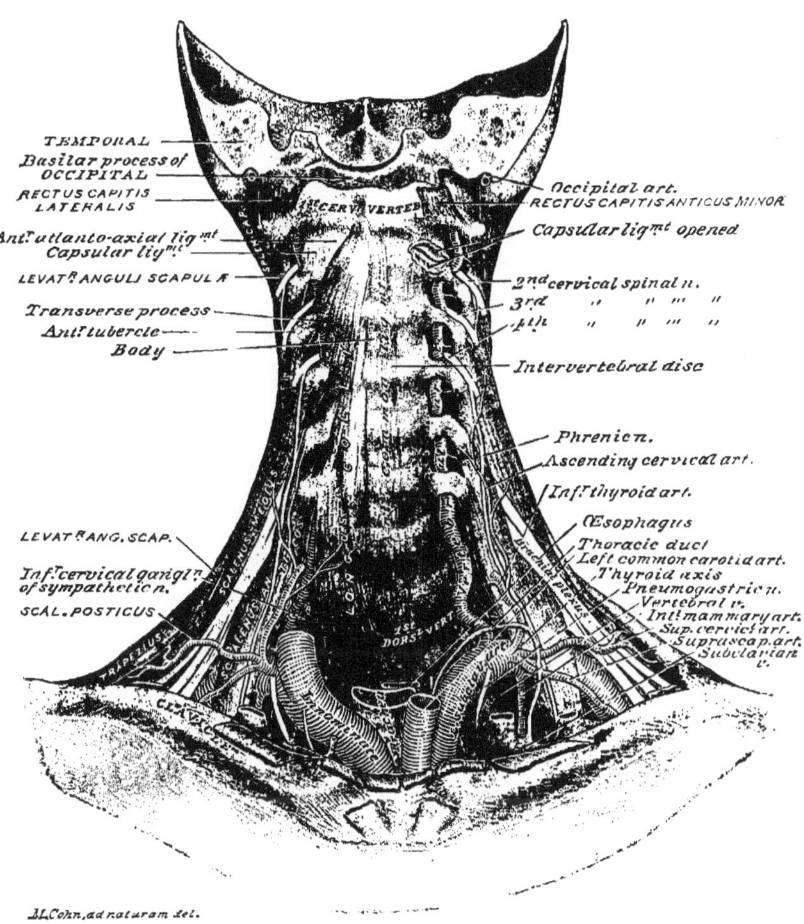
TEMPORAL
Basilar process of OCCIPITAL
RECTUS CAPITIS LATERALIS
Ant. atlanto-axial ligmt
Capsular ligmt
LEVATR ANGULI SCAPULÆ
Transverse process
Ant. tubercle
Body
Occipital art.
RECTUS CAPITIS ANTICUS MINOR
Capsular ligmt opened
2nd cervical spinal n.
3rd " " "
4th " " "
Intervertebral disc
Phrenic n.
Ascending cervical art.
Inf. thyroid art.
Œsophagus
Thoracic duct
Left common carotid art.
Thyroid axis
Pneumogastric n.
Vertebral v.
Int. mammary art.
Sup. cervical art.
Suprascap. art.
Subclavian v.
LEVATR ANG. SCAP.
Inf. cervical gangl. of sympathetic n.
SCAL. POSTICUS
TRAPEZIUS
1st CERV. VERTEB.
1st DORS. VERT.
Brachial plexus

M. Cohn, ad naturam del.

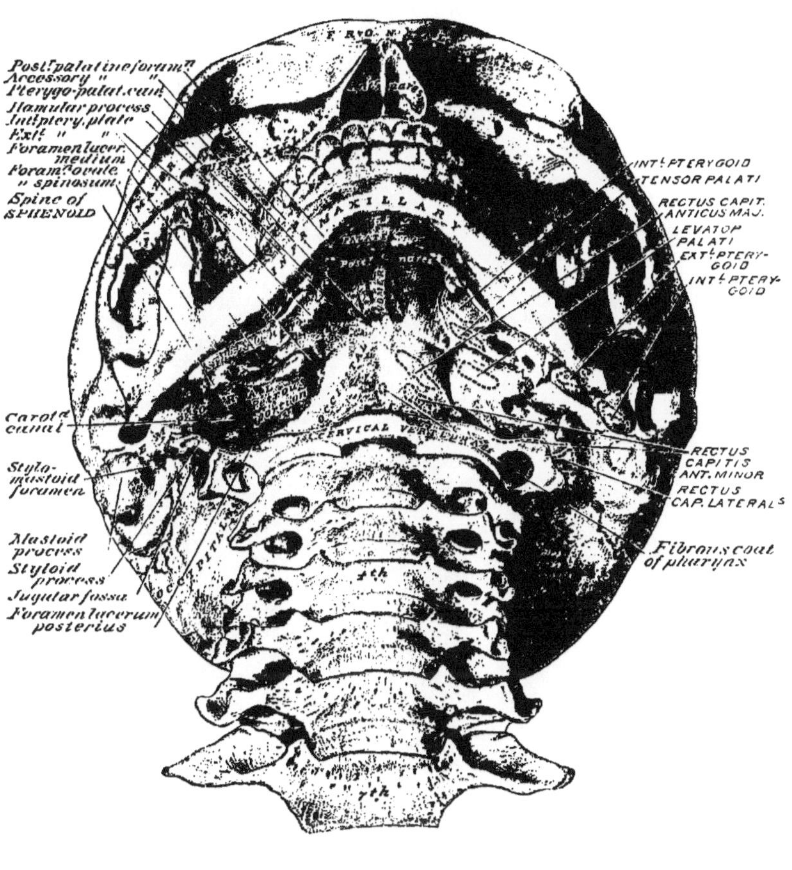
Postr. palatine foram.
Accessory " "
Pterygo-palat. can.
Hamular process
Intl. ptery. plate
Extl. " "
Foramen lacer. medium
Foram. ovale
" spinosum
Spine of SPHENOID
Carotd. canal
Stylo-mastoid foramen
Mastoid process
Styloid process
Jugular fossa
Foramen lacerum posterius
INTL. PTERYGOID
TENSOR PALATI
RECTUS CAPIT. ANTICUS MAJ.
LEVATOR PALATI
EXTL. PTERY-GOID
INTL. PTERY-GOID
RECTUS CAPITIS ANT. MINOR
RECTUS CAP. LATERALS
Fibrous coat of pharynx
MAXILLARY
CERVICAL VERTEBRAE
OCCIPITAL
7th

M. Cohn, ad naturam del.

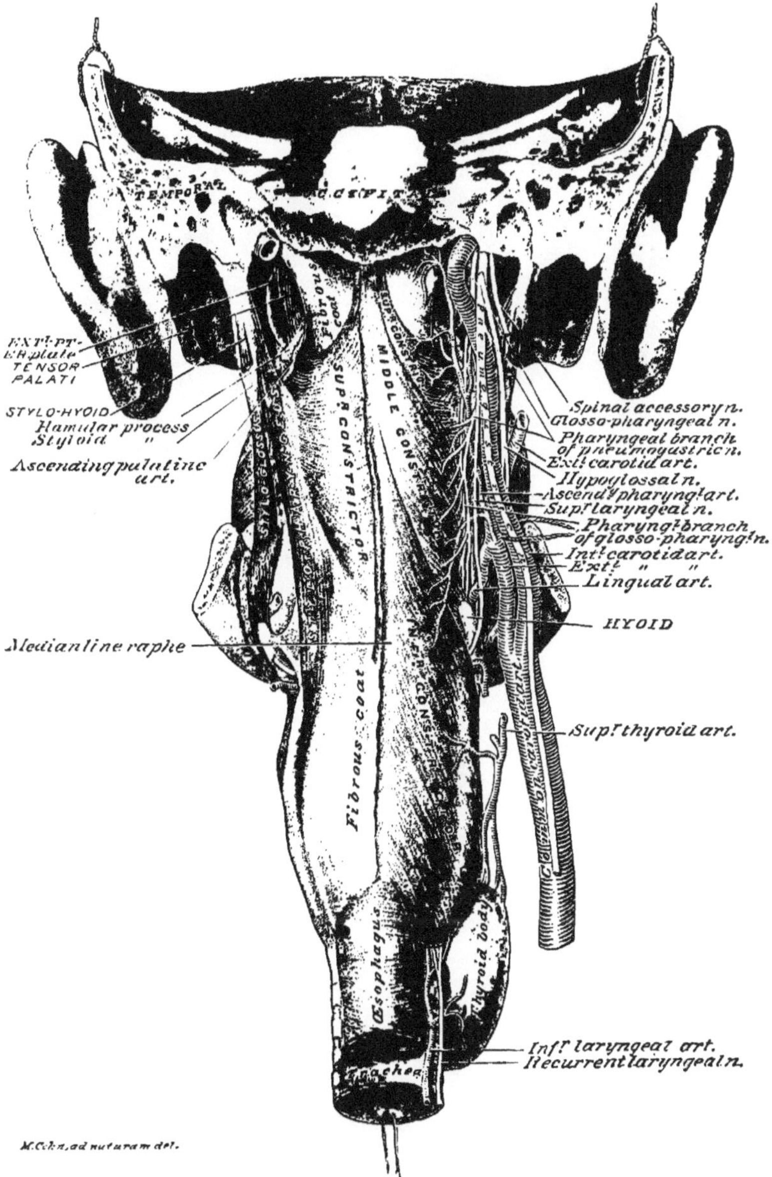
TEMPORAL
OCCIPIT
EXTl. PT-
Kplate
TENSOR
PALATI
STYLO-HYOID
Hamular process
Styloid "
Ascending palatine
art.
SUPR CONSTRICTOR
MIDDLE CONSTR
Spinal accessory n.
Glosso-pharyngeal n.
Pharyngeal branch
of pneumogastric n.
Extl. carotid art.
Hypoglossal n.
Ascendg pharyngl art.
Supr laryngeal n.
Pharyngl branch
of glosso-pharyngl n.
Intl. carotid art.
Extl. " "
Lingual art.
HYOID
Median line raphe
Fibrous coat
Supr thyroid art.
Œsophagus
thyroid body
Infr laryngeal art.
Recurrent laryngeal n.
Trachea
M. Cohn, ad naturam del.

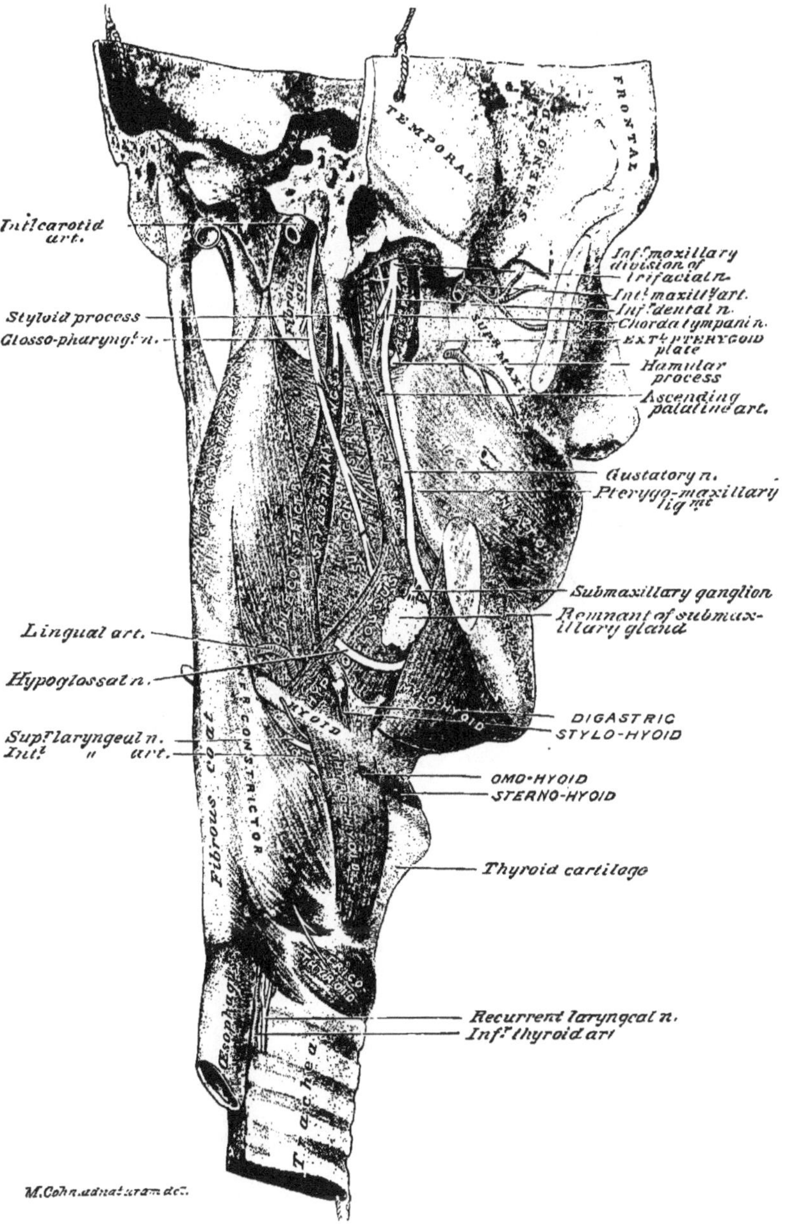
TEMPORAL
SPHENOID
FRONTAL
Intl carotid art.
Infr maxillary division of trifacial n.
Intl maxillry art.
Infr dental n.
Chorda tympani n.
EXTl PTERYGOID plate
Styloid process
Glosso-pharyngl n.
Hamular process
Ascending palatine art.
Gustatory n.
Pterygo-maxillary ligmt
Submaxillary ganglion
Remnant of submaxillary gland
Lingual art.
Hypoglossal n.
DIGASTRIC
STYLO-HYOID
Supr laryngeal n.
Intl " art.
OMO-HYOID
STERNO-HYOID
Fibrous coat
Thyroid cartilage
Recurrent laryngeal n.
Infr thyroid art
Œsophagus
Trachea
M. Cohn ad naturam del.

PLATE 202

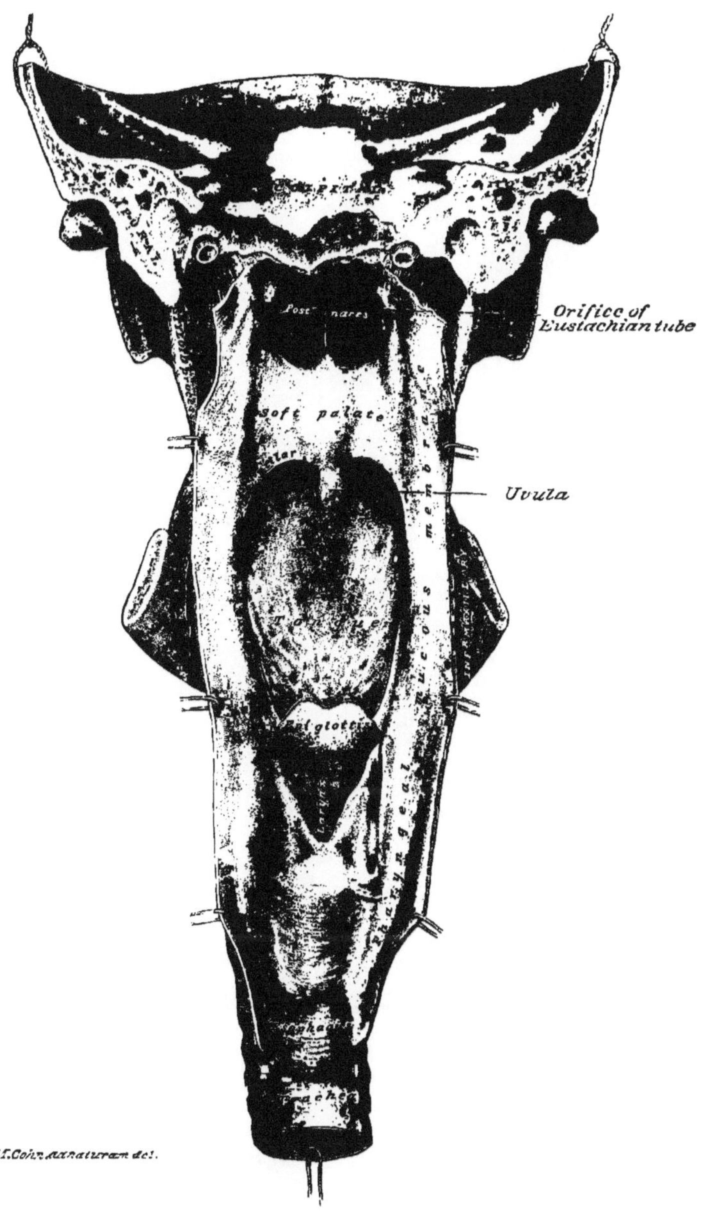
Orifice of
Eustachian tube
Uvula
Post. nares
Soft palate
M.Cohn ad naturam del.

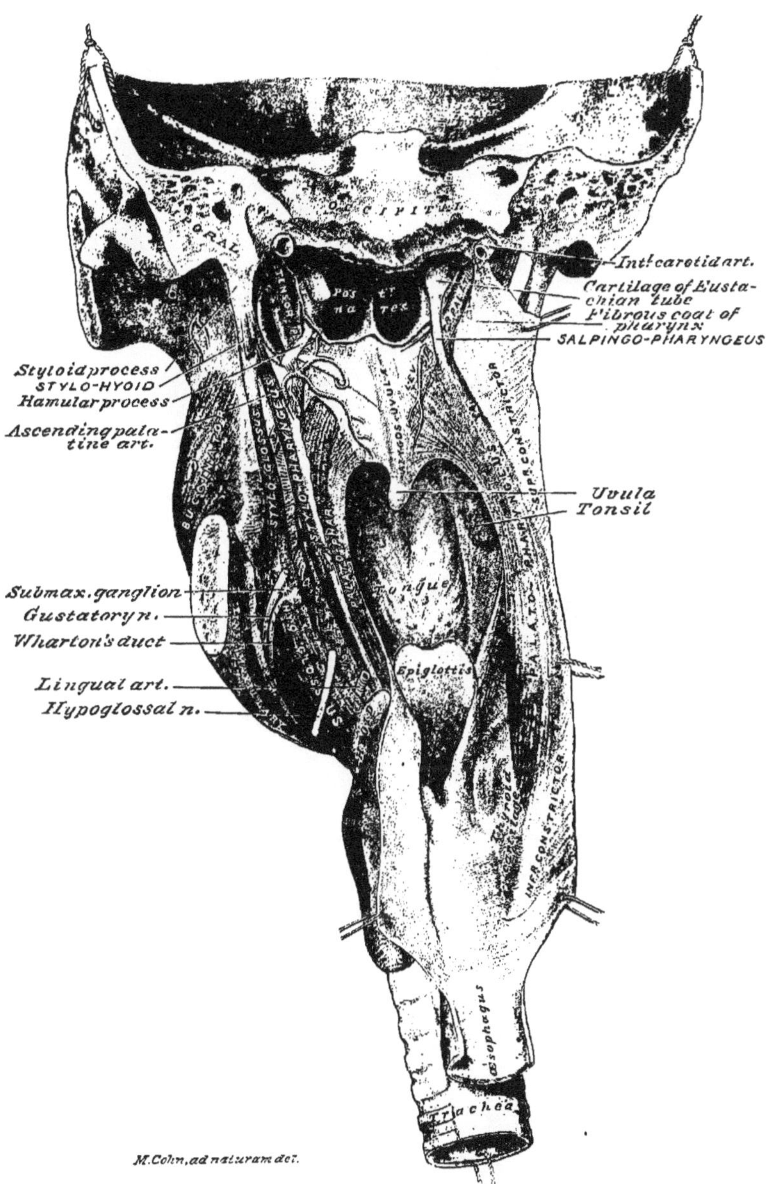
Int! carotid art.
Cartilage of Eustachian tube
Fibrous coat of pharynx
SALPINGO-PHARYNGEUS
Styloid process
STYLO-HYOID
Hamular process
Ascending palatine art.
Uvula
Tonsil
Submax. ganglion
Gustatory n.
Wharton's duct
Lingual art.
Hypoglossal n.
Epiglottis
Esophagus
Trachea
OCCIPITAL
TENSOR
SUP. CONSTRICTOR
INF. CONSTRICTOR

M. Cohn, ad naturam del.

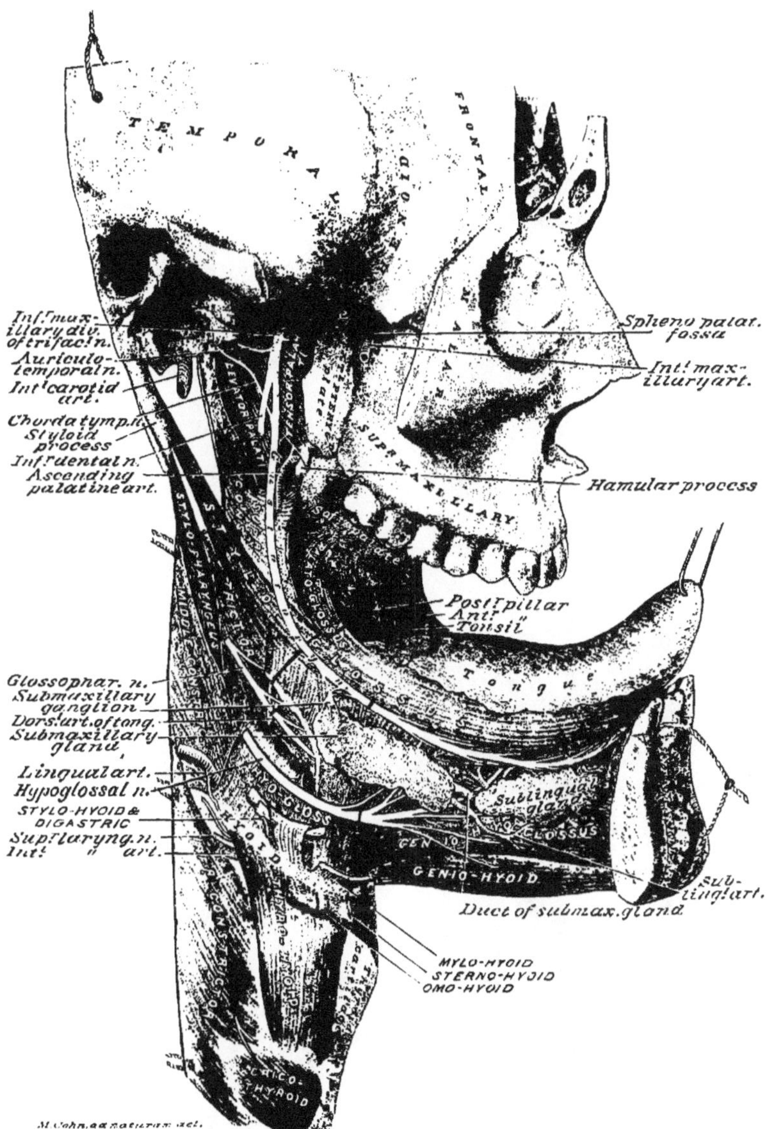
TEMPORAL
FRONTAL
SPHENOID
MALAR
SUPR MAXILLARY
Infr max-illary div. of trifac! n.
Auriculo-temporal n.
Int! carotid art.
Chorda tymp. n.
Styloid process
Infr dental n.
Ascending palatine art.
Spheno palat. fossa
Int! max-illary art.
Hamular process
Soft palate
Post! pillar
Antr "
Tonsil
Tongue
Glossophar. n.
Submaxillary ganglion
Dors! art. of tong.
Submaxillary gland
Lingual art.
Hypoglossal n.
STYLO-HYOID & DIGASTRIC
Sup! laryng. n.
Int! " art.
Sublingual gland
GENIO-HYOID
Sub-ling! art.
Duct of submax. gland
MYLO-HYOID
STERNO-HYOID
OMO-HYOID
CRICO-THYROID
M. Cohn, ad naturam del.

FIG. 1

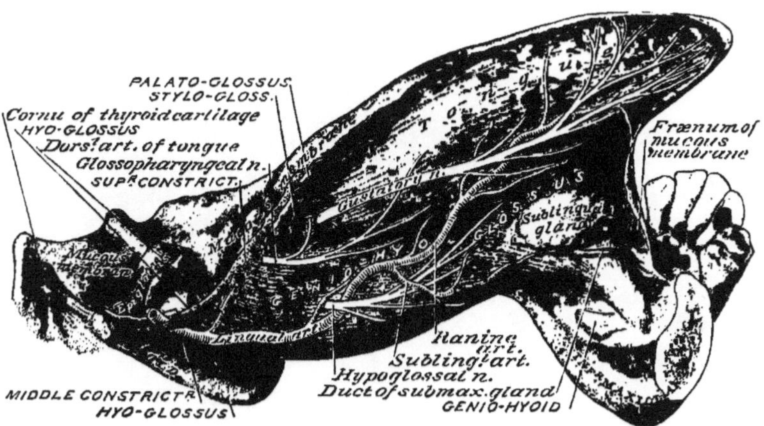
PALATO-GLOSSUS
STYLO-GLOSS.
Cornu of thyroid cartilage
HYO-GLOSSUS
Dors! art. of tongue
Glossopharyngeal n.
SUPR CONSTRICT.
Frænum of mucous membrane
Ranine art.
Subling! art.
Hypoglossal n.
Duct of submax. gland
GENIO-HYOID
MIDDLE CONSTRICTR
HYO-GLOSSUS

FIG. 2

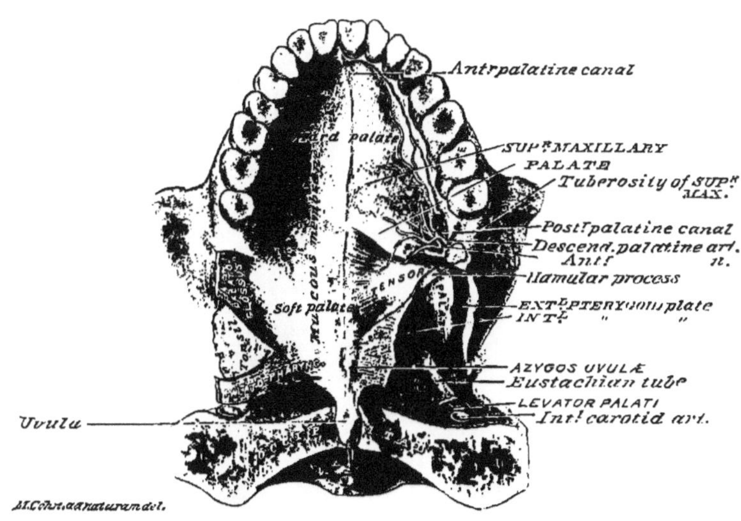
Antr palatine canal
SUPR MAXILLARY
PALATE
Tuberosity of SUPR MAX.
Post! palatine canal
Descend. palatine art.
Ant! n.
Hamular process
EXTL PTERYGOID plate
INTL " "
AZYGOS UVULÆ
Eustachian tube
LEVATOR PALATI
Int! carotid art.
Uvula
Soft palate
Hard palate
MUCOUS MEMBRANE
TENSOR

M. Cohn, ad naturam del.

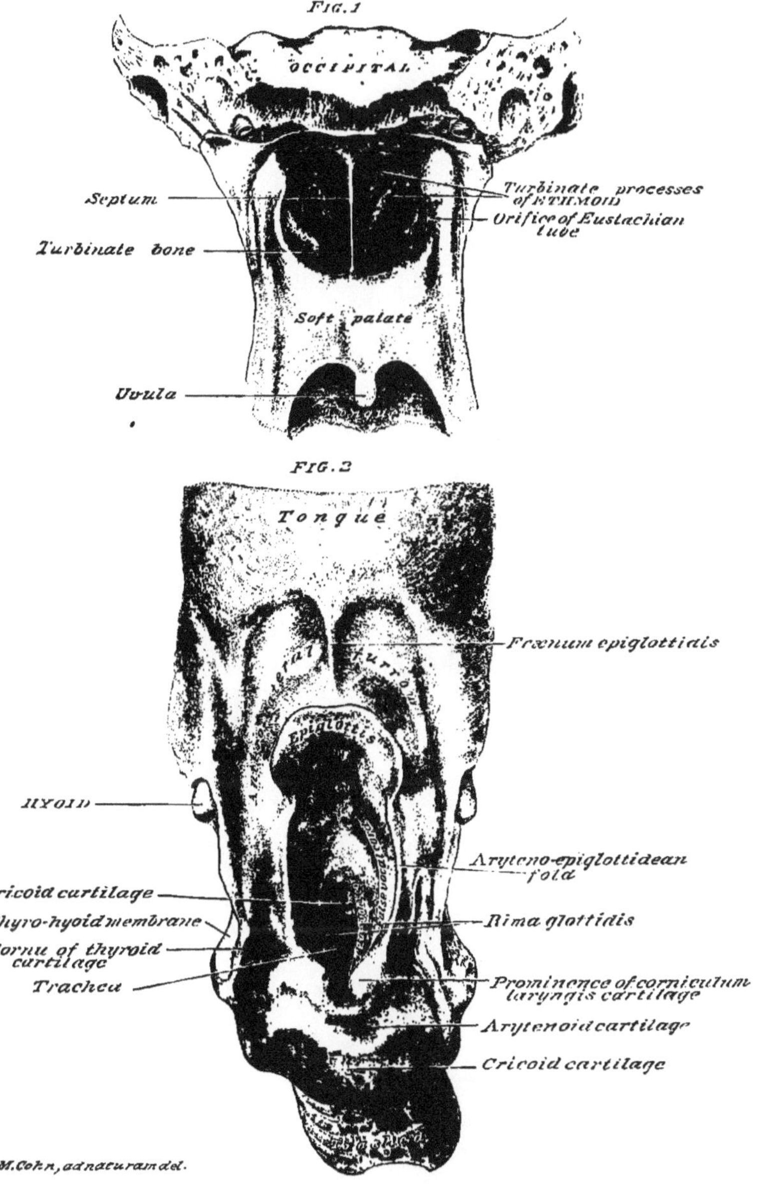
FIG. 1
OCCIPITAL
Septum
Turbinate bone
Turbinate processes of ETHMOID
Orifice of Eustachian tube
Soft palate
Uvula
FIG. 2
Tongue
Fræenum epiglottidis
Epiglottis
HYOID
Aryteno-epiglottidean fold
Cricoid cartilage
Thyro-hyoid membrane
Cornu of thyroid cartilage
Trachea
Rima glottidis
Prominence of corniculum laryngis cartilage
Arytenoid cartilage
Cricoid cartilage

M. Cohn, ad naturam del.

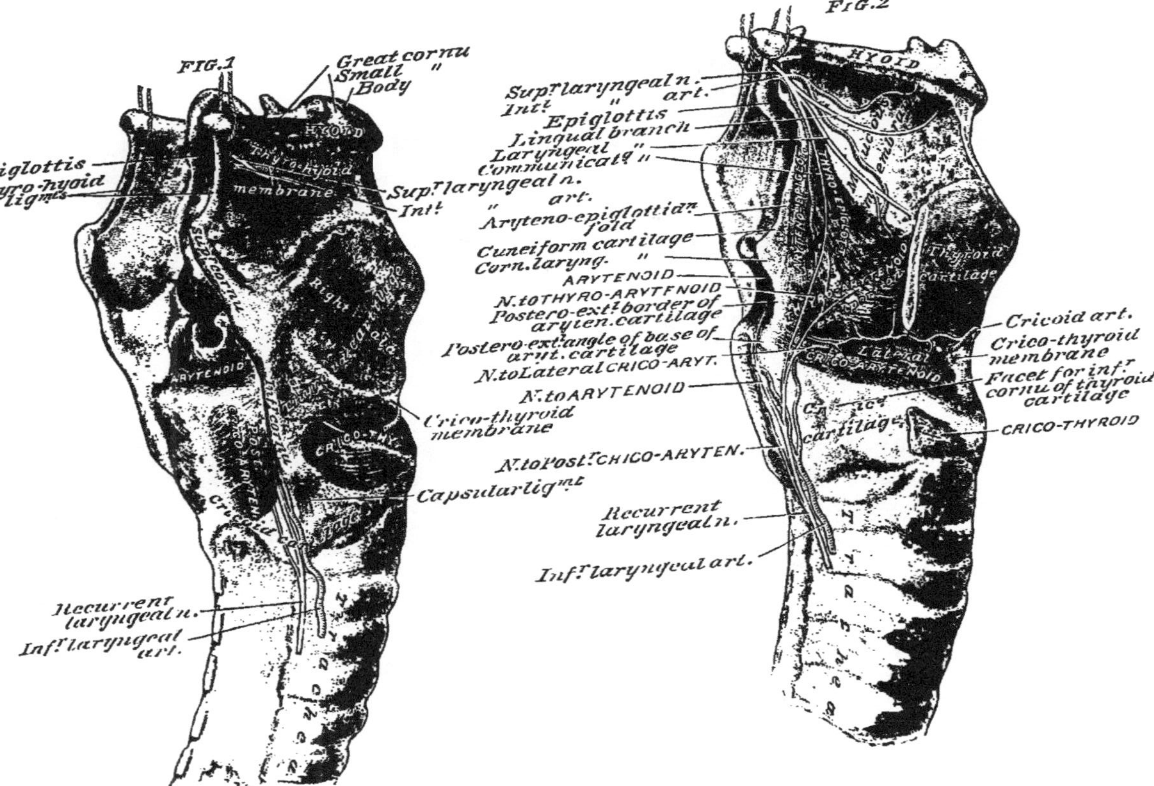
FIG.1
Great cornu
Small "
Body
HYOID
Thyro-hyoid
membrane
Epiglottis
Thyro-hyoid ligmts
Supr laryngeal n.
Intl
ARYTENOID
Crico-thyroid membrane
CRICO-THY.
Capsularligmt
Recurrent laryngeal n.
Infr laryngeal art.
M.Cohn, ad naturam del.
FIG.2
HYOID
Supr laryngeal n.
Intl " art.
Epiglottis
Lingual branch
Laryngeal "
Communicatg "
Supr laryngeal n.
" art.
Aryteno-epiglottidn fold
Cuneiform cartilage
Corn. laryng. "
ARYTENOID
N. to THYRO-ARYTENOID
Postero-extl border of aryten. cartilage
Postero-ext. angle of base of aryt. cartilage
N. to Lateral CRICO-ARYT.
N. to ARYTENOID
N. to Postr CRICO-ARYTEN.
Recurrent laryngeal n.
Infr laryngeal art.
Thyroid cartilage
Cricoid art.
Crico-thyroid membrane
Facet for infr cornu of thyroid cartilage
CRICO-THYROID
THYRO-EPIGLOTT.
Lateral CRICO-ARYTENOID
cartilage

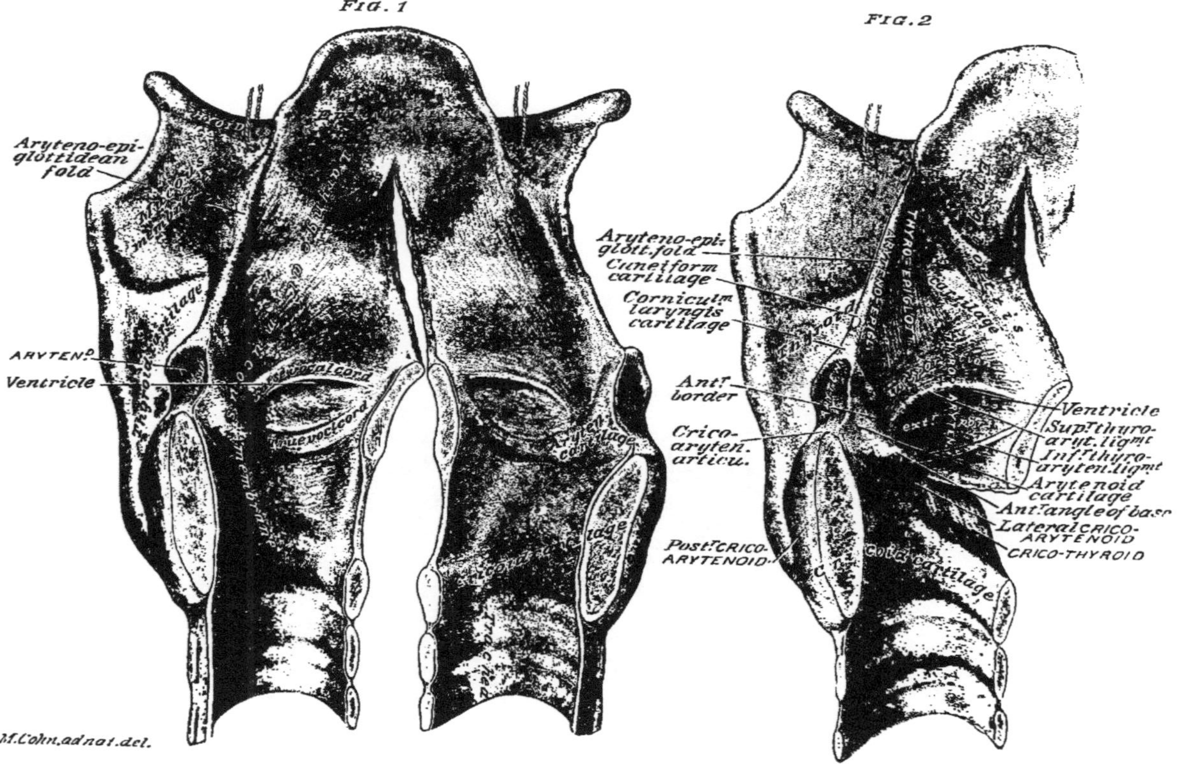
FIG. 1
Aryteno-epi-
glottidean
fold
ARYTEN.D
Ventricle
M.Cohn,ad nat.del.
FIG. 2
Aryteno-epi-
glott.fold
Cuneiform
cartilage
Corniculm
laryngis
cartilage
Antr.
border
Crico-
aryten.
articu.
Postr.CRICO-
ARYTENOID
Ventricle
Supr.thyro-
aryt.ligmt
Infr.thyro-
aryten.ligmt
Arytenoid
cartilage
Antr.angle of base
Lateral CRICO-
ARYTENOID
CRICO-THYROID
THYRO-EPIGLOTT
ARYTENO-EPIGL
Cricoid cartilage

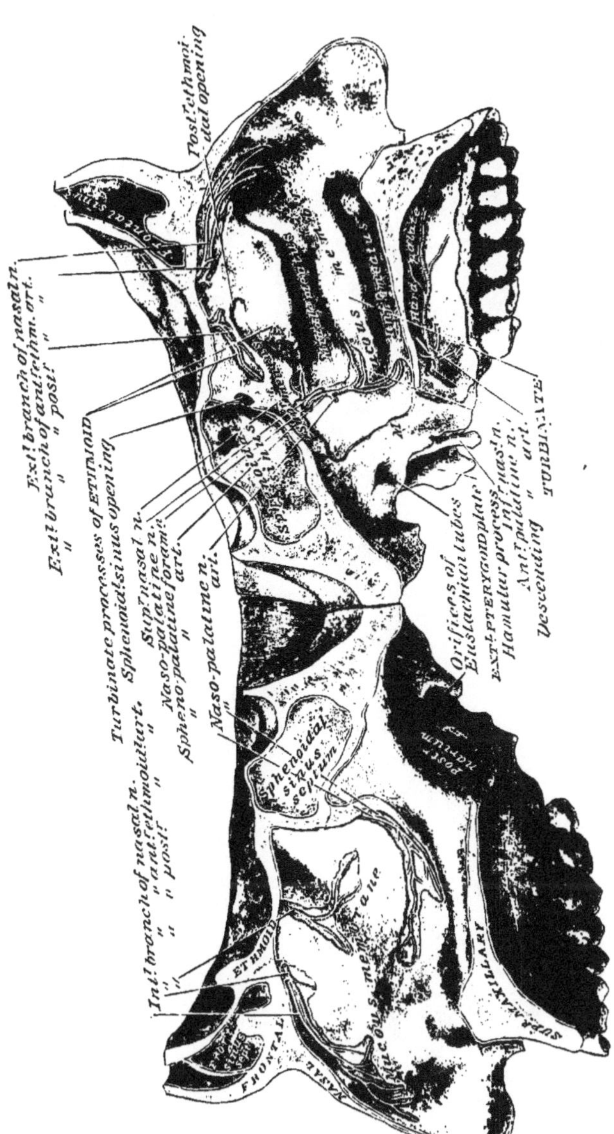
Ext'l branch of nasal n.
Ext'l branch of ant'r ethm. art.
Post'r ethmoidal opening
Turbinate processes of ETHMOID
Sphenoidal sinus opening
Sup'r nasal n.
Naso-palatine n.
Spheno-palatine foram'n
Naso-palatine n.
art.
Int'l branch of nasal n.
Orifices of Eustachian tubes
EXT'L PTERYGOID plate
Hamular process
Ant'r palatine n.
Descending
TURBINATE
Sphenoidal sinus septum
Mucous membrane
FRONTAL
ETHMOID
NASAL
SUP'R MAXILLARY
post nares

M. Cohn. ad naturam del.

FIG. 1

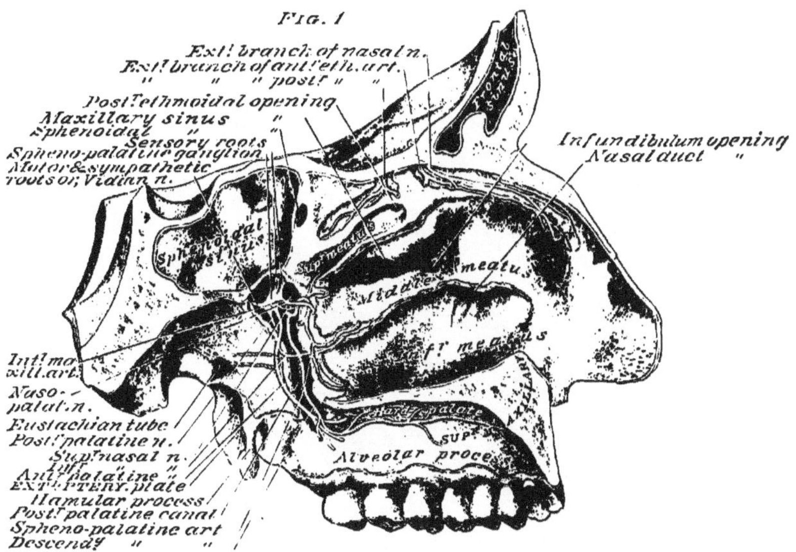
Ext! branch of nasal n.
Ext! branch of ant! eth. art.
" " " post! " "
Post! ethmoidal opening
Maxillary sinus "
Sphenoidal " "
Sensory roots
Spheno-palatine ganglion
Motor & sympathetic
roots or, Vidian n.
Frontal sinus
Sphenoidal sinus
Sup. meatus
Middle meatus
Infundibulum opening
Nasal duct "
Int! max. ill. art.
Naso-palat. n.
Eustachian tube
Post! palatine n.
Sup! nasal n.
Inf! " "
Ant! palatine "
EXT! PTERY. plate
Hamular process
Post! palatine canal
Spheno-palatine art
Descend! " "
Hard palate
SUP
Alveolar proce

FIG. 2

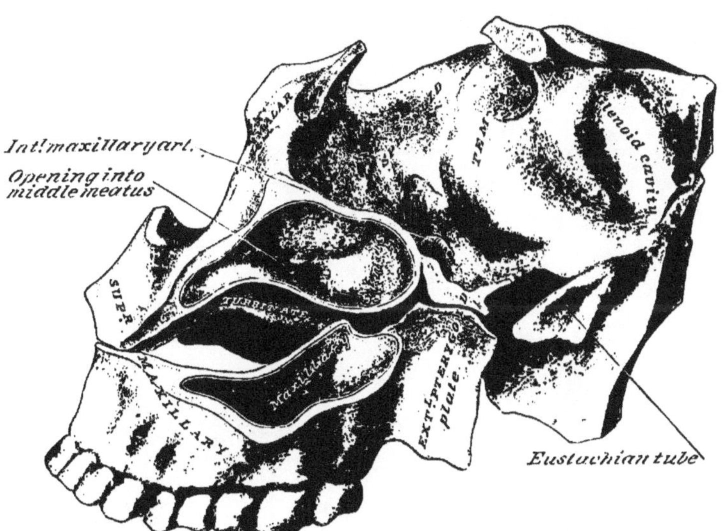
Int! maxillary art.
Opening into middle meatus
MALAR
TEM
Glenoid cavity
SUP.R
TURBINATE
MAXILLARY
EXT! PTERY plate
Eustachian tube

M. Cohn, ad naturam del.

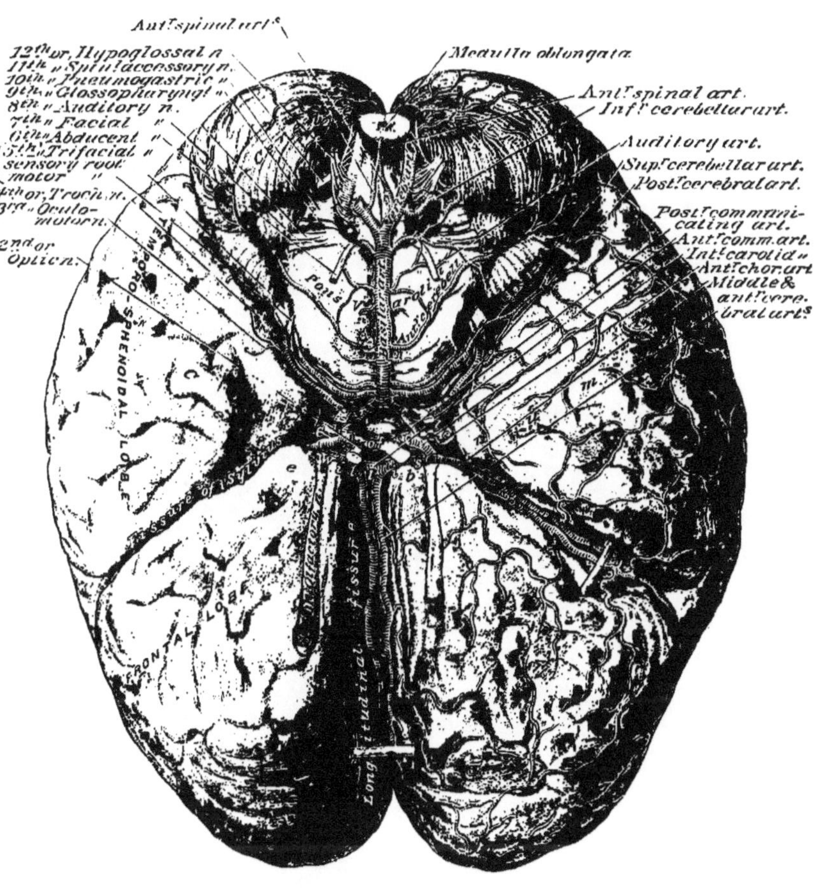
Antr spinal artˢ
12th or, Hypoglossal n.
11th " Spinal accessory n.
10th " Pneumogastric "
9th " Glossopharyngl "
8th " Auditory n.
7th " Facial "
6th " Abducent "
5th " Trifacial "
sensory root
motor "
4th or, Trochl n.
3rd " Oculo-motor n.
2nd or Optic n.
Medulla oblongata.
Antr spinal art.
Infr cerebellar art.
Auditory art.
Supr cerebellar art.
Postr cerebral art.
Postr communicating art.
Antr comm. art.
Intl carotid "
Antr chor. art
Middle & antr cerebral artˢ
Pons
TEMPORO-SPHENOIDAL LOBE
Fissure of Sylvius
FRONTAL LOBE
Longitudinal fissure

M. Cohn. ad naturam del.

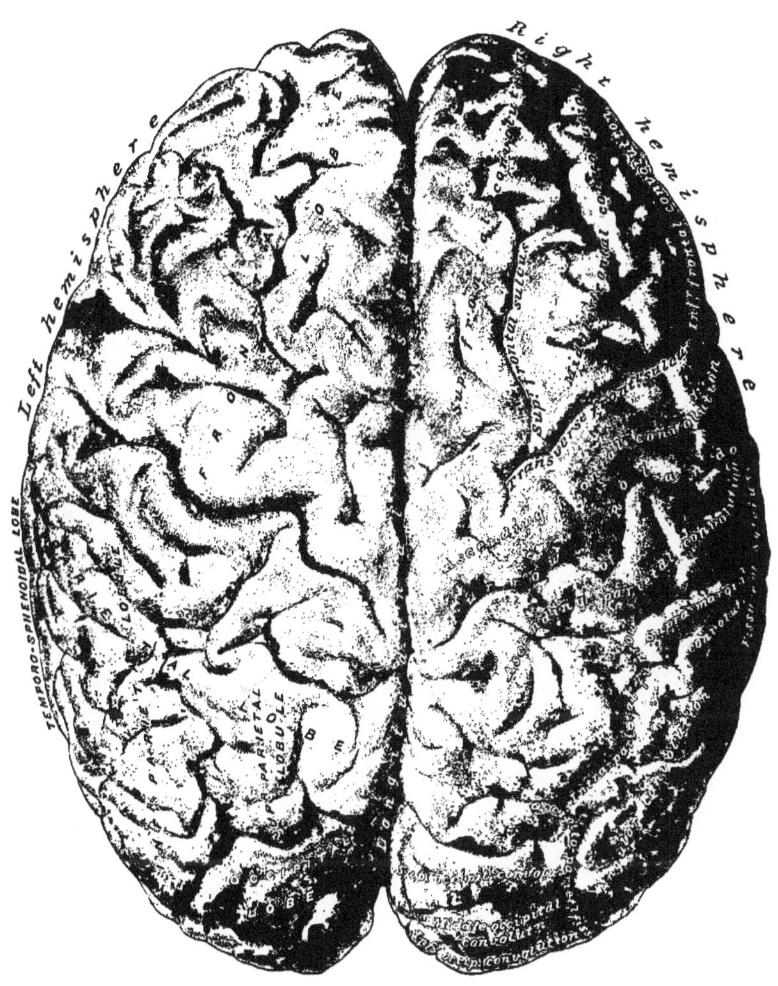
Left hemisphere
Right hemisphere
TEMPORO-SPHENOIDAL LOBE
Sup. frontal sulcus
Transverse frontal sulcus
Middle occipital convolutn

M. Cohn, ad naturam del.

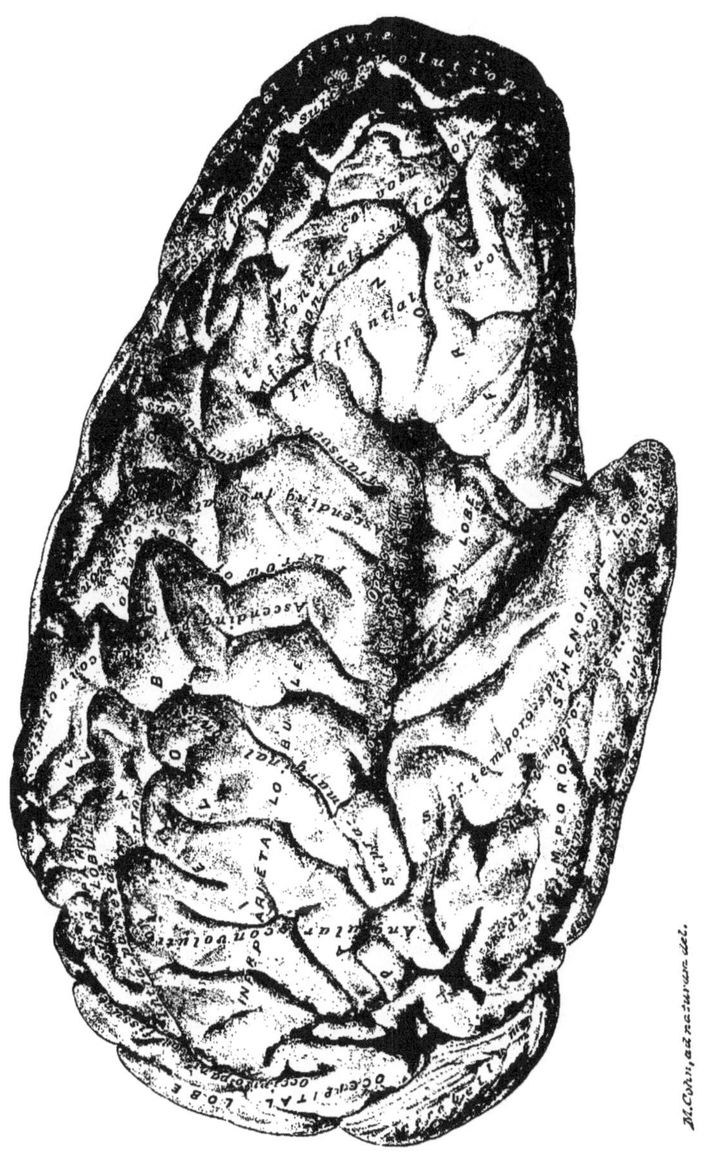
Infr. frontal convolution
Angular convolution
OCCIPITAL LOBE
Supr. temporo-sphenoidal

M. Cohn, ad naturam del.

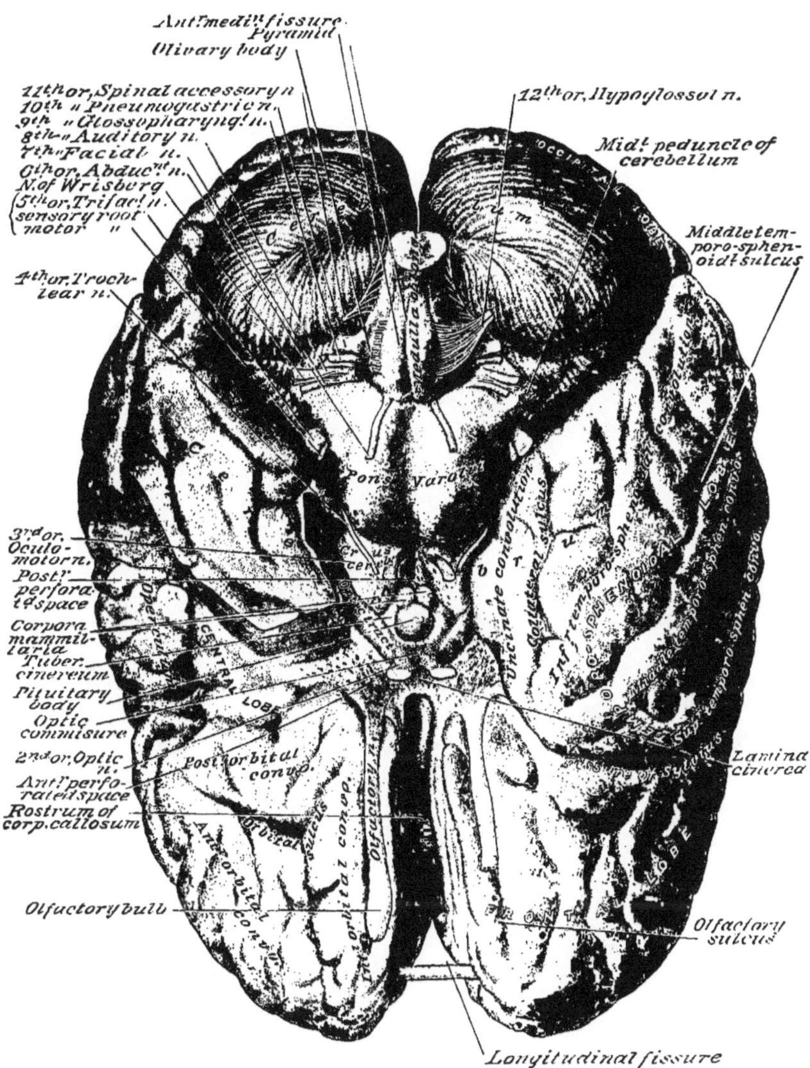

Ant! medin fissure
Pyramid
Olivary body
11th or, Spinal accessory n.
10th " Pneumogastric n.
9th " Glossopharyng! n.
8th " Auditory n.
7th " Facial n.
6th or, Abduc.nt n.
N. of Wrisberg
5th or, Trifac! n. sensory root
motor "
4th or, Trochlear n.
3rd or, Oculo-motor n.
Post.r perforat.d space
Corpora mammillaria
Tuber cinereum
Pituitary body
Optic commisure
2nd or, Optic n.
Ant.r perforated space
Rostrum of corp. callosum
Olfactory bulb
12th or, Hypoglossal n.
Mid! peduncle of cerebellum
Middle temporo-sphenoid! sulcus
Lamina cinerea
Olfactory sulcus
Longitudinal fissure
Pons Varolii
Medulla oblongata
Crus cerebri
Uncinate convolution
Collateral sulcus
Central lobe
Post. orbital convo.
Ant. orbital convo.
Orbital sulcus
Olfactory n.
Fissure of Sylvius

M. Cohn, ad naturam del.

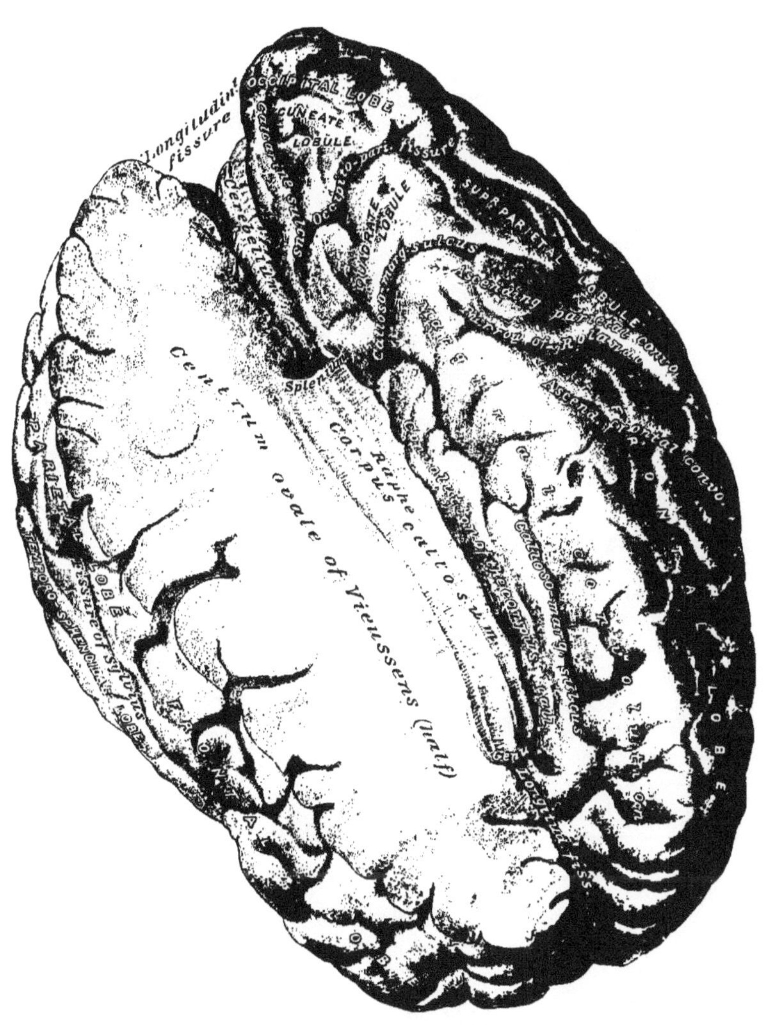
Longitudinal fissure
OCCIPITAL LOBE
CUNEATE LOBULE
Occipito-par. fissure
SUPRAPARIETAL LOBULE
Ascending parietal convo.
Fissure of Rolando
Ascend. frontal convo.
Splenium
Calloso-marg. sulcus
Centrum ovale of Vieussens (half)
Corpus callosum
Raphe
FRONTAL LOBE
Longitud. fiss.
Fissure of Sylvius
TEMPORO-SPHENOIDAL LOBE
PARIETAL LOBE

M. Cohn, ad naturam del.

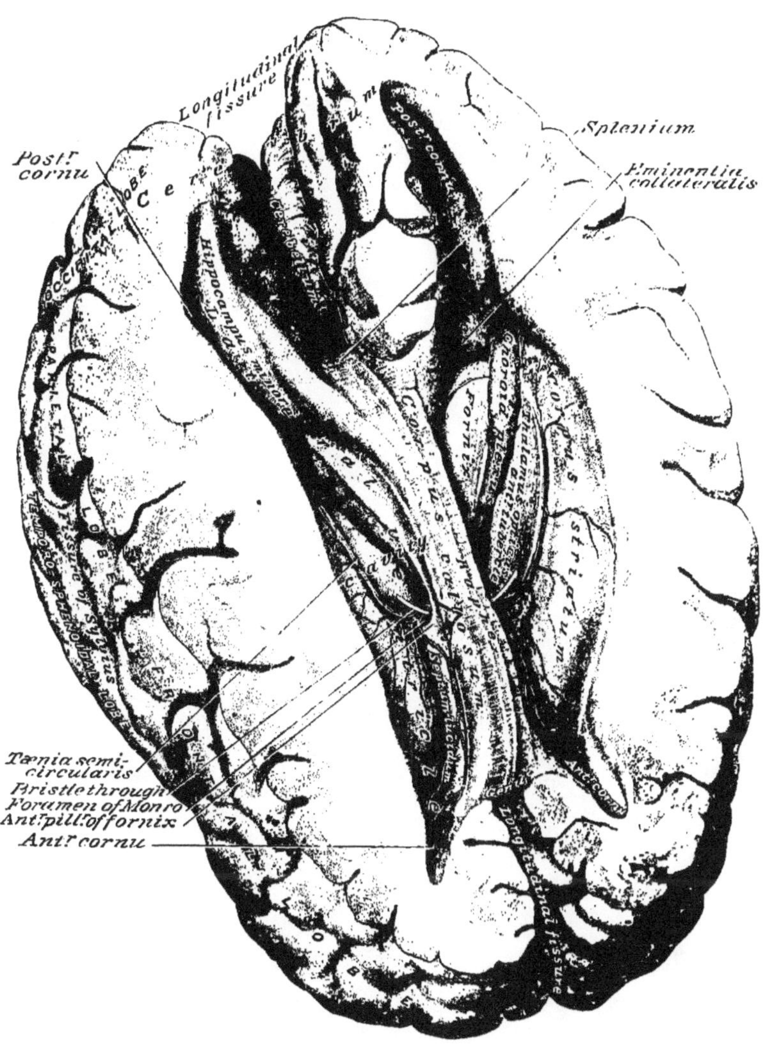
Longitudinal fissure
Splenium
Post.r cornu
Eminentia collateralis
Taenia semi-circularis
Bristle through
Foramen of Monro
Ant.r pill.r of fornix
Ant.r cornu
OCCIPITAL LOBE
Hippocampus minor
Corpus callosum
Fornix
Corpus striatum
Septum lucidum
Ant.r cornu
Longitudinal fissure
Post.r cornu
Fissure of Sylvius
TEMPORO-SPHENOIDAL LOBE

M. Cohn ad naturam del.

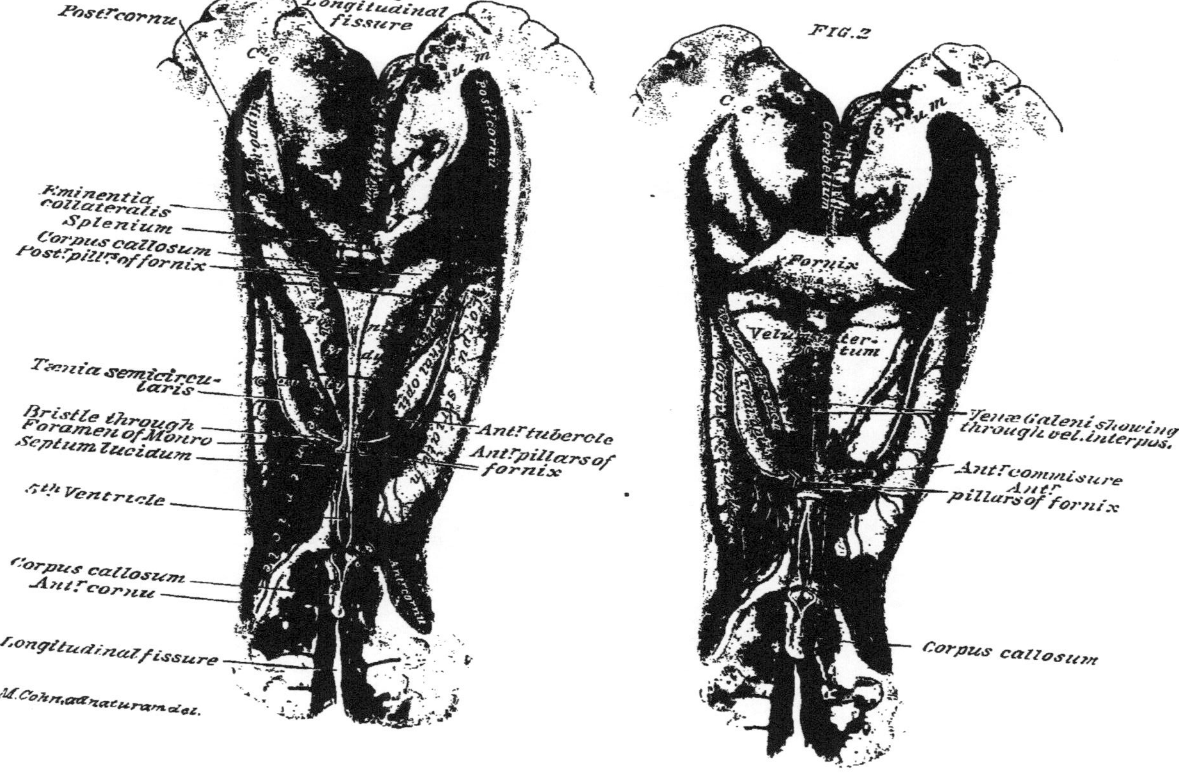

FIG. 1
Longitudinal fissure
Post. cornu
Eminentia collateralis
Splenium
Corpus callosum
Post. pill. of fornix
Tænia semicircularis
Bristle through Foramen of Monro
Septum lucidum
5th Ventricle
Corpus callosum
Ant. cornu
Longitudinal fissure
Ant. tubercle
Ant. pillars of fornix
M. Cohn, ad naturam del.
FIG. 2
Cerebellum
Fornix
Velum interpositum
Venæ Galeni showing through vel. interpos.
Ant. commisure
Ant. pillars of fornix
Corpus callosum

PLATE 217.

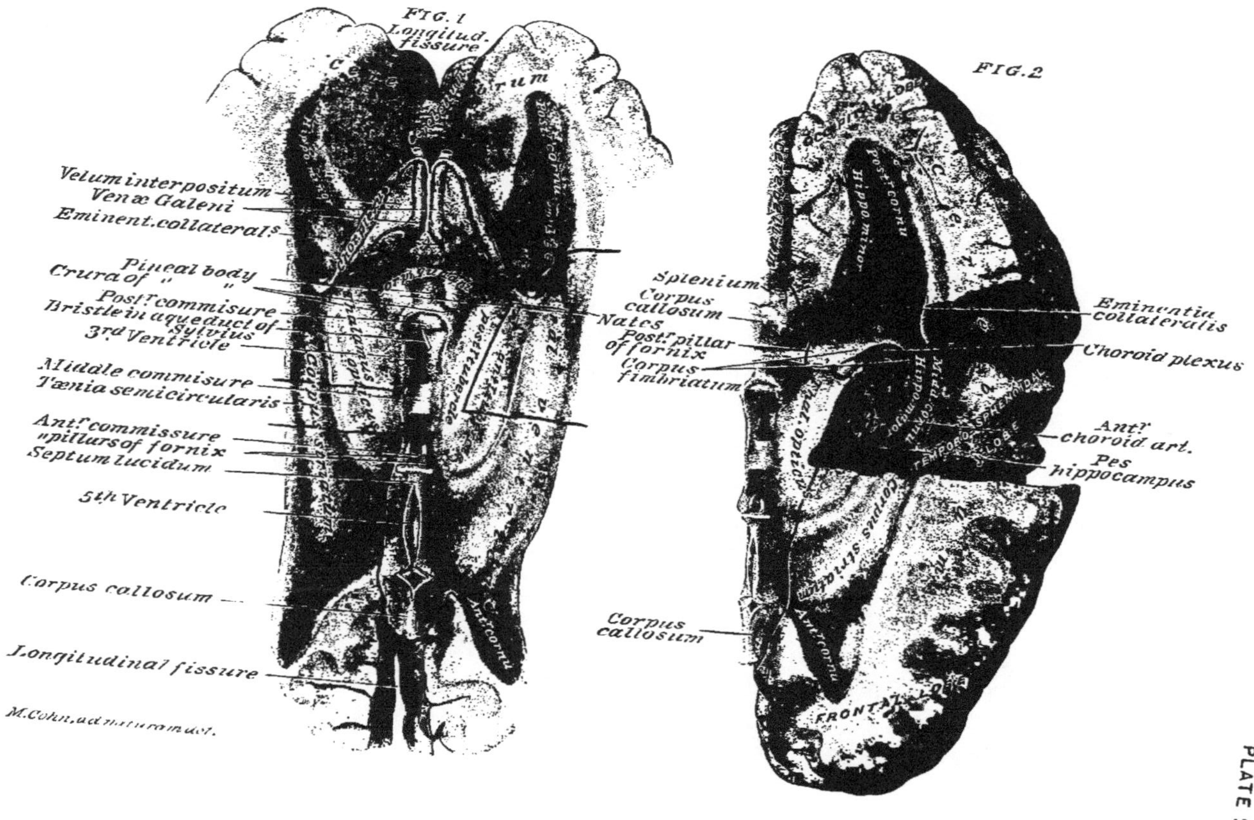
FIG. 1
Longitud. fissure
Velum interpositum
Venæ Galeni
Eminent. collateral.s
Pineal body
Crura of " "
Postr. commisure
Bristle in aqueduct of Sylvius
3rd Ventricle
Middle commisure
Tænia semicircularis
Antr. commissure
" pillars of fornix
Septum lucidum
5th Ventricle
Corpus callosum
Longitudinal fissure
Post. cornu
Ant. cornu
Thalamus opticus
Corpus striatum
Hippo. minor
Post. tubercle
Nates
FIG. 2
Splenium
Corpus callosum
Postr. pillar of fornix
Corpus fimbriatum
Corpus callosum
Eminentia collateralis
Choroid plexus
Antr. choroid art.
Pes hippocampus
Post. cornu
Hippo. minor
Midd. cornu
Hippo. major
Thal. optica
Corpus striatum
Ant. cornu
TEMPORO-SPHENOIDAL LOBE
FRONTAL LOBE
M. Cohn, ad naturam del.

PLATE 219.

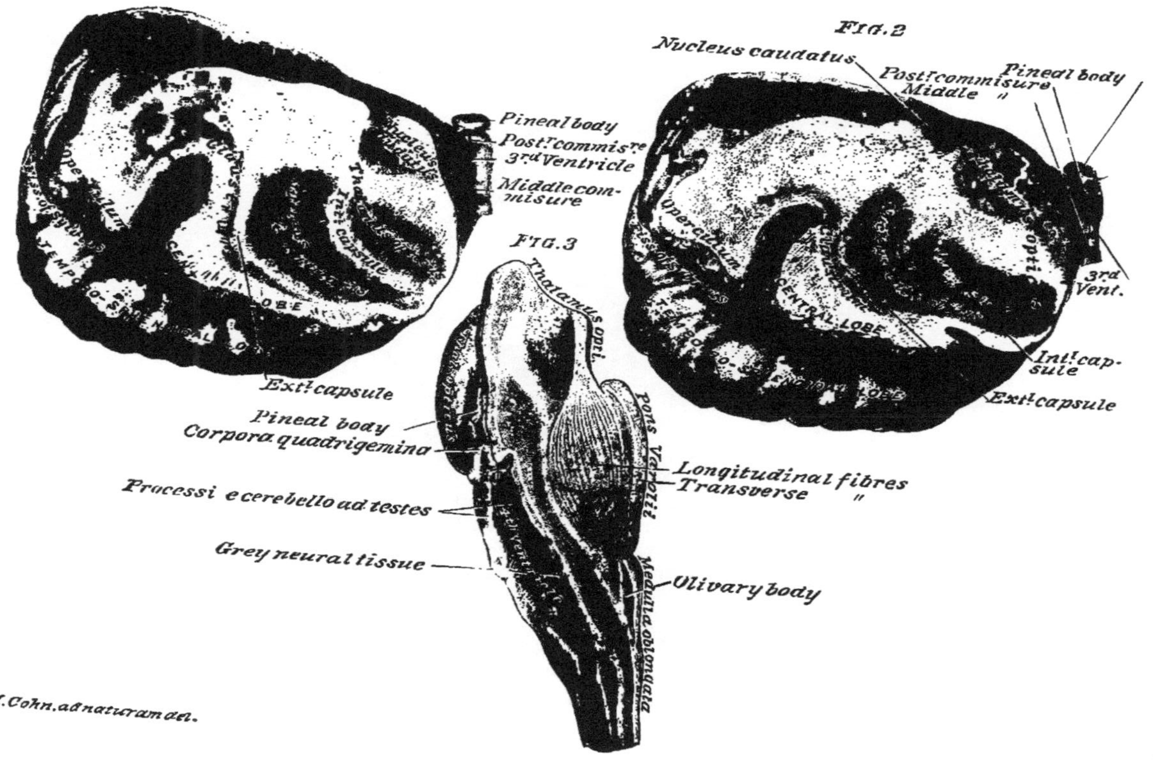

FIG. 1
Pineal body
Post. commisre
3rd Ventricle
Middle commisure
Ext. capsule
FIG. 2
Nucleus caudatus
Post. commisure
Middle "
Pineal body
3rd Vent.
Int. capsule
Ext. capsule
CENTRAL LOBE
FIG. 3
Thalamus opti.
Pons Varolii
Pineal body
Corpora quadrigemina
Longitudinal fibres
Transverse "
Processi e cerebello ad testes
Grey neural tissue
Olivary body
Medulla oblongata
M. Cohn. ad naturam del.

FIG. 1

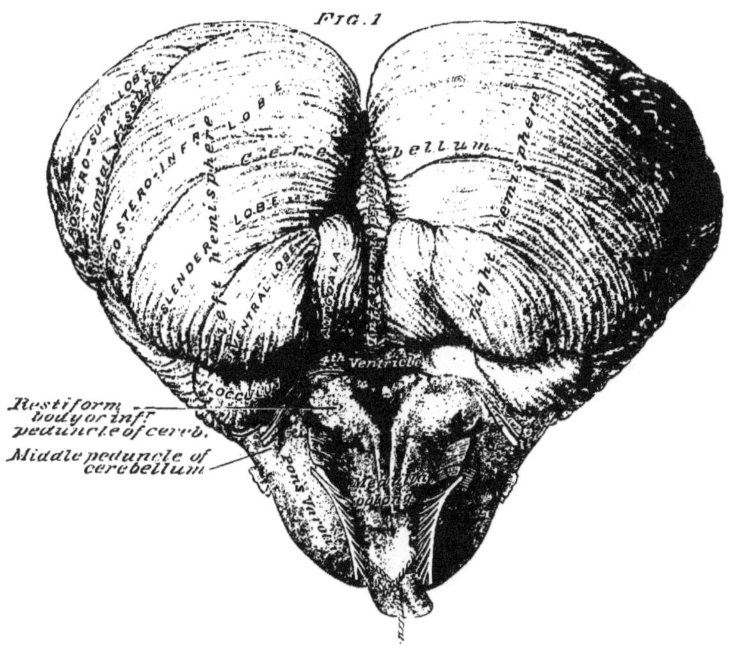
POSTERO-SUPR LOBE
POSTERO-INFR LOBE
SLENDER LOBE
Left hemisphere
Cerebellum
Right hemisphere
4th Ventricle
FLOCCULUS
Restiform body or infr. peduncle of cereb.
Middle peduncle of cerebellum
Pons Varolii

FIG. 2

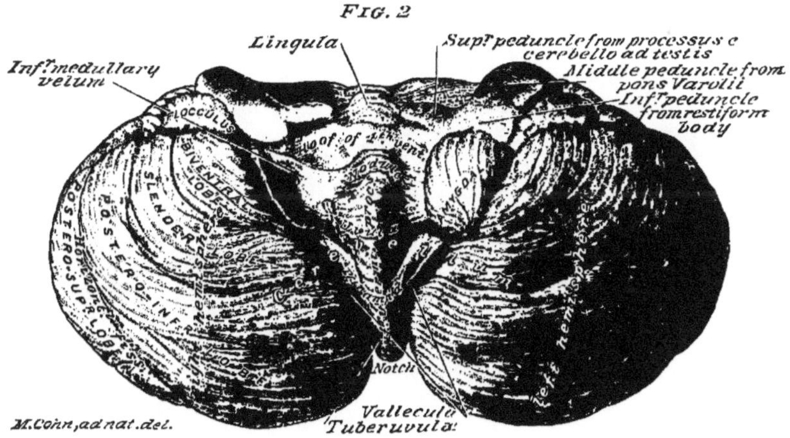
Lingula
Supr. peduncle from processus e cerebello ad testis
Infr. medullary velum
Middle peduncle from pons Varolii
Infr. peduncle from restiform body
FLOCCULUS
BIVENTRAL LOBE
SLENDER LOBE
POSTERO-INFR LOBE
POSTERO-SUPR LOBE
Left hemisphere
Notch
Vallecula
Tuberuvula

M.Cohn, ad nat. del.

FIG. 1

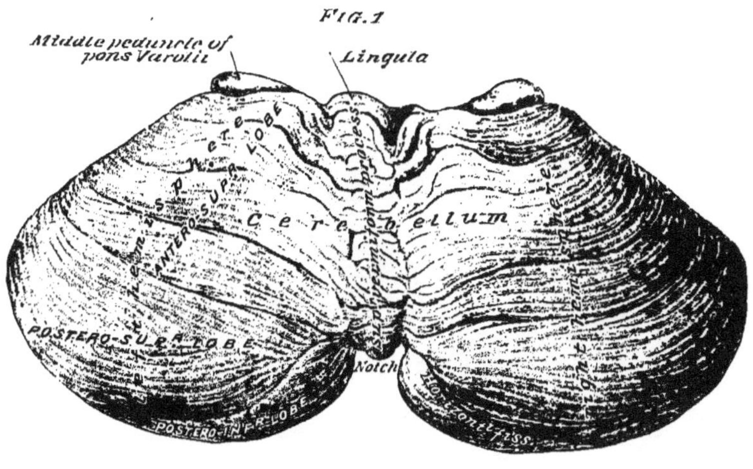
Middle peduncle of pons Varolii
Lingula
Cerebellum
ANTERO-SUP.R LOBE
POSTERO-SUP.R LOBE
POSTERO-INF.R LOBE
Notch

FIG. 2

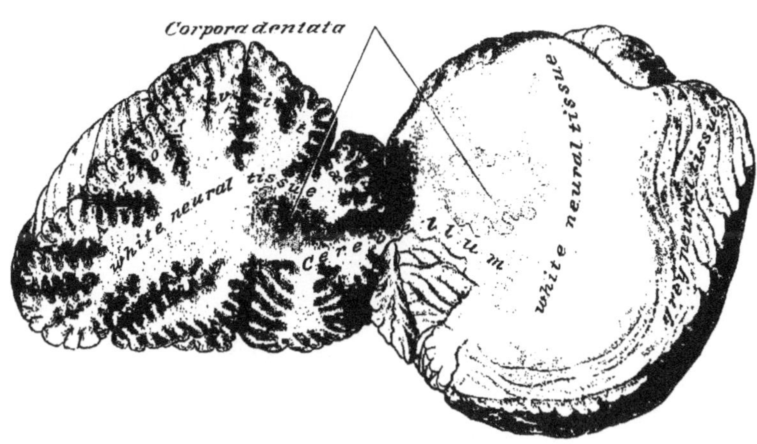
Corpora dentata
white neural tissue
Cerebellum
white neural tissue
grey neural tissue

M. Cohn, ad nat. del.

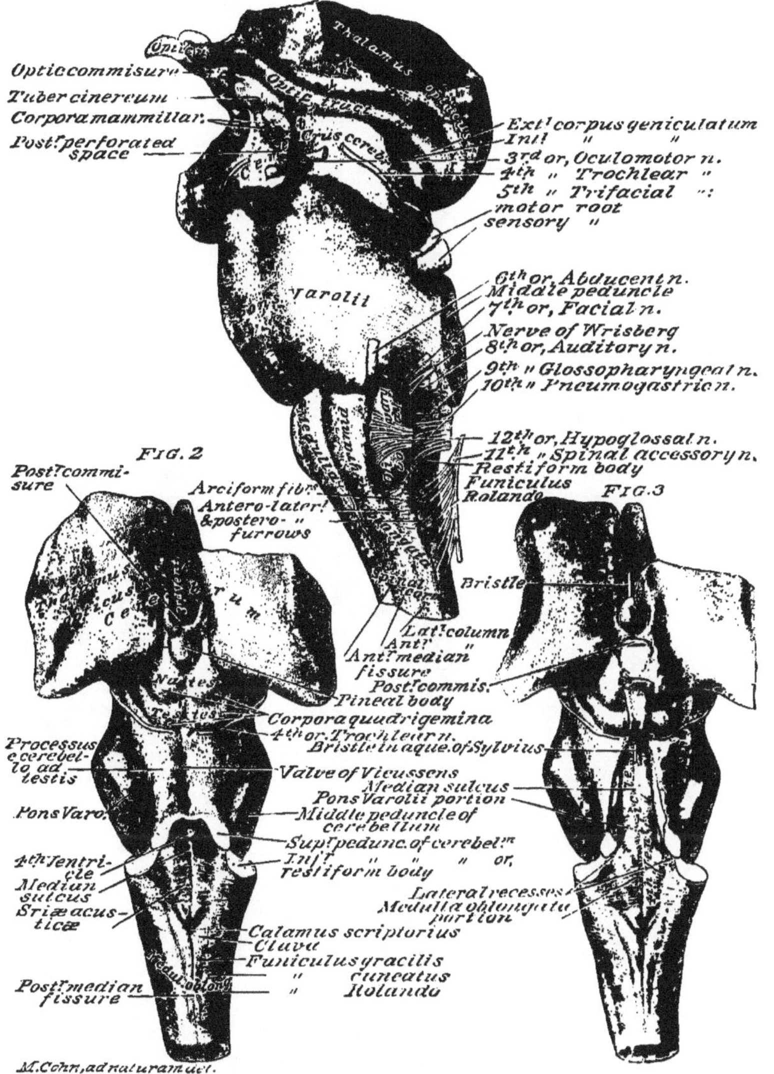
FIG. 1
Thalamus opticus
Optic tract
Crus cerebri
Varolii
Optic commisure
Tuber cinereum
Corpora mammillar.
Post! perforated space
Ext! corpus geniculatum
Int! " "
3rd or, Oculomotor n.
4th " Trochlear "
5th " Trifacial ":
motor root
sensory "
6th or, Abducent n.
Middle peduncle
7th or, Facial n.
Nerve of Wrisberg
8th or, Auditory n.
9th " Glossopharyngeal n.
10th " Pneumogastric n.
12th or, Hypoglossal n.
11th " Spinal accessory n.
Restiform body
Funiculus Rolando
Arciform fib.
Antero-later! & postero- " furrows
Lat! column
Ant! "
Ant! median fissure
FIG. 2
Post! commisure
Cerebrum
Naties
Testes
Pineal body
Corpora quadrigemina
4th or, Trochlear n.
Processus e cerebello ad testis
Valve of Vieussens
Pons Varol.
Middle peduncle of cerebellum
Sup! peduncle of cerebel.m
4th Ventricle
Median sulcus
Striæ acusticæ
Inf! " " " or, restiform body
Calamus scriptorius
Clava
Funiculus gracilis
" cuneatus
" Rolando
Post! median fissure
M. Cohn, ad naturam del.
FIG. 3
Bristle
Post! commis.
Bristle in aque. of Sylvius
Median sulcus
Pons Varolii portion
Lateral recesses
Medulla oblongata portion

INDEX.

☞ *Introduction text not indexed. The figures refer to the plates. To avoid complication, references to the figures of the plates are omitted, the plate reference being considered sufficient. A plate reference, like this—* 192 *to* 197—*means that the part referred to is illustrated in the plates from* 192 *to* 197, *inclusive.*

www.ingramcontent.com/pod-product-compliance
Lightning Source LLC
Chambersburg PA
CBHW031932220326
41598CB00062BA/1711

* 9 7 8 3 3 3 7 3 6 9 7 9 8 *